AF531359

BIOLOGY
OF
COELENTERATA

BIOLOGY
OF
COELENTERATA

By

Dr. D.R. Khanna
Reader in Zoology
Gurukul Kangri University
Haridwar (Uttaranchal)
&
Dr. P.R. Yadav
Reader
Department of Zoology
D.A.V. College
Muzaffarnagar (U.P.)

DISCOVERY PUBLISHING HOUSE
NEW DELHI-110002

First Published-2005

ISBN 81-8356-021-0

Published by

DISCOVERY PUBLISHING HOUSE
4831/24, Ansari Road, Prahlad Street,
Darya Ganj, New Delhi-110002 (India)
Phone: 23279245 • Fax: 91-11-23253475
E-mail:dphtemp@indiatimes.com

Printed at
Arora Offset Press
Laxmi Nagar, Delhi–92

Preface

The present title "Biology of Coelenterata" is a compilation work and embodies a fairly comprehensive treatment of the fundamental facts and aspects of morphology, anatomy, and physiology of various types of coelenterates. It is an indispensable text meeting complete requirement of undergraduate and post-graduate students of Biology. The purpose of this book is to provide students an authoritative and uptodate text in a very simple way, easy to grasp by those who do not have strong background of this subject. Much emphasis has been given on the anatomical studies of selected coelenterates. Ecological aspects have also been dealt with in sufficient details.

To make the work more comprehensive and informative, the author has consulted many authoritative books, research journals, abstracts, monographs etc. He is grateful to all those great scholars whose work are cited or substantially reproduced.

There can be no claim to originality except in the manner of treatment and much of the information has been obtained from the books and scientific journals available in the different libraries.

The author express his thanks to his friends and colleagues whose continue inspirations have initiated him to bring out this book.

The author is painfully aware of the shortcomings, errors and misprints that have crept in, and shall be grateful to receive suggestion for improvement of the next edition from all the readers.

The author expresses his gratitude to Mr. Wasan and staff of M/s Discovery Publishing House for their whole hearted co-operation in the publication of this book.

—Author

Contents

1

INTRODUCTION

The phylum Coelenterata, also called *Cnidaria*, includes the polyps, hydroids, jelly-fishes, sea-anemones and corals. The coelenterates show the following characters:

1. Most of the coelomates are marine. They are abundant on the sea-shore and also extend to great depths. A few polyps, like *Hydra*, and a few jelly-fishes, like *Craspedacusta*, are fresh-water. All are carnivorous. Some are fixed, others free-floating. Some grow on other animals and few permit other animals to live in or near them as commensals. Power of regeneration is well marked.
2. The body shows two main forms, the *polyp* and the *medusa*, which can be derived from each other. The polyp is cylindrical and usually fixed, but may be solitary or colonial. It is an asexual form. The medusa is umbrella-like and generally free swimming, but always solitary. It is a sexual form. Either or both the forms may occur in a species. Both forms may have a number of morphological varieties, several of which may exist in a single animal (polymorphism). When both the polypoid and medusoid forms are found in the same animal, the two forms alternate in the life-cycle. This phenomenon is termed the *alternation of generations*. Many authorities now hold that the medusa is adult coelenterate while the polyp is merely a juvenile (young) stage. Thus, there is no alternation of generations.
3. They are first multicellular animals to prossess true tissues, a digestive cavity and two distinct germ layers. They are, thus, the lowest Metazoa. They have developed all the four basic tissues found in all higher animals, namely, epithelial tissue for covering

surfaces, muscular tissue for contractility, connective tissue for support and transportation, and nervous tissues for irritability and coordination, but lack organs. They are, therefore, described as the *Metazoa with tissue grade of constructions*.

4. The body-wall consists of two distinct layers of cell: the outer *epidermis* and the inner *gastrodermis*, each containing several types of cells. Between these layers is a thin or thick sheet of gelatinous material, the *mesogloea*, which may be structureless or may have fibres and free cells, but is never formed of cells. Hence, the coelenterates are *diploblastic*. Mesogloea with free cells may be considered the beginning of a third layer.
5. The body encloses a single cavity, the *coelenteron*, which provides the phylum its name. The coelenteron is the digestive cavity. It may be sac-like or branched or divided into compartments by septa. It communicates with the exterior by a single aperture, the *mouth*, there being no anus. The mouth serves for both the ingestion of food and the egestion of faeces. Such a digestive cavity is said to be *incomplete*. In Actinozoa, the epidermis forms a stomodaeum which is a fore-runner of the pharynx in the higher phyla.
6. The soft and delicate body may be supported by horny or calcareous exoskeleton or by endoskeleton.
7. Symmetry is usually *radial* about an oral-aboral axis. In some forms, however, it is *biradial*.
8. There is no head in any of the two forms.
9. The oral end bears a number of slender, flexible processes, the *tentacles*.
10. Digestion is partly *intercellular* that occurs in the coelenteron and partly *intracellular*, which takes place in vacuoles within some gastrodermal cells. Besides being a digestive cavity, the coelenteron also serves for the distribution of food. It is, therefore, also known as the *gastrovascular cavity*.
11. There are no special organelles for respiration, circulation and excretion. Respiration simply occurs by diffusion of gases through the general body surfaces. Elimination of nitrogenous waste product, namely ammonia, also takes through the body surfaces by diffusion.
12. There are special stinging cells, the *cnidoblasts*, which produce in them peculiar organelles, the *nematocysts*, for offence and defence. Seventeen different types of nematocysts are known so far. There also occur free undifferentiated cells, the *interstitial cell*, for producing sex cells and cnidoblasts.

13. Sexes may be separate or united. Reproduction is usually asexual (budding) in the polyp form and sexual in the medusa form. Gonads are simple and have no ducts. Segmentation is holoblastic. Development includes a free-swimming ciliated larva, the *planula*. The long axis of a larva becomes the oral-aboral axis of the adult and its blastoporal end forms the oral end of the adult.
14. A muscular system consisting of contractile processes of epithelial cells or independent muscle cells or both is present to make the animals motile.
15. There is no separate body-cavity or coelom so that the coelenterates are *acoelomates*.
16. A primitive type of nervous system consisting of a diffuse network of unpolarised nerve-cells and their processes occurs in both the epidermis and gastrodermis. Numerous sensory cells are present in the epidermis and some also in the gastrodermis. Medusae have complicated sense organs. All this enables the coelenterates to behave as a coordinated unit and consequently makes them much more efficient than the sponges.

Advancement over Sponges

The coelenterates show advancement over the sponges in having—

1. definite form and symmetry,
2. well-organized layers of cells in the body-wall,
3. muscular tissue,
4. nervous system and sense organs,
5. digestive cavity, and
6. true germ layers.

Phylum-Coelenterata includes about 10,000 known species. About 5000 are known as fossils. They are grouped into three classes:

Class—I Hydrozoa.

Class—II Scyphozoa.

Class—III Anthozoa.

Class-I—Hydrozoa

(Gr. *Hydro* = water, *zoon* = animal).

1. Mostly colonial and marine, a few solitary and fresh-water, sessile or free-living.
2. They are *tetramerous symmetrical* or *polymerous symmetrical*.
3. Members of this class are medusoid of polypoid or show both forms in their life-cycle.

4. Mesoglea is acellular.
5. Nematocysts occur only in the epidermis.
6. Gametes develop in the epidermis.
7. Hydromedusae are usually small and plaktonic.
8. The asexual polypoid form arises from sexual medusoid form and the latter arises from the former, thus exhibiting metagenesis in the life-cycle of some hydrozoan e.g., *Obelia*.
9. Naked solitary species e.g., hydras probably stem from early polypoid forms which were not colonial.
10. Associated with colonial organization have been the evolution of a skeleton (support) and division of labour (Polymorphism).
11. Zygote undergoes complete cleavage and a hollow blastula is formed which changes into a solid *stereogastrula* whose ectodermal cells acquire cilia and then this stereogastrula changes into free-swimming *planula*. Planula settles down, attaches by its anterior end to a substratum and by budding forms the branched colonial polypoid stage.

The class Hydrozoa is divided into five order :

Order (i) Hydroida

1. Solitary or colonial.
2. The polyps are more developed.
3. Medusoid stage present or absent.
4. Sense organs of medusae are exclusively ectodermal.

It is divided in following sub-orders :

Suborder (a) Anthomedusae or Athecata

1. Skeletal covering, when present, does not surround hydrotheca. Freshwater or marine.
2. Polyps are naked.
3. Medusae are bud or bell-shaped. The gonads situated on the manubrium. Statocysts absent.

Examples—*Hydra, Protohydra, Bougainvillea, Tubularia, Hydractinia* etc.

Suborder (b) Leptomedusae or Thecata

1. Usually marine
2. Hydranth surrounded by a theca.
3. Medusa flattened, bowl shaped. Gonads are present on the radial canals. Eye-spot and statocysts are present.

Examples—*Obelia, Sertularia, Plumularia, Aglaophenia* etc.

Suborder (c) Limnomedusae

1. Polypoid stage without perisarcal skeleton.
2. Some only with medusoid stage.

   Example—*Graspeda custa.*

Order (ii) Hydrocorallina

1. Polypoid generation forming a colony which secretes a massive calcareous skeleton having minute pores at the surface through which the polyps protuade.
2. Polyps dimorphic, nutritive gastrozooids and dactylozooids.

Suborder (a) Milleporina

1. Skeleton covered by only a thin epidermal layer.
2. Dactylozooids are long, hollow and with tentacles.
3. Mature medusae lead an independent life.

   Example—*Millepora.*

Suborder (b) Stylasterina

1. Dactylozooids are small, solid and without tentacles.
2. Polyps are symmetrically arranged.
3. Medusa develop in special spore-sac and never free.

   Example—*Stylaster.*

Order (iii) Trachylina

1. *Medusoid* forms only, polypoid forms are absent or reduced.
2. Medusa with a velum beneath the margin of the bell and the tentacles inserted above the margin.
3. Statocysts and tentaculocysts present.
4. The planula give rise to the adult through an *actinula larva.*

   Order Trachylina is divided into two suborders :

Suborder (a) Trachymedusae

1. Sensory tentacles in pits or vesicles.
2. Gonads on radial canals.
3. Margins of bell are smooth.

   Examples—*Gonionemus, Petasus.*

Suborder (b) Norcomedusae

1. Sensory tentacles not enclosed.
2. Gonads on the floor of stomach.
3. Margins of bell scalloped by tentacle bases.

   Examples—*Cunina, Polycopa.*

Order (iv) Siphonophora

1. Free floating or free swimming colonies, showing maximum polymorphism.
2. Individuals attached to a linear stem or a circular disc.
3. A *pneumatophore* or float filled with air is present at the top in some cases.
4. Oral tentacles absent.
5. Nematocysts large and powerful.
6. Medusae incomplete, never free.

This order is divided into two suborders :

Suborder (a) Calycophora

1. Pneumatophore is absent.
2. One or more swimming zooids are present in the upper part of the colony.

Examples—*Abyla, Diphyses, Praya.*

Suborder (b) Physophorida

1. A large pneumatophore is present at the upper end of the colony.
2. Zooids are polypoid and medusoid.

Examples—*Physalia, Halistemma.*

Order (v) Chandrophora

1. Pelagic, polymorphic, polypoid colony with a gas filled float.
2. There is a central gastrozooid surrounded by many gonozooids.
3. Dactylozoids arranged in a marginal circlet.

Examples—*Porpita, Velella.*

CLASS-II—SCYPHOZOA

1. Class Scyphozoa include large jelly fishes.
2. Polyp stage is usually reduced or absent.
3. Medusa are large, umbrella-shaped without velum. Perisarc is absent.
4. The sense organs are in the form of hollow *tentaculocysts* having *statolith*.
5. The gastral tentacles are endodermal.
6. Mesogloea enlarged and usually cellular.
7. The gonads are endodermal and release the gametes in stomach.
8. Stomodaeum is absent in the gastro-vascular system but gastric tentacles are present and the cavity is divided into interradial pockets by four ridges or septa.

9. Alternation of generation is seen in some.
10. They are exclusively marine.

The class is divided into 5 orders.

Order (i) Lucernarida

1. Mostly found in cold littoral waters.
2. Sessile and sedentary, attached to the substratum by the aboral stalk.
3. Body is goblet or trumpet shaped.
4. Mouth is four cornered with small oral lobes and a short manubrium.
5. Tentacles are present in some, they are per-radial and inter-radial.
6. The umbrella is cup-shaped.
7. Gonads are long band-shaped lying on faces of septa.
8. Larva is planula with cilia.

Examples—*Lucernaria, Haliclystus.*

Order (ii) Coronatae

1. Free swimming forms living in deep sea.
2. Body conical or dome-shaped divided into an upper cone and lower crown by a coronary groove.
3. Solid tentacles are per-radial and ad-radial, 4 interradial *rhopalina* present.
4. Four to sixteen tentaculocysts.

Examples—*Pericolpa, Atolla.*

Order (iii) Cubomedusae

1. Free-swimming found in warm and shallow waters.
2. The umbrella is four-sided and cup-shaped.
3. Four hollow inter-radial tentacles and four per-radial tentaculocysts are present. Each tentaculocyst with one or more ocelli and a lithocyst.
4. A true velum is absent.
5. Gonads are leaf-like and there is no alternation of generation in life-cycle.

Examples—*Chiropsalmus, Tamoya.*

Order (iv) Semaeostomae

1. Common free-swimming animals found all over the world.
2. The umbrella is disc-like.
3. Mouth surrounded by four oral arms.

4. Eight or more tentaculocysts are present.
5. Gastric pouches and septa are absent.

Examples—*Aurelia, Cyanea.*

Order (v) Rhizostomae

1. Free-swimming forms found is shallow waters of tropical and subtropical oceans.
2. The mouth is obliterated by the growth across it of 8 very large and branched oral arms.
3. The stomach is continued into the canals opening by funnel shaped apertures on the edges of the arms.
4. Umbrella is saucer or bowl-shaped.
5. Marginal tentacles absent.
6. Tentaculocysts are 8 or more in number.

Examples—*Rhizostoma, Cassiopeia.*

Class-III—Anthozoa or Actinozoa

1. Exist only in the polyp form, no medusa stage.
2. Differ from hydrozoan and scyphozoan polyp in possessing a stomodaeum, paired mesenteries (their free ends bear coiled mesenteric filaments like the gastric filaments of scyphozoan but are partially endodermal in origin).
3. Body wall consists of ectoderm and endoderm separated by a strong mesogloea having fibres and cells.
4. Stomodacum consists of the same layers reversed i.e., its lining membrane is ectodermal. The mesenteries are formed by a double layer of endoderm with a supporting plate of mesogloea.
5. More complex nematocysts than in Hydrozoa and Scyphozoa are found in the tentacles, body-wall, stomodacum and mesenteric filaments.
6. Muscular system is well-developed containing both ectodermal and endodermal fibres and endodermal muscle processes.
7. Nervous system is a typical nerve net.
8. Gonads develop in the mesenteries, sex cells are located in the endoderm, sperms and ova are discharged into the coelenteron.
9. The zygote develops into a planula which after swimming freely for sometimes settles down and metamorphoses into the adult form.
10. Except in one doubtful instance there is no alternation of generations. The class-Anthozoa is divided into two sub-classes:

Subclass-1 Hexacorallina or Zoantharia

1. Solitary or colonial marine forms.
2. Tentacles and mesenteries very numerous, arranged in multiple of six.
3. Tentacles are simple, unbranched, hollow cones.
4. Two siphonoglyphs are present and two pairs of directive mesenteries; the remaining mesenteries are generally arranged in couples with the longitudinal muscles of each couple facing one another.

Order (i) Actiniaria

1. The animals are commonly called sea-anemones. They are solitary or colonial, brightly coloured.
2. Tentacles and mesenteries are numerous.
3. Skeleton is absent.
4. One or more siphonoglyphs.

   Examples—*Metridium, Adamsia, Edwardsia.*

Order (ii) Madreporaria

1. The animals are usually colonial.
2. Skeleton is external and calcareous.
3. Siphonoglyph is absent.
4. Polyps are very small.
5. These are stony corals.

   Examples—*Corallium, Madrepora, Fungia, Meandra, Favia.*

Order (iii) Zoanthidea

1. Mostly colonial and some are solitary forms.
2. The tentacles are unbranched
3. The skeleton is usually of a calcareous nature but in few cases there is a horny axial skeleton.
4. Polyps are small and often united by basal stolons.
5. Mesentries paired. Each pair with a complete and an incomplete mesentery.
6. They have single siphonoglyph.

   Example—*Zoanthus.*

Order (iv) Antipatharia

1. These are compound tree-like animals. Commonly known as *black corals*.
2. The tentacles and mesenteries are comparatively few (6-24) in number.

3. The skeleton is present in the forms of a branched chitinoid axis.
4. Two siphonoglyphs.

   Example—*Antipathes.*

Order (v) Ceriantharia

1. The individuals are long and solitary.
2. Skeleton is absent.
3. The mesenteries are incomplete.
4. Numerous, simple tentacles are arranged in two whorls-oral and marginal.
5. There is a single dorsal siphonoglyph.

   Examples—*Cerianthus, Pachycerianthus.*

Order (vi) Corallimorpharia

1. Solitary or colonial polyps without skeleton.
2. Radially arranged tentacles.

   Example—*Corynactis.*

Subclass-2 Octocorallina

1. Colonial marine forms.
2. Only polyps no medusae.
3. Tentacles and mesenteries always eight in number.
4. The tentacles pinnate; i.e., produced into a symmetrical branchlets.
5. Never more than one siphonoglyph, which is ventral in position, i.e., faces the proximal end of the colony.
6. The arrangement of mesenteries is not always in couples (or pairs) and all their longitudinal muscles are directed ventrally i.e., towards the same side as the siphonoglyph.

   The subclass is divided into following orders :

Order (i) Stolonifera

1. Found in shallow waters.
2. Polyps arising form a creeping stolon.
3. Skeleton either absent or of calcareous tubes or separate calcareous spicules.
4. Found on coral-reefs in Old and New World.

   Examples—*Tubipora, Clavularia.*

Order (ii) Telestacea

1. Colony formed of simple or branched stems arising from a creeping base.

2. Skeleton of calcareous spicules.

   Example—*Telesto.*

Order (iii) Alcyonacea

1. The skeleton usually consists of calcareous spicules or the spicules become aggregated so as to form a coherent skeleton.
2. Some forms may be dimorphic
3. Body is branched with a central axis.
4. Polyps embedded in fleshy coenochyme.

   Examples—*Alcyonium, Gersemia.*

Order (iv) Coenothecalia

1. Blue or brown corals found on coral reefs in the Indo-Pacific waters.
2. Skeleton massive, calcareous and blue-green formed from iron salts.

   Example—*Heliopora.*

Order (v) Gorgonacea

1. The individuals are compound tree-like.
2. Skeleton is calcareous or horny.
3. Spicules are present in mesogloea.
4. Siphonoglyph is absent.
5. A central skeletal axis composed of a horn-like substance, the *gorgonin.*

   Examples—*Gorgonia, **Corallium.***

Order (vi) Pennatulacea

1. The colony is usually elongated.
2. One end of the colony remains embeded in the mud at the sea-bottom while other end bears the polyps.
3. The rachis is the axial polyp bearing numerous dimorphic polyps laterally.
4. The skeleton is calcareous or horny.

   Examples—*Pennatula, Renilla, Pteroeides.*

SERTULARIA

Phylum	—	Coelenterata
Class	—	Hydrozoa
Order	—	Hydroida
Suborder	—	Leptomedusae
Genus	—	*Sertularia*

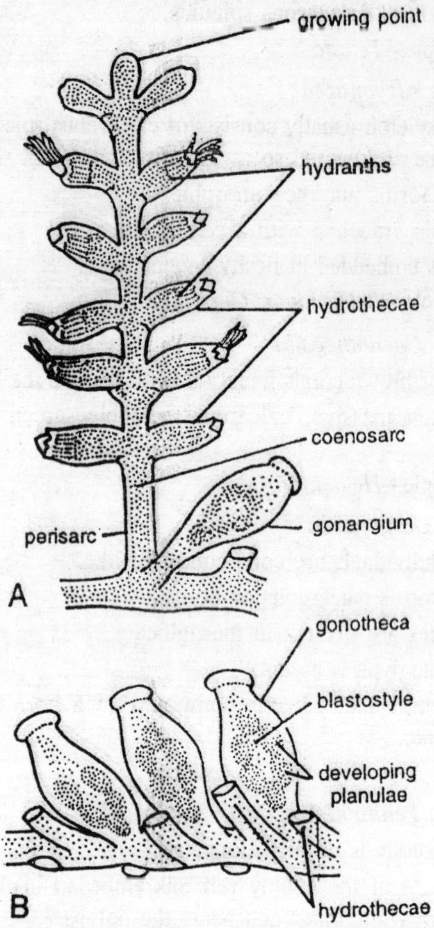

Fig. 1.1. Sertularia. A—Branch with hydranths. B—Branch with gonangia.

It is a small branching colony occurs in shallow marine water on submerged objects or rocks. Hydrotheca sessile i.e., not stalked, with *opercula* in pair along the stem being exactly opposite to each other. Hydrothecae are large, polyps retract within themselves. *Gonangia* are of simple form, much larger than hydrothecae, a few in a colony and present only in a certain period of three years. Blastostyles produce *planulae* and no free medusae.

Plumularia

The classification is the same as that of *Sertularia*. It is a feather-like colony occurs in shallow marine water attached to some substratum. The plume-like colony comes out from creeping hydrorhiza. A series

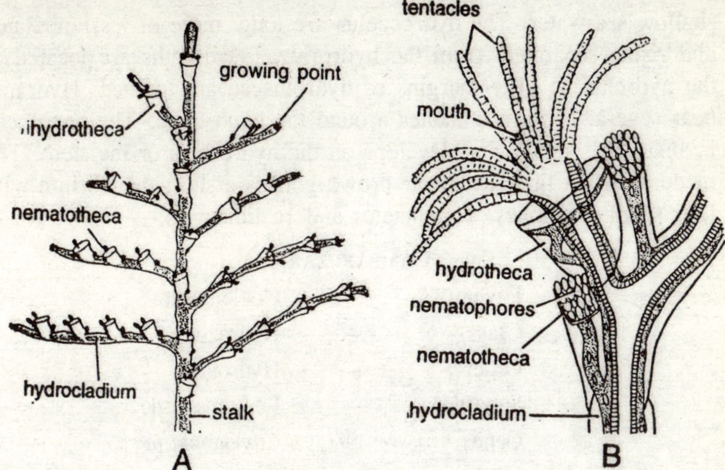

Fig. 1.2. Plumularia. A—A portion of colony. B—A polyp (magnified).

of hydrothecae are found on one side of the branches of the colony. Hydrothecae are small and without stalk and the polyps can be partly retracted within them. Special mouthless polyps known as *nematophores* are found which have long amoeboid projections. The gonangia are long and superficially resembling the spadix inflorescence.

CLYTIA

The systematic position is same as that of *Sertularia*. It forms a simple or sparsely branched colony attached to docks, algae etc. in

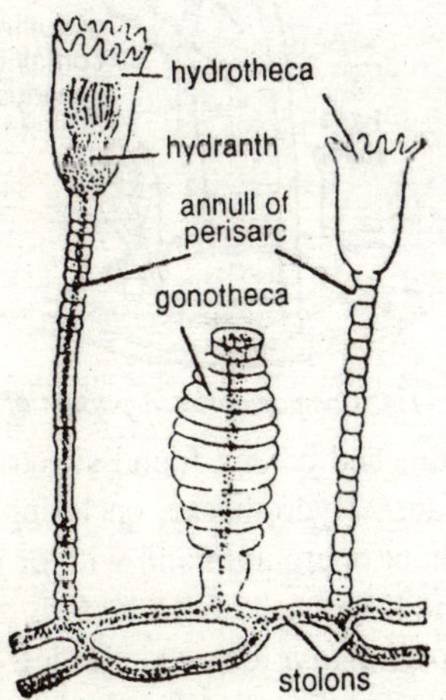

Fig. 1.3. Clytia.

shallow sea water. The hydrocaulus are long, more or less annulated and rising separately from the hydrorhiza. Hydranths are located in the hydrothecae. The margins of hydrothecae are toothed. Hydranth bear several filiform tentacles around the hypostome. The gonotheca is annulated, sessile and develops on the hydrorhiza or the stem. The medusae when liberated, bear brown gonads a short manubrium with four small oral lobes, 16 tentacles and 16 lithocysts.

CAMPANULARIA

Phylum	—	Coelenterata
Class	—	Hydrozoa
Order	—	Hydroida
Suborder	—	Leptomedusae
Genus	—	*Campanularia*

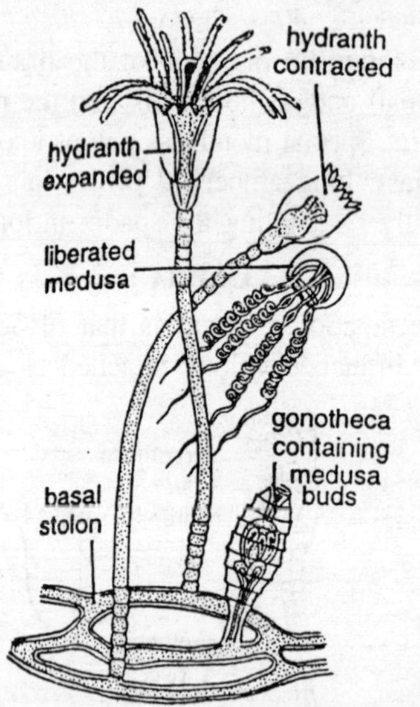

Fig. 1.4. Companularia. A portion of colony.

It is a simple branched colony found attached to submerged marine objects in shallow water. Hydrothecae, enclosing polyps, are bell-shaped usually stalked, without operculum and with or without marginal teeth. Annulated stem at the base of the branch. *Blastostyle* produces *planulae* and not medusae. *Gonanagium* long and slender. Hypostome of hydranth or polyp tumpet-shaped surrounded by tentacles.

HYDRACTINIA

Phylum	—	Coelenterata
Class	—	Hydrozoa
Order	—	Hydroida
Suborder	—	Anthomedusae
Genus	—	*Hydractinia*

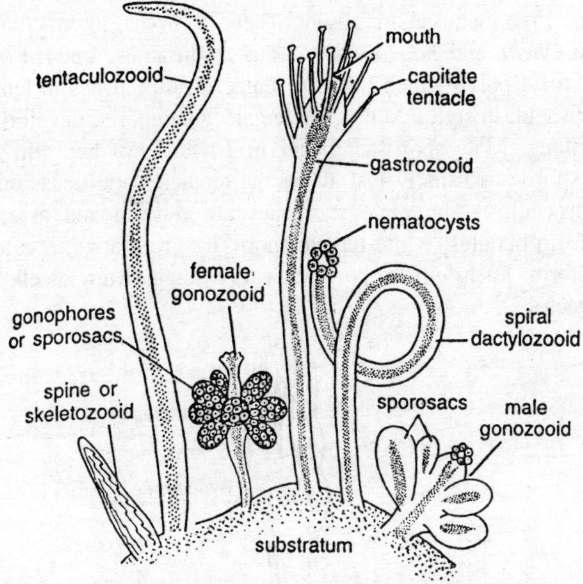

Fig. 1.5. Hydractinia. A polymorphic colony.

It is a small, marine hydroid found in shallow water attached to stones, sea-weeds and shells. It is commonly found on Atlantic coast, U.S.A., Europe and also found in Sanjuan Island. Zooids are of four types *gastrozooid* (feeding polyp), *gonozooid* (reproductive zooid), *dactylozooid* (protective zooid) and *tentaculozooid* (sensory polyp). The hydranths are simply mouthless and are called as dactylozooids. The dectylozooids have very short tentacles abundantly supplied with nematocysts. The dactylozooids are capable of very active movements. With the help of massive coenosarc consisting of a number of branches, *Hydractinia* form a firm brownish crust on the surfaces of dead gastropod shells inhabited by hermit-crabs. The association between *Hydractinia* and hermit crabs is known as *commensalism*. The hydroid feed upon minute fragments of the hermit crab's food and the hermit crab is protected from its enemies by the presence of the inedible stinging hydroid.

EUDENDRIUM

The systematic position is same as that of *Hydractinia*. *Eudendrium* forms a branching colony arising from a reticulated hydrorhiza. Perisarc is well distinct and variously annulated. Hydranth has a trumped-like hypostome. The hypostome is surrounded by a whorl of filiform tentacles. Free medusae are absent. There occurs an intermediate stage between *Obelia* and *Bougainvillea*. The medusae are budded off from the stalk of a polyp which loses its tentacles, diminishes in length and becomes a blastostyle. Male and female gonophores develop on the same colony. Male sporosacs occur in a whorl just beneath, and the female sporosacs usually just above the tentacles and occasionally on the hydrocaulus. The male sporosacs are reddish and arranged in monoliform bundles, while female sporosacs are orange in colour and are pyriform. Each female sporosac bears a single ovum attached above the tentacles.

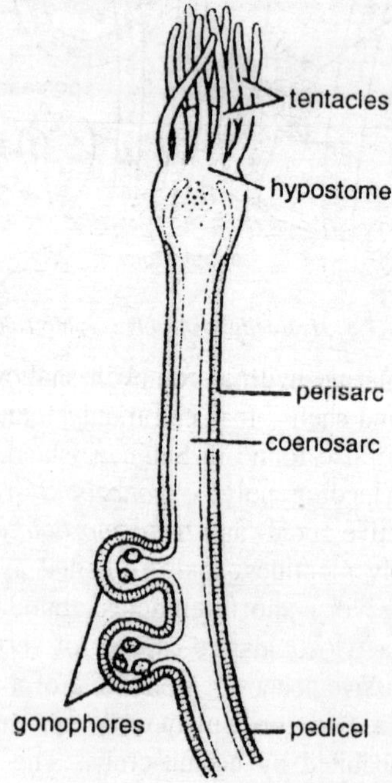

Fig. 1.6. Eudendrium.

CERATELLA

Classification is the same as that of *Hydractinia*. It is a colonial, sedentary hydroid coelenterate. It is tree-like highly branched animal. It consists of a branching axis composed of many inter-twisting and anastomosing tubes. Hydrotheca covers polyp while gonotheca covers medusa. The zooids have scattered capitate tentacles. The medusae bear gonads on the manubrium and devoid of lithocytes. Eye spots are present.

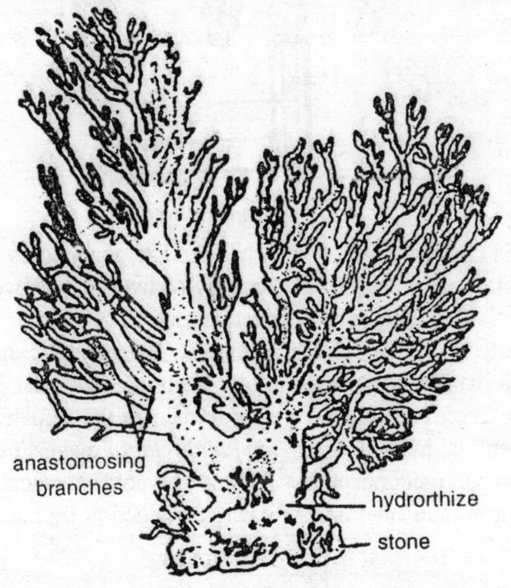

Fig. 1.7. Ceratella.

TUBULARIA

The systematic position of *Tubularia* is same as that of *Hydractinia*. It is a marine hydroid occuring solitary or in large pink colonies attached to substratum. The colony consists of tufts of irregularly branched stems bearing hydranths arising from hydrorhiza. The hydranths are large in size and bright in colour, with a flower-like appearance. Each hydranth consists of a proximal narrow neck attaching it to the stem and a distal larger conical region bearing two whorls of solid, filiform tentacles. The distal smaller tentacles arise from the hypostome surrounding the mouth. The basal tentacles are large and surround the base of hydanth. All the base of hydranth, the stalk forms a swelling where the perisarc comes to an end without forming a protective hydrotheca around the hydranth. The stalk or stem contains

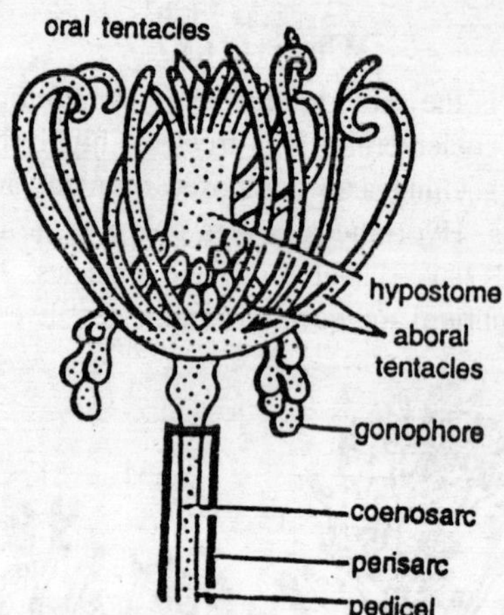

Fig. 1.8. Tubularia. A hydranth.

perisarc and coenosarc and is transparent. The gonophores or sporosacs arise as hollow branches from the body of hydranth between two sets of tentacles. Gonophores produce medusae. The colony is dioecious with separate male and female members. The medusae are deep bell-shaped with manubrial gonads and four knob-like tentacles. The medusae are not set free. The fertilized eggs remain in the female gonophore. Development is indirect with a *planula larva* never becomes free swimming, but metamorphoses into a polypoid, tentaculate *actinula larva* which is then set-free. Asexual reproduction by budding.

Bougainvillea

The classification is the same as that of *Hydractinia*. It is a plant-like hydroid colony inhabiting in sea waters. Two types of zooids are present the *polyp* or *hydranth* and *medusa*. Polyps are vegetative responsible for nutrition and food catching, while medusa is reproductive in function. Hydrotheca and gonotheca coverings are absent on polyp and medusa respectively. Polyp is beset with a single circlet of filiform tentacles which surround the hypostome. Medusa bears four radial canals and a wide velum. On the manubrium of medusa develops the gonads.

Millepora

Phylum	—	Coelenterata
Class	—	Hydrozoa
Order	—	Hydrocorallina
Suborder	—	Milliporina
Genus	—	*Millepora*

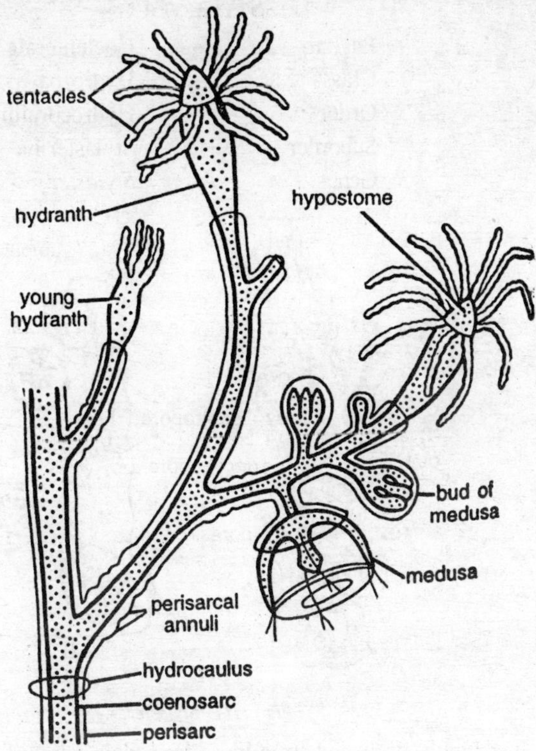

Fig. 1.9. Bougainvillea. A portion of colony.

Marine colonial form found in tropical seas and associated with other corals. Ectoderm secretes a calcareous yellowish or white skeleton of perisarc. The dried colony is perforated by numerous pores. These are of two different sizes; large *gastropores* and small *dactylopores* which surround the gastropores. In living specimens two types of zooid—*gastrozooids* and *dactylozooids* which protrude out of these pores. Gastrozooids are feeding zooids, having 4-5 short knobbed tentacles. The dactylozooids are protective zooids with capitate tentacles having nematocysts. Pores lead into canals which forms network in coenosarc. Medusae with 4 or 5 rudimentary tentacles. These are free, simple and originate from coenosarc.

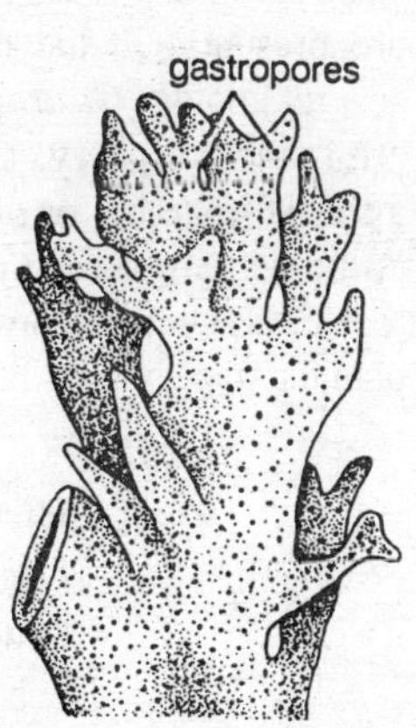

Fig. 1.10. Millepora. A portion of dried colony.

STYLASTER

Phylum	—	Coelenterata
Class	—	Hydrozoa
Order	—	Hydrocorallina
Suborder	—	Stylasterina
Genus	—	*Stylaster*

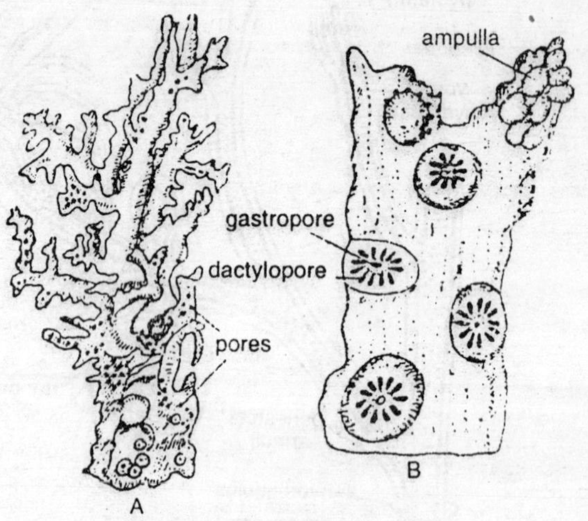

Fig. 1.11. Stylaster.

It is highly branched, tree-like, deep pink colony found in tropical and subtropical seas. Numerous upright branches of calcium carbonate are present. On the branches are cup-shaped projections formed of several zooids. *Gastrozooids* situated in the centre, while *dactylozooids* are in the periphery. *Gonophores* or reproductive zooids are lodged in a special chamber or *ampulla* of the coral. Form the horizontal partition at the bottom of each cup projects a calcareous projection called *style*, hence the generic name *Stylaster* is given. The young sets free in the planula stage which later on metamor-phoses into a new colony.

GONIONEMUS

Phylum	—	Coelenterata
Class	—	Hydrozoa
Order	—	Trachylina
Suborder	—	Trachymedusae
Genus	—	*Gonionemus*

It is a medusa found in bottom of sea water. It is cosmopolitan in distribution. It is like the medusa of *Obelia* and generally bell-like or

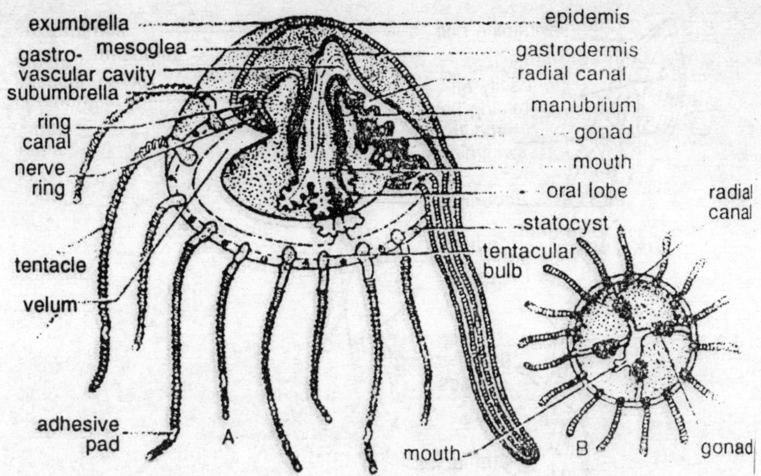

Fig. 1.12. Gonionemus. A—Section to show internal structure; B—Sub-umbrellar side.

saucer shaped. The convex outer surface of the bell is known as exumbrella while the concave inner surface is subumbrella. From the centre of the concave or subumbrella surface hangs down a short, hollow and quardrangular process known as the *manubrium* bearing four brief frilled oral lobes surrounding the mouth. The rim of the margin of the medusa bears numerous 16 to 18 highly contractile tentacles provided with belts of nematocysts and adhesive pads. The adhesive pads are generally situated at the bending on each tentacle and helps in anchoring to marine plant at rest. The gonads are situated on four radial canals and medusae are unisexual. Planula develops into small polyp which has mouth tentacles and is known as *haleremite*. The haleremite reproduces asexually and gives rise to planula-like buds which are nonciliated and are called *frustules*. The frustule develop into new polyps which from external side but off gonophores. These naked gonophores give rise to medusae.

Diphyes

Phylum	—	Coelenterata
Class	—	Hydrozoa
Order	—	Siphonophora
Suborder	—	Calycophora
Genus	—	*Diphyes*

It is a polymorphic colonial, marine and pelagic coelenterate. It has cosmopolitan distribution. Pneumatophore is absent. The colony consists of a coenosarcal stem bearing groups of zooids called *cormidia*

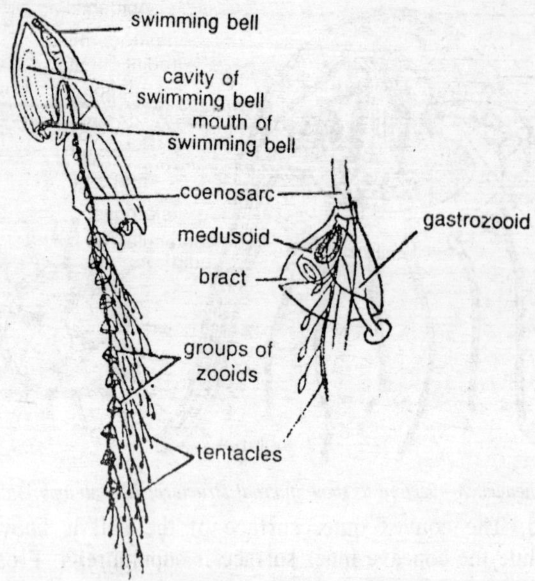

Fig. 1.13. Diphyes. A—Entire colony; B—Single group of zooids.

and two large swimming bells or *nectophores*. The bells are present at the proximal end of colony. Below the nectophores, the stem bears widely separated groups of zooids. These can be retracted into a groove in nectophores. Other zooids attached to the bells are tentacles bearing gastrozooids, a medusa and a large bract of protective nature. Each group of zooid (Cormidium) consists of a hydrophyllum, a gastrozooids with its tentacles and a gonozoid. The stem often breaks at the internodes. Detached zooids are called *endoxia*, which swim about like independent organisms and become sexually mature.

Halistemma

Phylum	—	Coelenterata
Class	—	Hydrozoa
Order	—	Siphonophora
Suborder	—	Physophorida
Genus	—	*Halistemma*

It is a polymorphic colony found floating on the surface of Mediterranean sea. Uppermost end of the stem possesses an ovid bubble-like body, known as *float* or *pneumatophore* containing air. Below the penumatophore, there is a series of unsymmetrical medusae, called *nectocalyces* each having a deep bell-like body with a velum and without manubrium. *Swimming bells* or *nectocalyces* contract, drawing water

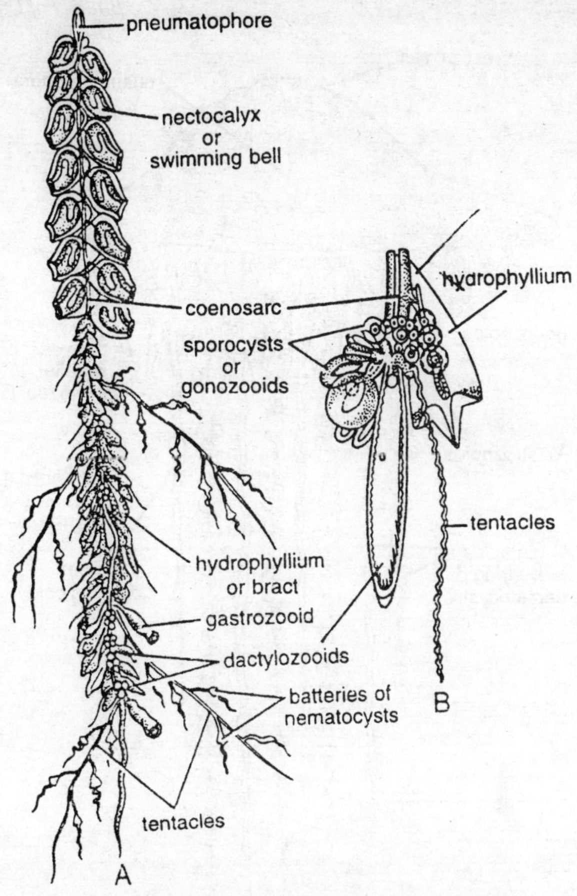

Fig. 1.14. Halistemma. A—Entire colony; B—Cormidium.

into their cavities and pumping it out to propel the whole organism through water. Stem underneath divides into nodes and internodes. Below certain nodes arise *polyps* with a long, branched tentacle and bearing batteries of "*stinging capsule*". In the remaining nodes, *dactylozooides* or *feelers* take the place of polyps. Below these dactylozooids are *sporosacs*, some are males and others females. Delicate leaf-like, transparent bodies, the bracts or *hydrophylia*, spring from internodes.

Physalia

The classification is the same as that of *Halistemma*. It is a marine, pelagic, polymorphic hydroid. It is found in the gulf stream from Florida to Vineyard Sound and occasionally to the Bay of Funday.

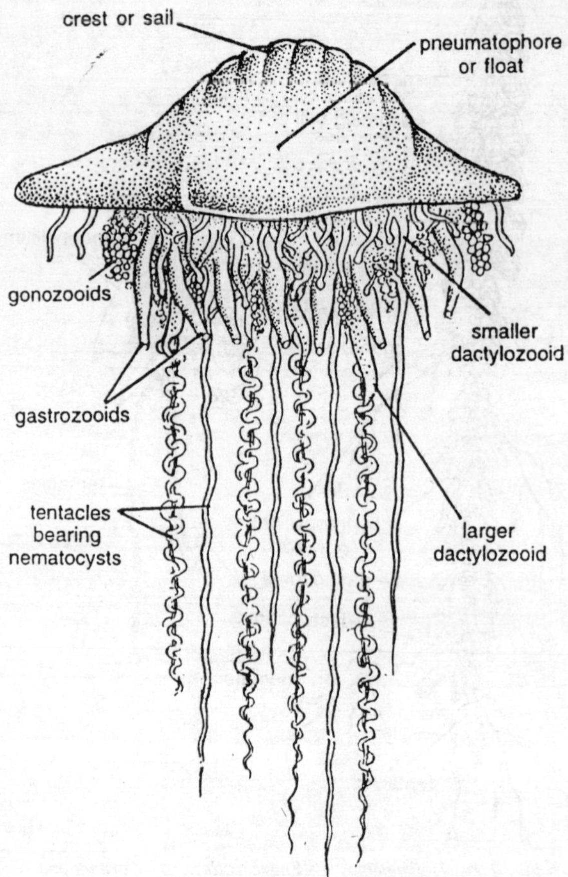

Fig. 1.15. Physalia. A colony.

It is commonly called '*Portuguese man of war*', because it has a large, brilliant coloured pneumatophore or float which is like the cap of Great Napolean. The dorsal side of float is called *crest* or *sail*. The float is filled with a gas. The composition of the gas is 85-91% nitrogen, 1.5% argon and 7.5-13.5% oxygen. The float or pneumetophore has a pore called *pneumatopore*. The swimming bells or nectocalyces are absent. Hanging from the underside of pneumatophore are three types of zooids: (i) *Gastrozooids* are nutritive zooids without tentacles. (ii) *blastostyles* are reproductive zooids containing served medusae. (iii) *Dactylozooids* are the protective zooids with tentacles and nematocysts. The female gonophores are medusoid and swim free, while male ones are small and remain attached. The sting of *Physalia* is very poisonous and the nematocysts are so highly poisonous as to cause danger to man.

PORPITA

The classification is the same as that of *Physalia*. *Porpita* is a marine coelenterate found usually on South Atlantic Coast and occasionally near U.K. Coast. The float is very large, circular and disc like (like the float of *Physalia*). The colony consists of disc-like body enclosing a chambered chitinoid shell. It exhibits the remarkable phenomenon of *polymorphism*. Long dactylozooids or tentacles are present around the disc. The under-surface is beset with gonozooids or blastostyles provided with mouth bearing the medusae. The single gastrozooid is present in the centre of disc on ventral side as in *Velella*. Crest is wanting.

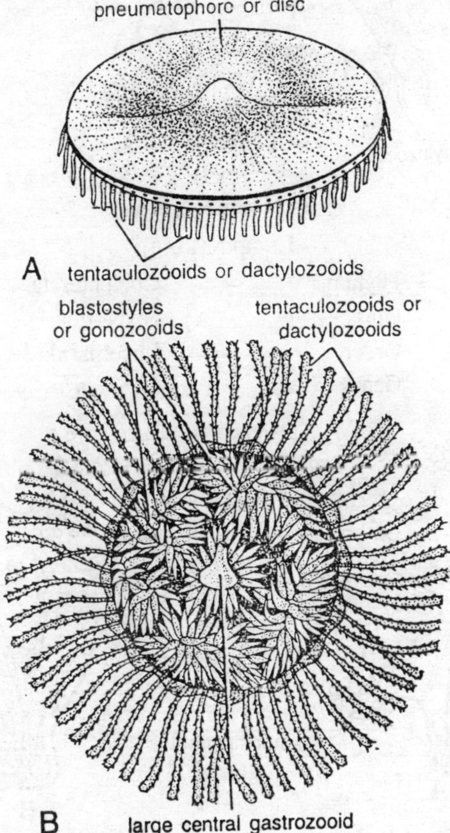

Fig. 1.16: Porpita. A—Colony in dorsal view; B—Colony in oral or ventral view.

Velella

The classification is the same as that of *Physalia*. It is a beautifully coloured pelagic form found in warm water. It is common in Pacific Coast and South Atlantic Coast. Body rhomboidal with a medusa-like appearance. Pneumatophore dorsal, in the form of a chambered chitinous disc and bears a vertical crest like ridges. Hanging from the centre of the under surface is a single large gastrozooid which is surrounded by numerous gonozooids which give off medusa-buds. The gonozooids are devoid of tentacles. Dactylozooids hang down from the rim of the disc forming a circular fringe. It is incapable of sinking down by altering the gas contents of the float.

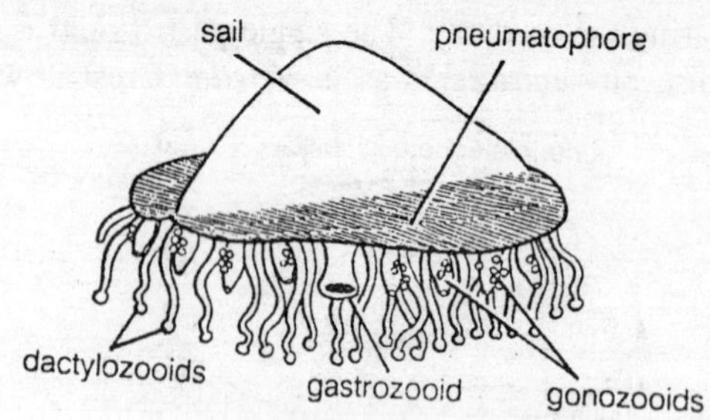

Fig. 1.17. Velella. A colony.

Lucernaria

Phylum	—	Coelenterata
Class	—	Scyphozoa
Order	—	Lucernarida
Genus	—	*Lucernaria*

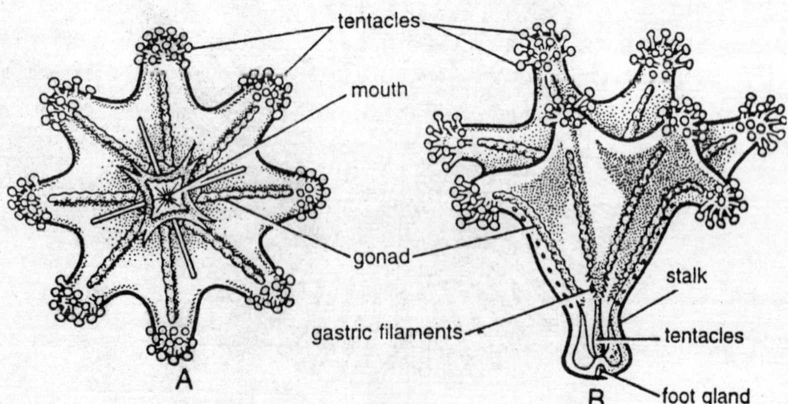

Fig. 1.18. Lucernaria. A—Oral view; B—Side view.

It is a sedentary, sessile animal found on British Coast. Body is trumpet-shaped or conical differentiated into aboral exumbrella and oral subumbrella surface. Exumbrella surface drown out into a short, cylindrical stalk used for attachment to sea-weeds. Margin of umbrella drawn into 8 short and hollow adhesive lobes or adradial arms. Each arm bears a cluster of knobbed adhesive tentacles. Mouth is cruciform (four cornered) with small oral lobes and a short manubrium. Gastrovascular system consists of central stomach and four perradial pouches divided by four interradial septa. Gastric filaments numerous. Gonads are band-like brone on septa. Marginal adhesive gastric pits and velum are absent.

Charybdaea

Phylum	—	Coelenterata
Class	—	Scyphozoa
Order	—	Cubomedusae
Genus	—	*Charybdaea*

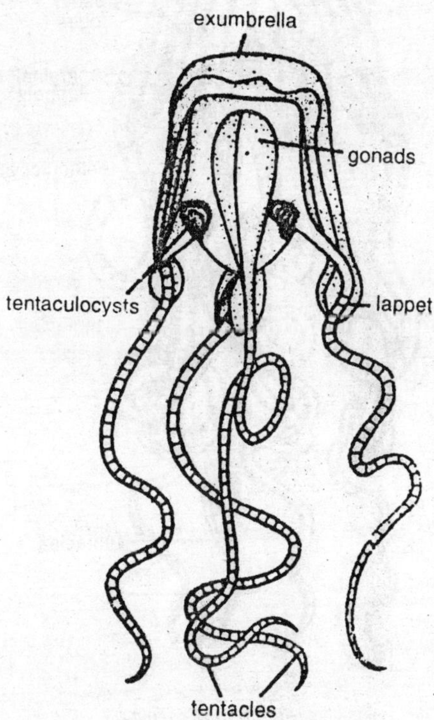

Fig. 1.19. Charybdaea.

It is a marine animal. It is commonly found in the warm shallow waters of tropical and subtropical region. It is an active swimmer and swims through the water appearing beautiful. It resembles a deep bell with somewhat flattened top and square in transverse section. The margin of the umbrella bears 4 tentacles and 4 tentaculocysts. The tentacles are inter-radial and tentaculocysts are per-radial. The tentaculocysts are set in deep marginal notches and the tentacles arise from gelatinous lobes. The tentaculocysts are very complex, each bearing a lithocyst and several eye-spots. The margin of the umbrella is produced into a false velum like the velarium of *Aurelia*. The nervous system is present in the form of nerve ring round the margin of bell.

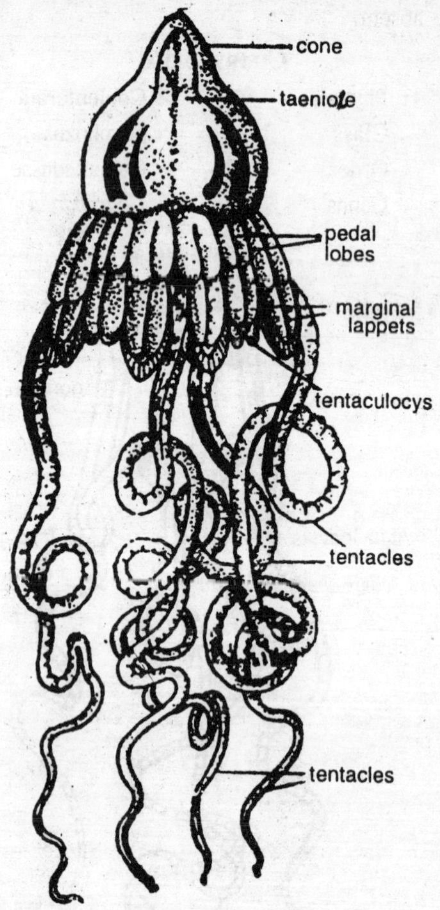

Fig. 1.20. Pericolpa.

Pericolpa

Phylum	—	Coelenterata
Class	—	Scyphozoa
Order	—	Coronatae
Genus	—	*Pericolpa*

It is a beautiful marine solitary medusoid animal. It is cosmopolitan is distribution, but abundent found in Greenland. Umbrella is divided into apical cone and marginal crown by a furrow. Marginal crown is divided into a series of pedal lobes and series of marginal lappets by a second horizontal furrow. Marginal lappets and pedal lobes are present in same radii. Tentaculocysts are present on four interradial pedal lobes. Large mouth opens into the stomach by manubrium.

Cyanea

Phylum	—	Coelenterata
Class	—	Scyphozoa
Order	—	Semaeostomae
Genus	—	*Cyanea*

It is the largest, solitary, bioluminescent and marine medusa. It is found in the costal waters of America, Pacific Coast and Polar Regions. It is commonly called as '*sun-jelly*' or '*sea-blubber*'. Umbrella

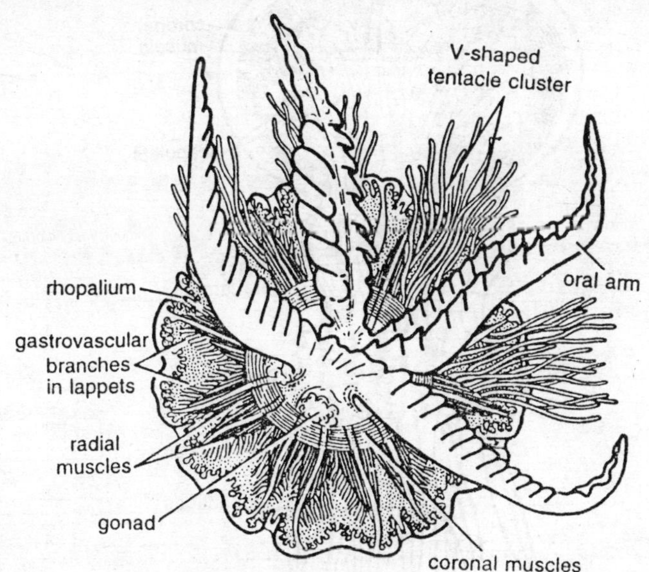

Fig. 1.21. Cyanea.

dome-shaped or saucer-shaped with margin scalloped into eight lappets having eight rhopalia. Subumbrella carries four long oral arms and eight V-shaped bundles or tentacles, the marginal tentacles. Four gonads, placed between oral arms and tentacles. Circular canal absent. Radial canals branch profusely and extend into rhopalia and tentacles are present. It produces burning sensation.

RHIZOSTOMA

Phylum	—	Coelenterata
Class	—	Scyphozoa
Order	—	Rhizostomae
Genus	—	*Rhizostoma*

It is an inhabitant of shallow water in tropical and subtropical zones, sometimes, in temperate zones. These are also found in Indo-Pacific Region and North Carolina. The colour varies. The colour of umbrella is pale green with a deep reddish margin, arms bright blue. The marginal tentacles are absent. The original four arms become

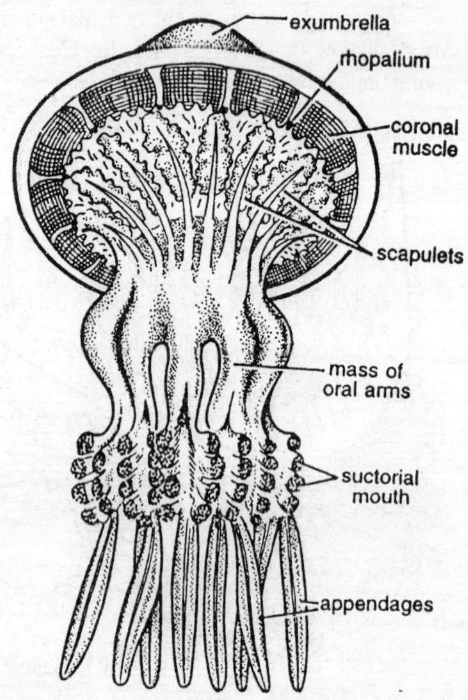

Fig. 1.22. Rhizostoma.

divided longitudinally into eight. The mouth is one is young but it is replaced by numerous small "sucking mouths" in adult which lie along the lips. Lips act as organs for external digestion. It feeds upon small animals like fishes etc. The mesogloea of the medusae of *Rhizostoma* is neither gelatinous nor mucilaginous. The entire medusa contains 91 to 96% water content. The umbrella is about 2 feed in length and specimen is about 4 feet.

EDWARDSIA

Phylum	—	Coelenterata
Class	—	Anthozoa
Subclass	—	Hexacorallina
Order	—	Actiniaria
Genus	—	*Edwardsia*

It is a small, solitary, marine animal, burries in sand. It is found in U.S.A., Southern California and North of Cape Cod. Body is

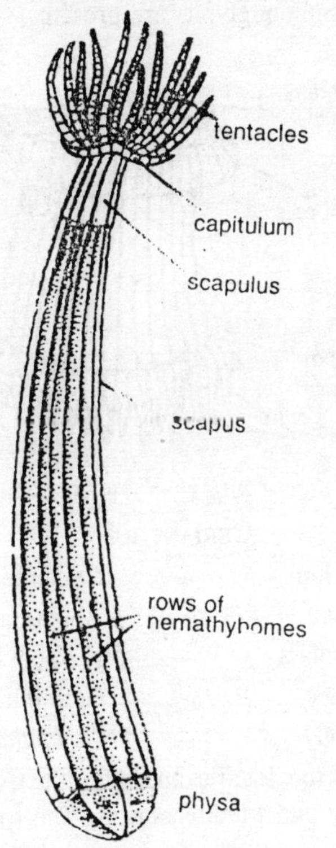

Fig. 1.23. Edwardsia.

elongated and is differentiated into oral disc and column. Oral disc is short and carries centrally placed mouth and a circlet of 16 tentacles arranged in two rows. The cylindrical column has three parts, the *capitulum*, *scapulus* and *scapus*. The body surface has 8 longitudinal ridges and the posterior half of the scapus contains rows of nematocysts called as *nemathybomes*. The basal part or *physa* is demarcated by limbus from scapus. Siphonoglyphs and mesenteries are 8 in number are present. Septa are in primitive condition consisting of 8 macrosepta and morc than 4 microsepta. The young of *Edwardsia* is parasitic on Ctenophore.

MINYAS

The systematic position is same as that of *Edwardsia*. *Minyas* is a colonial, pelagic floating sea-anemone. It is found in tropical regions and most abundant in shallow and costal waters. It is a blue coloured anthozoan. The oral disc bears a large number of tentacles. The column is short. Skeleton is absent. The basal disc contains an air-filled sac, due to which the animal floats. The mesenteries are arranged in the multiple of six. The siphonoglyphs are present.

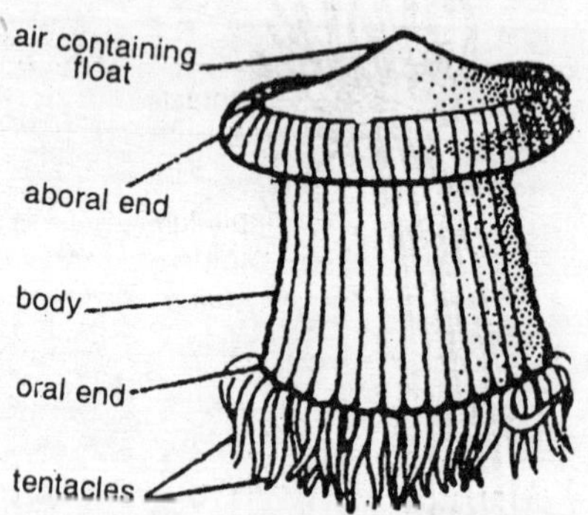

Fig. 1.24. Minyas.

CERIANTHUS

Phylum	—	Coelenterata
Class	—	Anthozoa
Subclass	—	Hexacorallina
Order	—	Ceriantharia
Genus	—	*Cerianthus*

It is found along the Pacific and Atlantic Coasts, Cape Cod to Florida, Bay of Fundy and Mediterranean. The form resembles a sea

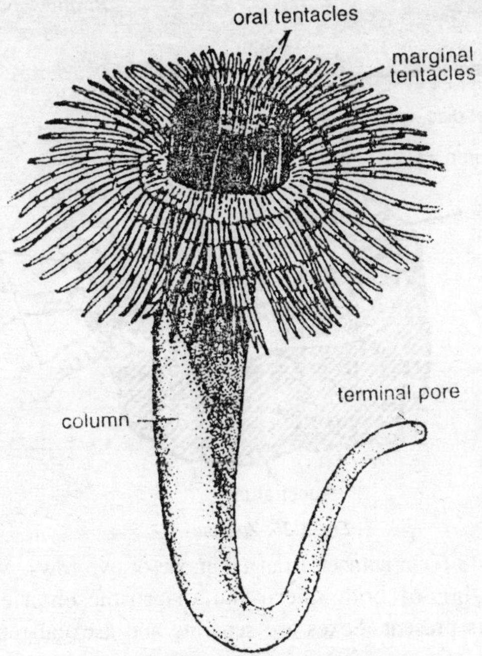

Fig. 1.25. Cerianthus.

anemone and has cylindrical elongated body but without pedal disc. The oral region has numerous tentacles about 130 arranged in 2 rows, the marginal whorl of tentacles and inner circlet of tentacles. The lower end is round and has terminal pore. Pharynx has only one siphonoglyph. The ectoderm secretes in long tube of mucus in which the animal lives and the tube is fixed in soft, sandy or muddy sea bottom. On the tube sand particles are deposited so it becomes hard.

ZOANTHUS

Phylum	—	Coelenterata
Class	—	Anthozoa
Subclass	—	Hexacorallina
Order	—	Zoantheria
Genus	—	*Zoanthus*

It is a colonial form found attached to rocks or corals, sponges, shells of molluscs etc. It is found in West Indies. Body is differentiated into column and oral disc having mouth and marginal tentacles. Cylindrical, small polyps arise singly from a network of stolon. Tentacles

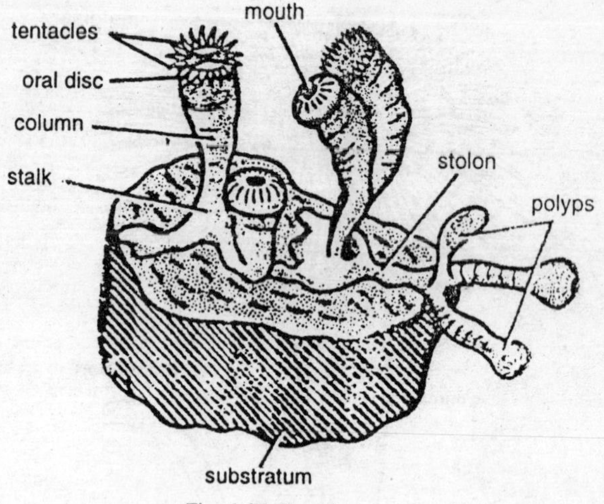

Fig. 1.26. Zoanthus.

are numerous, 48-60 in number arranged in one or two rows. Mesenteries paired consisting of both micro-and macro-mesenteries. Single siphonoglyph is present. Sexes are separate and asexual reproduction by budding.

MADREPORA

Phylum	—	Coelenterata
Class	—	Anthozoa
Subclass	—	Hexacorallina
Order	—	Madreporaria
Genus	—	*Madrepora*

It is a colonial, symbiotic and marine coral found in West Indies and Florida. It is commonly called as '*horn coral*' and plays an important role in coral-reef formation. Colony is branched with numerous, small crowded polyps in elevated cylindrical cups separated by coenosteum. Terminal polyps contain 6 tentacles and lateral polyps with 12 tentacles. Internally mesentries are bilaterally arranged and coenosarc contains a network of canals. Sometimes small crustaceans are found in association with horny corals.

MEANDRINA

The classification is the same as that of *Madrepora*. Colonial form inhibiting the coastal waters of West India, Asia and America. It is popularly known as *brain-coral* since resembles the human brain due to the presence of long curved grooves and ridges. The colony is

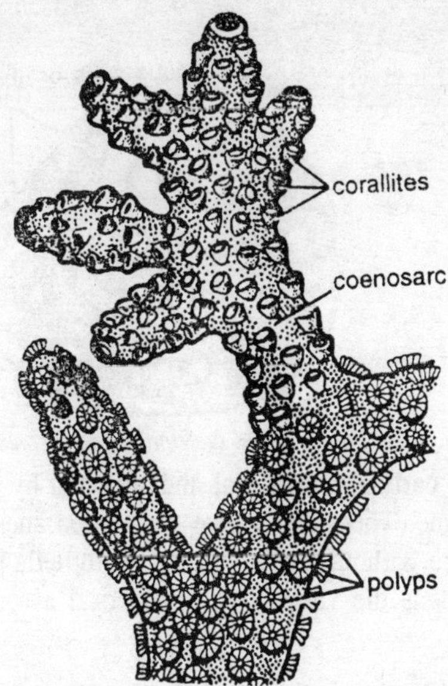

Fig. 1.27. Madrepora.

large and is composed of lime stone secreted by ectoderm. Polyps bear separate mouth, tentacular fringe, septa and mesenteries. Polyps remain confluent in the living condition.

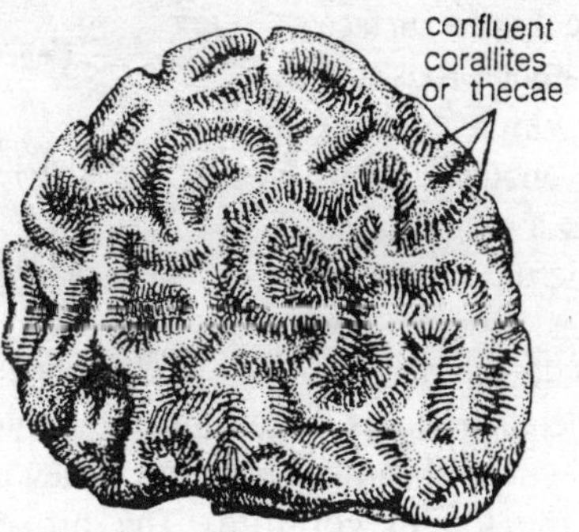

Fig. 1.28. Meandrina. Dried skeleton.

Favia

The classification is the same as that of *Madrepora*. It is commonly found in Florida and West Indies. *Favia* has a compact and huge colony produced by buds. The surface of the colony has closely placed polygonal cases or cups. The cups are so near to each other as to have common boundaries. The skeleton which is very hard, is made

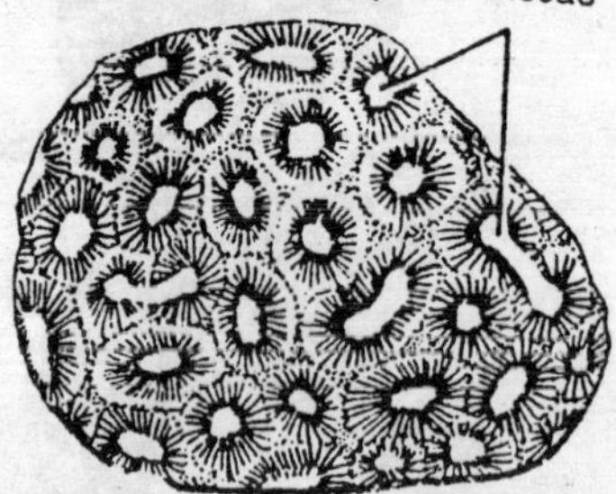

Fig. 1.29. Favea. Dried skeleton showing continuous cups.

up of calcium carbonate material and secreted by ectoderm for support of delicate tissue. The polyp is like that of sea anemones which contract in cups and are without siphonoglyph. Columella present. The *Favia* is aporose corals as the various parts of coral are solid and stony.

FUNGIA

The classification is the same as that of *madrepora*. It is a solitary and marine coral found in warm sea, generally in Gulf of California. It is commonly called '*mushroom coral*'. The coral is flat and discoidal, the theca is confined to the lower surface and small calcareous rods and synapticulae, which connect septa with one another. Adult animal contains a single large polyp with numerous tentacles. Siphonoglyph is absent. The life-history includes planula larva which metamorphoses into adult. Reproduction by monodisc strobilation which are plate-like structures. They are separated off one by one from the top of corallite. The broken discs are called *anthocyanthus* and develop into adults. The remaining broken part is called *anthocaulus* which grows again into new disc and is set off.

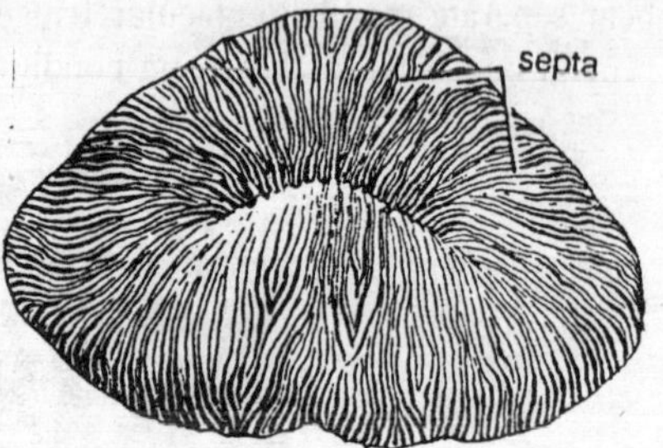

Fig. 1.30. Fungia. Skeleton.

FLABELLUM

The systematic position is the same as that of *Madrepora*. It is a solitary coral, remains embeded in the sand or remains attached to some object. The corallite forms a short conical cup or *theca*. The theca is laterally compressed. It consists of dense stony calcium carbonate, rough and brownish externally but smooth and white internally. The tapering base forms a short and fragilc stalk for peduncle which is attached in young but may become detached when adult. The theca is covered by a calcarious layer called *epitheca*. Several radiating partations (Septa) proceed from the inner surface of theca towards the axis of the cup. These are primary, secondary and tertiary, like the mesenteries. The primary septa meet in the centre forming the *columella* which is an irregular mass.

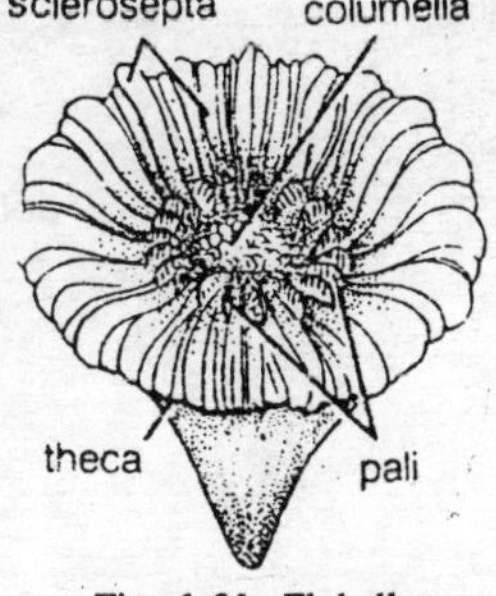

Fig. 1.31. Flabellum.

Astrangia

The classification is the same as that of *Medrepora*. It forms small encrusting colonies and found on rocks and shells—*Astrangia* is found along the coasts of Atlantic and America. Colony consists of theca and polyps. Polyps are white pinkish or greenish in colour. Zooids are more or less isolated with six septa of the first cycle, six smaller ones of the second and fourth cycles. Mouth is present on the oral disc, which is surrounded by double rings of tentacles. Bigger polyps have three cycles of tentacles mostly 12 larger tentacles alternating with 12 smaller tentacles. It feeds on crustaceans and small fishes.

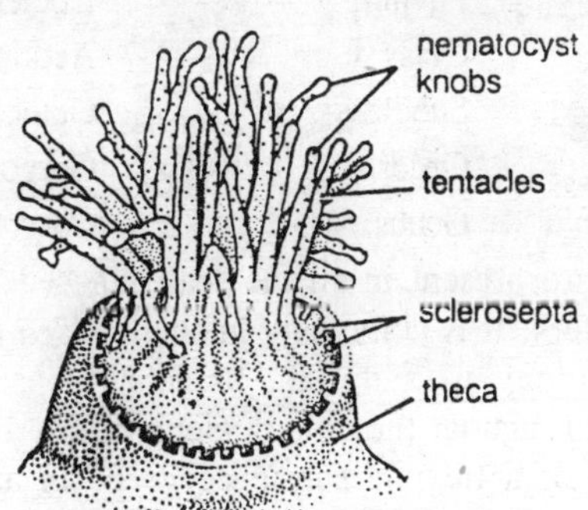

Fig. 1.32. Astrangia.

Dendrophylia

The systematic position is same as that of *Madrepora*. It is a marine, colonial and compound coral. It is found in tropical and subtropical waters. It appears to be a tree-like anthozoan. The corallites (Polyps) do not lie in close contact with each other but are tree-like. The corallites arise from a common calcareous stem or coenechyme. The stem is formed by calcification of coenosarc. Each polyp possesses

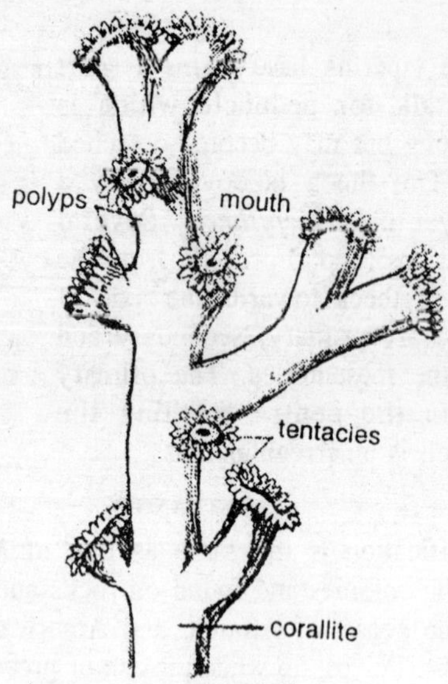

Fig. 1.33. Dendrophylia.

a centrally placed mouth which is surround by a whorl of small tentacles. Reproduction by budding.

GORGONIA

Phylum	—	Coelenterata
Class	—	Actinozoa
Subclass	—	Octocorallina
Order	—	Gorgonacea
Genus	—	*Gorgonia*

There are present in all the seas attached to rocks and stones in shallow waters. It is commonly known as '*sea-fan*'. It is yellowish or brownish colony with upright branches consisting of an axial rod extending throughout the colony along all the branches. The skeleton is made up of a flexible substance *gorgonin* and calcareous spicules embedded in mesogloea. Polyps or anthrocodia are retractile and present in rows on both the sides of the branches. Sexes are separate. The dried skeletons of the colony are used in decorative art.

CORALLIUM

The classification is the same as that of *Gorgonia*. It is a colonial marine animal commonly found in Mediterranean at a depth of 10-30 fathoms, Cape Verde Island, Japan, Italy etc. It is commonly called

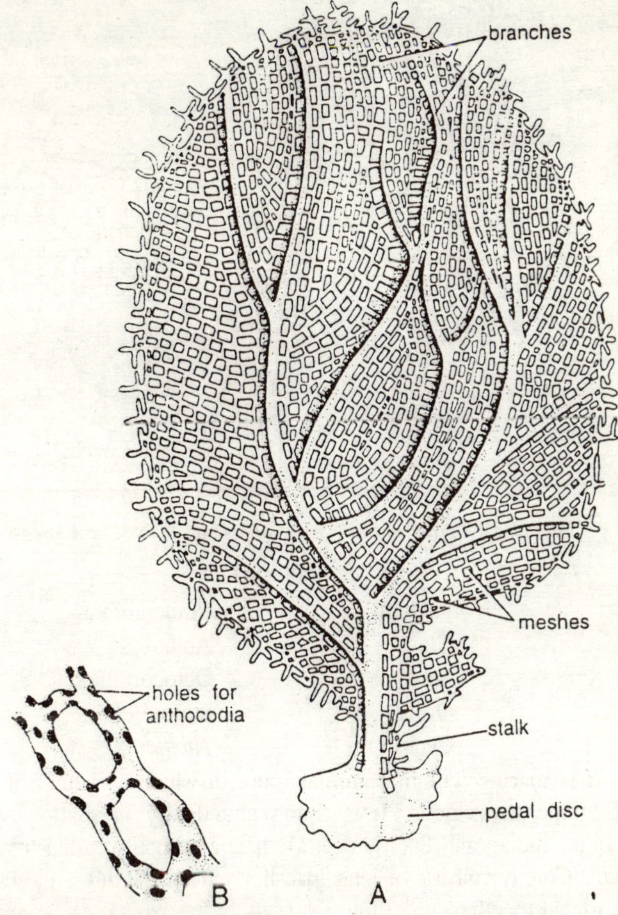

Fig. 1.34. Gorgonia. A—Entire colony; B—A portion (magnified).

'*red coral*'. The colony is upright and branched and is supported by calareous axis of fused spicules. The colony is dimorphic as there are two type of zooids (i) nutritive zooids, the *autozooids* and (ii) for circulation of water in the colony, the *siphonozooids*. The zooids are white in colour and the colony is dark red due to the prepesence of red calcareous spicules. Gonads are borne by siphonozooids. Planula develops inside the zooids and so the viviparity is found in *Corallium*. The *Corallium* is of great economic importance as the precious red coral of commerce known as red moonga in N. India which is obtained from this animal.

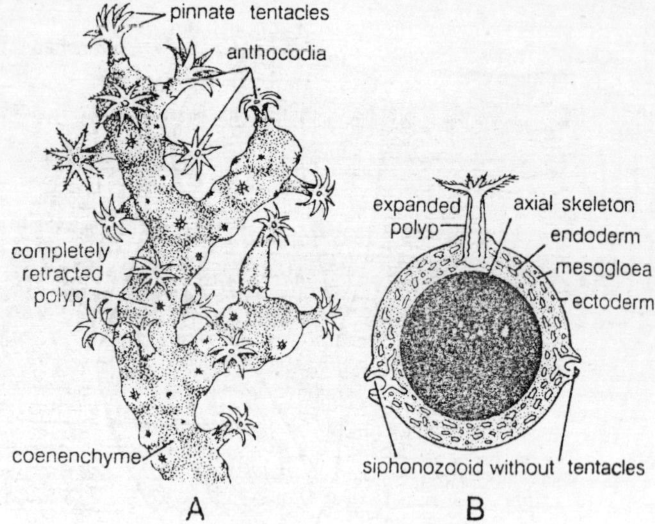

Fig. 1.35. Corallium. A—Portion of colony; B—T.S. of a branch.

TUBIPORA

Phylum	—	Coelenterata
Class	—	Anthozoa
Subclass	—	Octocorallina
Order	—	Stolonifera
Genus	—	*Tubipora*

It is marine, colonial animal found in shallow waters of tropical and temperate regions. It is found abundantly in shallow water of Atlantic, Indian and Pacific oceans. It is commonly called '*organ pipe coral*'. Colony consists of long, parallel and upright tubes closely fitted, and joined together at definite intervals by horizontal calcareous tubes. The polyps are bright green and secrete these tubes. The skeleton is internal and is covered by ectoderm in living condition. The mesogloeal spicules become closely fitted together and form a continuous tube for each polyp. Asexual reproduction by peculiar budding. The base of original polyp expands from which new polyps originate.

TELESTO

Phylum	—	Coelenterata
Class	—	Anthozoa
Subclass	—	Octocorallina
Order	—	Telestacea
Genus	—	*Telesto*

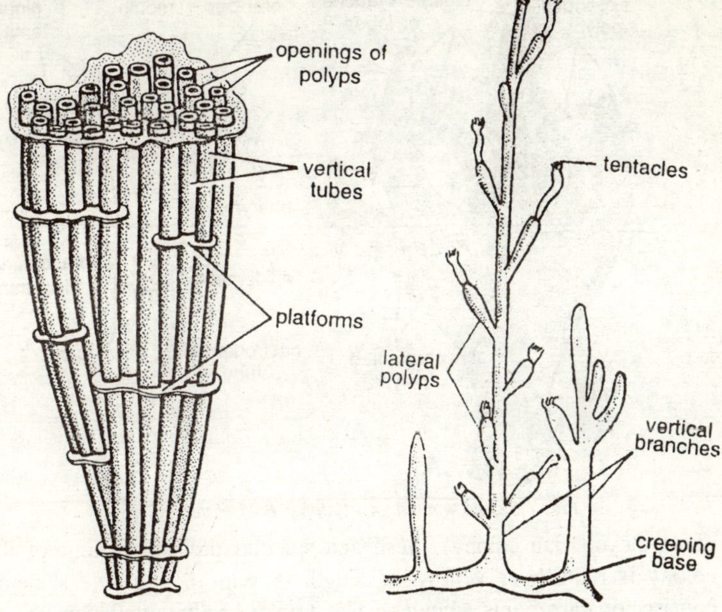

Fig. 1.36. Tubipora musica. *Fig. 1.37. Telesto.*

It is a sedentary, marine and colonial coelenterate. It is found along Atlantic coast. The colony comprises simple or branched stem, which arises from a creeping base. The stem bears lateral hydranths, which are arranged alternately. The stem is formed by the elongation of a single polyp and the lateral originate by way of solenial networks. The spicules may be somewhat united by calcareous and horny secretions.

ALCYONIUM

Phylum	—	Coelenterata
Class	—	Anthozoa
Subclass	—	Octocorallina
Order	—	Alyconacea
Genus	—	*Alcyonium*

Alcyonium or *Dead man's finger* is found attached to stones or rocks in sea water. It has cosmopolitan distribution, but mostly found in temperate and cold water of seas. Commonly found in long Island and St. Lawrence. Colony of short and thick leathery lobes attached to stones. Long polyps with tentacles except those of the outer end, can be contracted and everted. *Coenechcyme* (mesogloea or soft parts

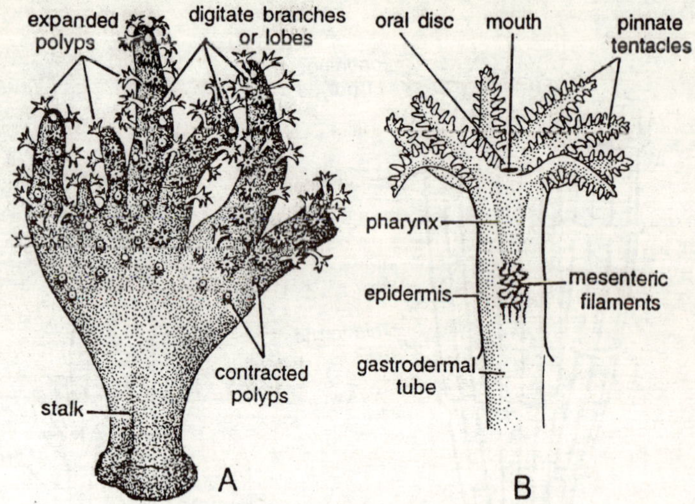

Fig. 1.38. Alcyonium. A—Colony; B—Isolated polyp.

of an alcyonarian colony), flesh and spicules present. Colour of the colony is reddish or yellowish; length is 4 to 10 cm. The skeleton consists of calcareous spicules. The food (*A. digitatum*) consists of various fine fragments of muscles of fish but they may reject many kinds of fish ova (according to *Pratt*). Gland cells occur in stomodaeum and it is probable that they secrete a fluid for digestion. The polyps are arranged on the upper part of the colony, and the lower part is sterile.

Hartea

The systematic position is same as that of *Alcyonium*. It is a marine solitary anthozoan found in shallow water. The body is elongated, cylindrical. The anterior end possesses eight pinnately branched tentacles. Skeleton is simplest, minute irregular deposits of calcium carbonate called spinules are deposited in the mesogloea. Mouth is surrounded by the tentacles. Reproduction by budding. It has a simple organisation and is supposed to be persistant larval form.

Heliopora

Phylum	—	Coelenterata
Class	—	Anthozoa
Subclass	—	Octocorallina
Order	—	Coenothecalia
Genus	—	*Heliopora*

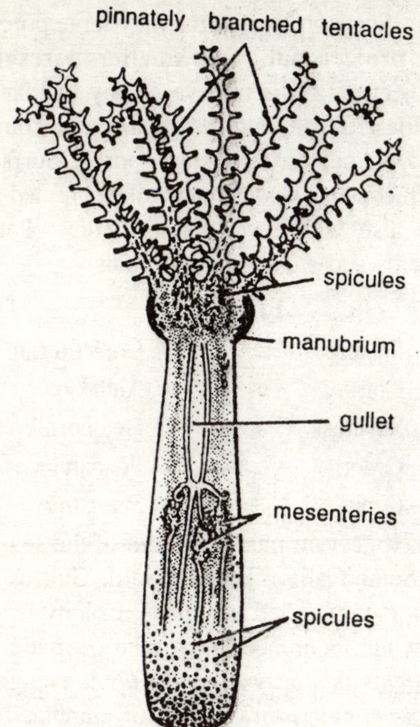

Fig. 1.39. Hartea.

It is a marine, sedentary coral found on coral reefs in the Indo-Pacific ocean. It is commonly called as '*blue coral*' because of its bright blue skeleton. The colour is due to iron salts. Its massive skeleton, composed not of spicules but of crystalline fibres of arragonite

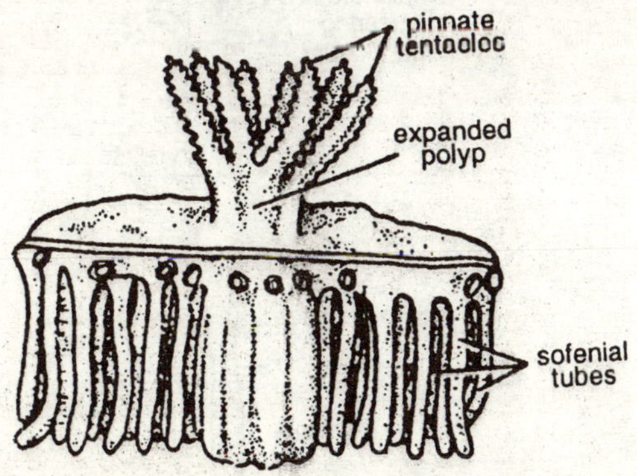

Fig. 1.40. Heliopora.

fused into lamellae. Skeleton is perforated by large pores through which ordinary polyps project out, and smaller pores through which siphonozooids project out. It is transversed by tubular cavities of two sizes, larger cavities-few in number open out through large pores. Smaller cavities open out through small pores. Surface contains flat coenechyme, which contains solonial network connected to middle region of the polyp and also with erect solonial tubes. Polyps dimorphic, large ordinary zooids and small siphonozooids.

PENNATULA

Phylum	—	Coelenterata
Class	—	Anthozoa
Subclass	—	Octocorallina
Order	—	Pennatulacea
Genus	—	*Pennatula*

It is found fixed deep in muddy bottom of the sea. It is commonly found in Europe, South California Northward, Gulf of St. Lawrence to Carolina. It is commonly called '*sea pen*'. Colony is divided into two parts : (a) stalk, which is embedded in sand or mud and (b) an upper part, called the *rachis* or *horny axis*. *Rachis* is elongated with paired lateral leaves or *pinnulae*. Lateral leaves or pinnulae long, from 20 to

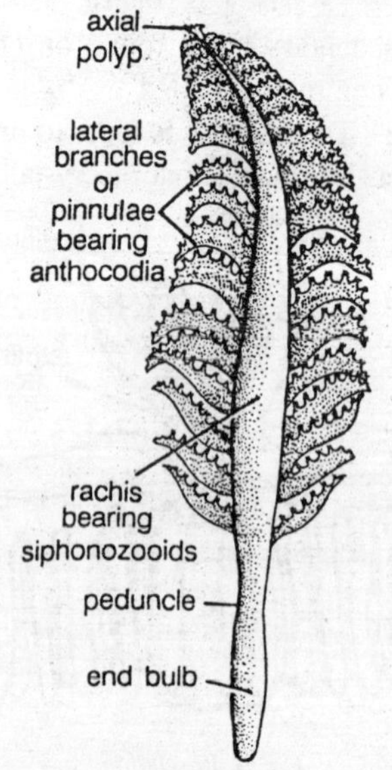

Fig. 1.41. Pennatula. Entire colony.

25 in number on each side. Polyps, arranged on horny axis, are of two types : (a) *Autozooids*, whose function is nutritive, are situated side-by-side as regular lateral branches giving an appearance of a feather. (b) *Siphonozooids*, which maintain circulation of water in the colony, are situated on the back of the axis.

Renilla

The classification is the same as that of *Pennatula*. It is a marine, colonial form found in shallow waters of Carolina Coast, California Coast and West Indies. Colony composed of a small *peduncle* which lies embedded in the mud and a kidney-shaped *rachis*. *Rachis* has a broad ventral surface devoid of polyps and a dorsal surface covered with autozooids and siphonozooids. *Autozooids* or *anthocodia* lie scattered on the dorsal surface are nutritive in function. *Siphonozooids* are arranged in groups and maintain the circulation of water within the colony. A median bare 'track' devoid of polyps extends from the peduncle to about the middle of rachis, where it terminates at a special exhalent siphonozooid. Axial skeleton absent.

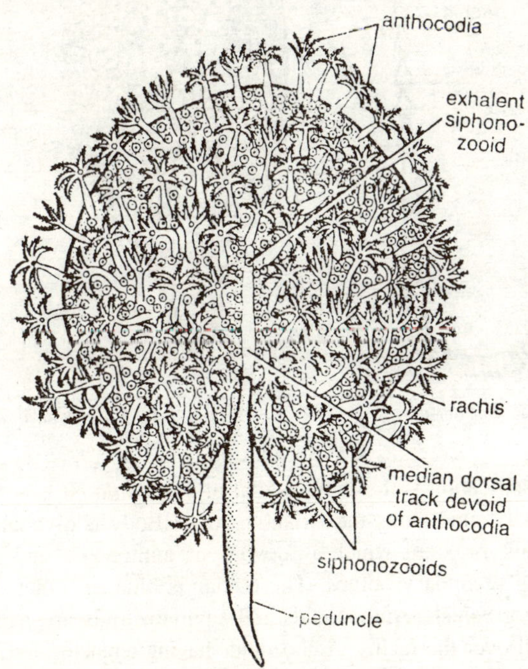

Fig. 1.42. Renilla.

Virgularia

The classification is the same as that of *Pennatula*. *Virgularia* or "walking stick" is found partly embedded in warmer water having a muddy and soft bottom. It is found along Pacific Coast. The rachis is long and elongated like a walking stick. The elongated body is divisible into two parts, a *stalk* and the *rachis*. The polyps arranged in transverse rows at regular intervals and are slightly fused and arise from the *rachis*. The stalk is without polyps. The stalk is thrust into the muddy bottom. The distal rachis stand straight outside water.

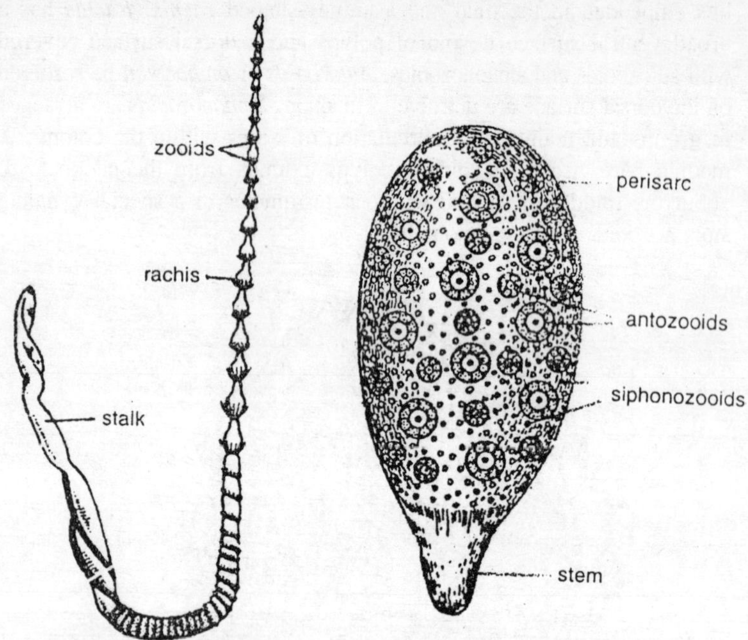

Fig. 1.43. Virgularia.

Fig. 1.44. Cavernularia.

Cavernularia

The classification is the same as that of *Pennatula*. It is a colonial, marine animal found along the Atlantic Coast. Body is divisible into a proximal stalk or stem which is devoid of anthocodia and a distal rachis having secondary polyps. The colour is blue or violet, may be yellowish or greenish. Anthozooids and siphonozooids are irregularly distributed all over the rachis. Anthozooids having tentacles and gonads. Coenosarc is fleshy and traversed by coenosarcal canals and contains long spicules. It is a bioluminescent, emits light.

2

First Appearance

During the exploration of the oceans it had become clear that reefs are distributed into three main types: fringing reefs growing close to a higher land, barrier reefs separated from the land by a lagoon, and atolls consisting of a coral rim around a lagoon without any land in it other than low islets. Other types are variations on these three main types, with the possible exception of ridge reefs. Fringing reefs seemed simple in origin, but barrier reefs and atolls require particular explanations for their genesis. The theories that have been proposed will be reviewed in this chapter, more or less in historical order, with more detail on the more recent ones, since they take account of modern research and techniques which were not available in the nineteenth and the first part of the twentieth century.

Darwin's theory, completed by DANA and development by Davis, As an explanation of fringing Reefs, Barriers, and Atolls

Darwin's theory is one of the most famous geomorphological theories ever proposed, and it may be interesting to recall how it was developed. As pointed out by Stoddart (1962b), the idea that fringing reefs became converted into barrier reefs, and then into atolls, by slow subsidence of an island foundation, did not arise from the observation of Pacific and Indian Ocean coral reefs during Darwin's voyage around the world on the *Beagle* (1832-1836). In his Autobiography, Darwin wrote: 'No other work of mine was begun in so deductive a spirit as this: for the whole theory was thought out on the west coast of South America before I had seen a true coral reef. I had therefore only to verify and extend my views by a careful examination of living reefs. But it should be observed that I had during the previous years been incessantly attending to the effects on the shores of South America of the intermittent elevation of the land, together with the denudation and the deposition of sediment. This necessarily led me to reflect on the effects of subsidence, and it was easy to replace in imagination the continued deposition of sediment by the upward growth of coral. To do this was to form my theory of the formation of barrier reefs, and atolls.

The observations themselves were made initially in November 1835, when sailing across what was then called the Low or Dangerous Archipelago, now the Tuamotus, and soon after visiting the Society Islands, where the examined particularly Tahiti and Eimeo, now Moorea. From making these observations, he wrote: 'If we imagine such an island, encircled by a coral reef separated from the shore by channels and basins of still water, after long successive intervals to subside a few feet, in a manner similar, but with a movement opposite to the continent of South America; the coral would be continued upwards, rising from the foundation of the encircling reef. In time the central land would sink beneath the level of the sea and disappear, but the coral would have completed its circular wall. Should we not then have a Lagoon Island (i.e. an atoll)? Under this view, we must look at a Lagoon Island as a monument raised by myriads of tiny architects, to mark the spot where a former land lies buried in the depths of the ocean.'

He next examined the Cocos or Keeling Atoll in the eastern Indian Ocean, which was considered by him as type of atoll, and prompted him to expound his theory in the first chapter of his book as follows: 'Let us in imaginative place within one of the subsiding areas, an island surrounded by a "fringing reef"—that kind, which alone offers no difficulty in the explanation of its origin. Let the unbroken lines and the oblique shading in the woodcut represent a vertical section through such an island; and the horizontal shading will represent the

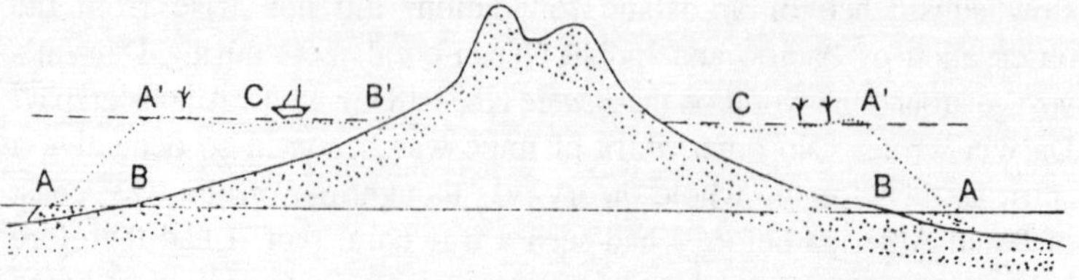

Fig. 2.1. Figure showing the formation of barrier reefs and atolls.

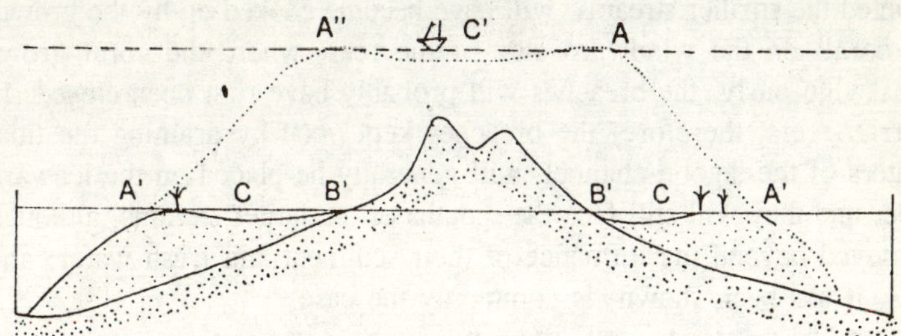

Fig. 2.2. ***A'A: outer edge of the barrier reef at the level of the sea. The coconut trees represent coral islets formed on the reef; CC: the lagoon-channel; B'B: the shores of the island, generally formed of low alluvial land and of coral detritus from the lagoon-channel; A"A": the outer edges of the reef now forming an atoll; C': the lagoon of the newly formed atoll.***

section of the reef. Now, as the island sinks down, either a few feet at a time or quite insensibly, we may safely infer from what we know of the conditions favourable to the growth of coral, that the living masses bathed by the surf on the margin of the reef, will soon regain the surface. The water, however, will encroach, little by little, on the shore, the island becoming lower and smaller, and the space between the edge of the reef and the beach proportionally broader. A section of the reef and island in this state, after a subsidence of several hundred feet, is given by the dotted lines: coral islets are supposed to have been formed on the new reef, and a ship is anchored in the lagoon-channel. This section is in every respect that of an encircling barrier-reef; it is, in fact, a section taken E and W through the highest point of the encircled island of Bolabola. The same section is more clearly shown in the following woodcut by the unbroken lines. The width of the reef, and its slope both on the outer and inner side, will have been determined by the growing powers of the coral, under the conditions (for instance the force of the breakers and of the currents) to which it has been exposed; and the lagoon-channel will be deeper or shallower, in proportion to the growth of the delicately branched corals within the reef, and to the accumulation of sediment, relatively, also, to the rate of subsidence and the length of the intervening stationary periods.

'As the space between the reef and the subsiding shore continued to increase in breadth and depth, and as the injurian effects of the sediment and fresh water borne down from the land were consequently lessened, the greater number of the channels, with which the reef in its fringing state must have been breached, especially those which fronted the smaller streams, will have become choked up by the growth of coral: on the windward side of the reef, where the coral grows most vigorously, the breaches will probably have first been closed. In barrier-reefs, therefore, the breaches kept open by draining the tidal waters of the lagoon-channel, will generally be placed on the leeward side, and they will still face the mouths of the larger streams, although removed beyond the influence of their sediment and fresh water; and this, it has been shown, is commonly the case.

'Referring to the following diagram, in which the newly-formed barrier-reef is represented by unbroken lines, intend of by dots as in the former woodcut, let the work of subsidence go on, and the doubly-pointed hill will form two small islands (or more, according to the number of the hills) included within one annular reef. Let the island continue subsiding, and the coral-reef will continue growing up on its own foundation, whilst the water gains inch by inch on the land, until the last and highest pinnacle is covered, and there remains a perfect atoll'.

Initially accepted with admiration, Darwin's theory, which received Dana's agreement, was subject to many discussions and controversy at the end of the nineteenth century, and various theories were presented at that time, summarized by Davis (1928) as follows:...whether the visible reefs have been formed by upgrowth as still-standing, aggraded submarine banks, as Rein first and Murray later suggested and as Wood-Jones still more recently argued; by outgrowth from still-standing foundations, the lagoon being excavated by solution back of the living reef, as Murray also proposed; by outgrowth on rising foundations, as Semper and Guppy believed, Semper having anticipated Murray in explaining lagoons by solution; by upgrowth on still-standing foundations completely truncated by abrasion, as Wharton proposed; or incompletely truncated, as Tyerman and Bennet long ago and as Guppy later suggested; or on platforms produced around still-standing foundations by subaerial erosion and submarine denudation, as Agassiz believed; or on still-standing foundations abraded by the lowered Glacial ocean and submerged by the rise of the Postglacial ocean, as Daly argues—for all these theories succeed in explaining the visible features of reefs that they were invented to explain. It is with regard to the invisible conditions and processes of the past that the theories differ.'

It must be pointed out that Darwin knew well, and said in his book, that many reefs around the world needed other explanations, generally complementary to his own general theory. He noted that in several areas there are elevated coral reefs. For example, at Savage Island, southeast of the Friendly Group (now called the Tonga Islands) (1842), 'an ancient lagoon is now represented by a central plain: here we cannot doubt that the elevatory forces have recently acted. The same conclusion may be extended, though with somewhat less certainty, to the islands of the Friendly Group...The surface of Tongatabou is low and level, but with some parts a hundred feet high; the whole consists of coral rock...On the New Hebrides, Mr. G. Bennett informs me he found much coral at a great altitude, which he considered of recent origin...In the East Indian archipelago, many authors have recorded proofs of recent elevation.' He accepted evidence on other raised reefs, as in Mauritius, Timor, New Guinea, and the Sandwich Islands (now Hawaii).

Concerning Tongatabou Island, Darwin's view has been supported by more recent and more accurate work. Taylor and Bloom (1977) described terraces at 40-80 metres and 110-140 metres, indicating uplift and tilting on this island. The uplift has occurred at an average rate of at least 3.3 metres/100 ky in post-Miocene times, but has now creased, or become very slow. In many respects, it would be wise for modern scientists to read Darwin's book carefully, for they would find many facts and opinions which are often presented as recent discoveries: for example, the fact that passages through barrier reefs and atoll rims occur on the leeward side, but are closed on the windward side, where they are thought to have formerly existed, is mentioned by Darwin (1842).

In the Red Sea also, it will be shown later in this book that essential features of coral reefs were mentioned by Darwin in 1842 on the basis of informations he had collected from sailors. It is curious, as Davis (1928) has pointed out, that Darwin failed to cite the embayed shorelines of islands surrounded by barrier reefs in support of his theory of subsidence. It was Dana who drew attention to that evidence, and so he must be mentioned with Darwin as a pioneer of the explanation of barrier reefs and atolls. Dana explored the Pacific reefs in 1839, just four years after Darwin, as a member of the United States expedition under Captain Wilkes. The radial valleys of Tahiti suggested to him that 'sunk to any level above that of five hundred feet, the erosion-made valleys of Tahiti would become deep bays, and above that one of one thousand feet, fiord-like bays, with the ridges spreading in the water like spider's legs.' It is also surprising that, in spite of the fact that Dana's ideas in this respect were documented in the three successive editions (1872, 1874, 1890) of his classic book *Coral and Coral Islands*, the value of embayments as evidence of subsidence was not appreciated by other scientists interested in coral reefs, such as Guppy (1888) and Agassiz (1903b). 'Darwin—Dana—Davis, the three Ds. The subsidence theory of coral barrier reefs and atolls will always be associated with these three men. Darwin, the originator, Dana, the strengthener, and Davis, the skillful champion. It is a curious fact that all three were born on February 12th, Darwin in 1809, Dana four years later, and Davis in 1850.'

The antecedent-platform theory, which was proposed by Hoffmeister and Ladd (1944), may be mentioned here as an extension of Darwin's theory with inclusion of further data. It had evolved progressively through the work of authors like Wharton, Murray, Vaughan, Gardiner, Semper, Rein, Wood Jones, details of their discussions being given by Davis (1928). According to these ideas, corals would have grown on any basement providing favourable conditions: old volcanic islands truncated by abrasion; submarine volcanoes not truncated; deep submarine foundations of any origin, raised by sedimentation until they would become shallow enough to allow coral growth; or uplifted or down faulted rocky basements affording the same possibilities. Later on, reef building would have continued during slow subsidence, as in Darwin's theory, particularly in Indonesia and in the Central Pacific. The subsidence may be tectonic, or isostatic, or due to compaction in the deep-sea floor. Such an evolution would have not occurred in the Western Pacific area, where reefs have been commonly uplifted. These differences in evolution between parts of the Pacific were suggested at a time when the notion of sea-floor spreading had not yet been developed, but the idea proved, later on, to be fruitful and even premonitory.

Sea-Level Shifts and Their Effects on Reefs

When the effects of Pleistocene glaciations and deglaciations on sea-level were discovered (and some authors mentioned them long before the end of the nineteenth century), it appeared necessary to take account of them in explanations of the coral reef types, especially as the more elaborate types apparently required a long time to be built. But knowledge of the Quaternary shift in sea-level has been considerably improved in the course of time, as the knowledge of the growth of corals during the last, Holocene transgression. Daly is the outstanding name in the early phase of these studies. His contributions to the study of coral reefs in relation to seal-level changes extended over a long period. His basic idea concerned the formation of the large barrier reefs and atoll lagoons, and may be summarized as follows. The depth of these lagoons is thought to be generally nearly uniform and flat at approximately 70-80 metres. It is suggested that they were abrasion platforms, now covered by a veneer of debris, and calcareous algae; they would have been, perhaps not completely shaped, but at least levelled, during a low, interglacial sea-level.

The fall of sea temperature at that time would have killed the majority of the corals, and even if some of them survived, their growth was hindered by increased turbidity. As temperature rose in Late-Glacial and Post-Glacial times, corals would have spread from their Late Pleistocene sanctuaries, and recolonized the platforms, their ensuing upward growth accompanying the rising sea-level. Nugent (1946) published a description of the coral reefs in the Gilbert, Marshall and Caroline Islands. Which was in support of Daly's theory. Beside the basic fact that sea-level variations have occurred, which is undoubted, Daly's theory has been heavily criticized. The assumption of widespread influence of turbidity during low sea-level phases is very contestable. As has been said turbid waters are unfavourable to coral growth, but in several instances corals (or species of corals) are able to grow in such conditions.

Shepard (1963) quoted Ladd and Hoffmeister 'who have called attention to the existence of reefs off Rewa River in the Fiji Islands in spite of the tremendous amounts of silt brought to the sea... Kuenen also refers in Indonesia to floods covering a reef with silt without killing it'. He mentions similar cases in Hawaii and elsewhere. Even more important is the fact that the depth of lagoons is not as uniform as Daly assumed. Its diversity was already recognized when Davis wrote his book (1928) and has been checked by more recent work. Most soundings show that lagoon depths are often very irregular. The fine charts of four to the Marshall Islands, published by Emery et al. (1954), are quite clear in this respect. From Japanese surveys, Tayama (1952) has concluded that average depths of the lagoons are 27 metres in the Palau Group, 41 metres in the West Caroline Group, and 63 metres in the East Caroline Group.

In addition, there are some extremely shallow lagoons, as at Christmas Atoll, Line Islands, and Elizabeth Atoll, between Australia and new Caledonia, each of which having a lagoon only a few metres deep. By contrast, a number of lagoons with depths of 150 and even

200 metres have been mentioned in the Moluccas. Even if the depth of most atoll lagoons would be around 70-80 metres, this figure would not be indicative of low sea-level bevelling, since most recent estimates of sea-level lowering during the coldest period of the Wichselian (Wisconsinian, Devensian) have been around 125 to 130 metres. There have also been oscillations of sea-level during the Weichselian, with a temporary rise to slightly below present datum around 35000 BP with a deep fall afterwards. Similar shifts must have occurred during earlier glaciations. It thus seems unlikely that there have been sufficiently long stillstands at low levels to permit the bevelling envisaged by Daly. A more valuable result from recent research is that the lowerings of the sea-level during Pleistocene glaciations have often determined a karstification of preexisting coral structures and a subaerial dissection of these structures by rivers, hence a diversification of their topography.

At Mayotte Island in the western Indian Ocean, the floor of the western part of the lagoon between the barrier reef and the central volcano has several enclosed depressions with steep sides. 60-70 metres deep, lying at approximately 20 metres below the surrounding lagoon floor. The largest one measures 1800 by 2800 metres, and includes pinnacles 15 to 20 metres high. These features have been interpreted as the results of a subaerial karstification, the pinnacles being needles of a small Kegel Karst, possibly coated by thin coral growth added sine the postglacial transgression.

At Clipperton Atoll, off Mexico in the Eastern Pacific, the southeastern part of the lagoon includes a narrow pit with vertical sides, more than 36 metres deep, surrounded by shallow flats. Diving in this pit by Cousteau's team, and further measurements by Ehrhardt (1976), revealed that its water has a high content of hydrogen sulphide, an evidence for its confinement. It is undoubtedly a result of subaerial karstification.

Off the Sudanese coast in the Red Sea, the Harvey Reef, 14 kilometres SSE of Port-Sudan, which runs north and south and rises to the surface, is surrounded by enclosed depressions, 50 to 60 metres deep, their margins being only 15 to 20 metres in depth. The general pattern is extremely irregular and consists of a system of pinnacles and sharp ridges 150 kilometres north of Port-Sudan (same author), the Baraja Reef also includes enclosed depressions and irregular pinnacles which rise to within 15-20 metres of the water surface across the bank. These landforms point to a pre-Holocene dissection of previous reefs, the pits approaching perhaps 60 per cent of the area covered.

The floors of the pits are now blanketed with recent sediments, while the walls of the intervening ridges have become sites for reef growth.

Drowned features that are presumably karstic were also mentioned from near Port-Sudan by Schroeder and Nasr (1983). They noted that the basins or pits provided environments of deposition for very fine grained material, including a relatively high proportion of non-carbonate particles, and that such very low energy environments were located in close vicinity to reefs exposed to strong surf. In the Bahamas, which may be classified as a special type of coral reef, drowned karst has been described by several authors, especially on and around Andros Island, and other karstic features occur on reefs off Belize in the Caribbean. At Andros, numerous pits, known as 'blue holes', are circular submerged dolines, connected beneath the reef with the eastern submarine scarp of the Great Bahama Bank. Some are small, others reach 60 to 80 metres in diameter, sometimes with overhanging sides.

The Holocene transgression resulted in muddy sedimentation with algal components in the shallow dolines. Off Belize, fossil stalactites were observed in caves down to 122 metres. These caves showed horizontal enlargements of joint patterns, which form tunnels at depths of 12, 18, 28, 40, 45, 55 metres, considered to indicate stillstands of sea-level during the Pleistocene regression. Sand waves are evidence that strong currents are now moving sediments down to 55 metres in open spaces. The calcareous rocks of Late Cretaceous to Eocene age which outcrop on the continental shelf off Northern Florida are not former coral reefs, but they show submarine dissolution features in the form of sinkholes, which are interesting in comparison with those found in reefs, since they help to understand the subaerial weathering on various limestones during phases of low sea-level. These sinkholes are incised abruptly into the sea floor, particularly the Red Snapper Sink, which occurs 40 kilometres offshore at a depth of 27 metres, and has been investigated down to 100 metres.

The major period of solution occurred in the Late Oligocene and Early Miocene, but it continued during Pleistocene low sea-level phases. As in the Red Sea off Sudan, these holes are being in filled with shelf sands, indicating that they are no longer being shaped by solution processes. The effects of processes at work during low sea-level phases on the general shape of atolls, particularly in the Pacific, will be discussed later in this chapter. Beside the subaerial karstic effects of water circulation resulting in the sinkholes presently submerged, the phases of low sea-level led to the cutting of valleys now drowned

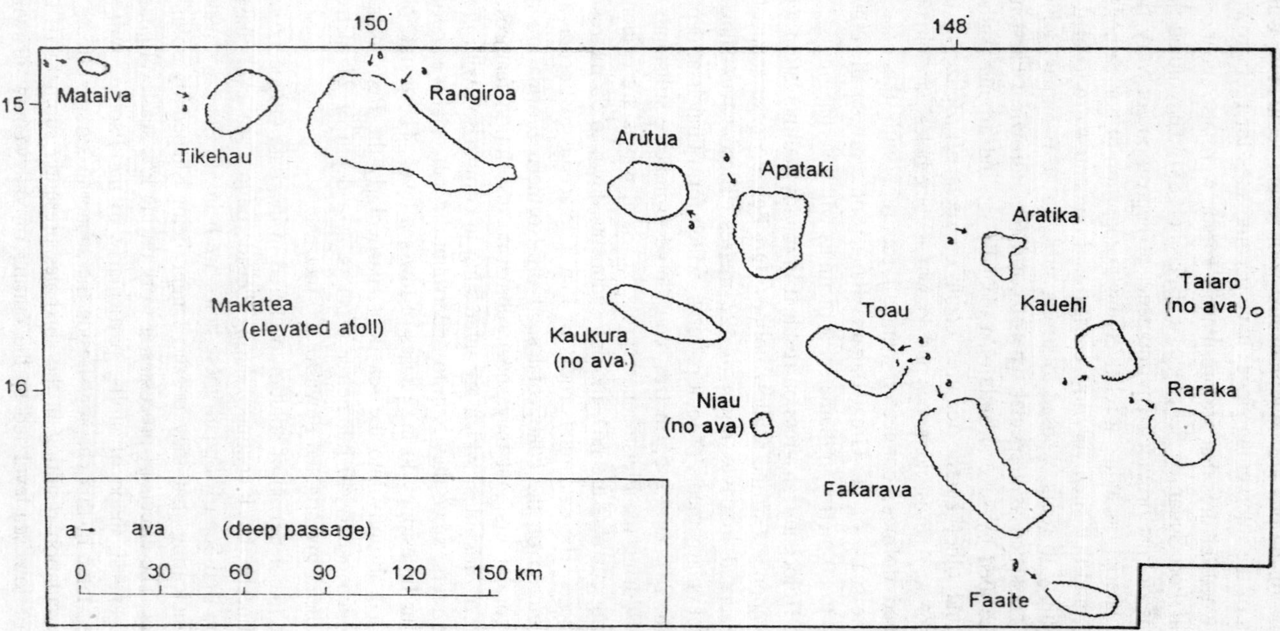

Fig. 2.3. Northwestern part of the Tuamotu Islands, showing the location of the avas. The avas are generally located on the leeward side of the atolls, except when one atoll is sheltered from the trade winds by another.

beneath lagoons and through atoll rims and barriers. Four such valley patterns were reported in the Mayotte lagoon, connected with passages across the barrier. The most remarkable one, in the northeastern part of the lagoon is 60 to 80 metres deep and ends in the Mzambourou Passage. Another one, 30 to 40 metres deep, is connected with the Longogori Passage in the east: this sinuous passage 4 kilometres long, only 300 to 400 metres wide, is 50 to 60 metres deep, and has only occasional, discontinuous sandy deposits on its bottom, and steep slopes on either side.

The unusual lack of sedimentation in such Pleistocene erosional landforms in the Mayotte lagoon seems due to strong scouring currents produced by the large tidal range, which reaches 4 metres at spring tides, a rather high figure for tropical countries. Similar features are found at Guadeloupe Island, in the Caribbean where passages across the barrier reef of Grand Cul-de-Sac Marin are connected with former valleys in the lagoon behind; although the depths are not so large as at Mayotte (35 metres across the barrier), the origin is the same.

The Tuamotu atolls in the Pacific are connected with the open sea by two types of passes: the shallow ones known as *hoa*(s) and the *ava*(s) which are much less numerous, deeper and navigable by ships. The latter are certainly remnants of fluviatile Pleistocene valleys that drained from the emerged lagoon into the lowered sea. They are generally located on the leeward side of atolls (or the barrier as in the case of Bora Bora in the Society Islands), in sectors not exposed to the trade wind swell, except when one atoll is sheltered by another, in which case they are found where the surf action is minimal. The location of the passages thus seems to be determined by swell, a feature independent of sea-level lowering.

The same leeward situation of passages may be seen in the Marshall and Caroline Islands, where they lie mostly in the southwest at the opposite side of the trade wind action in the northern hemisphere. Guilcher (1979b) thought that other valleys existed all around the atoll rims and barriers during low sea-level phases, but that the surviving ones e those which were sheltered from sedimentation which has closed the passages on the windward side. Such an explanation, previously suggested by Darwin (1842) on the windward side of the reef...the breaches will (probably have first been closed'), receives confirmation from recent high resolution seismic investigations carried out on the rim of Mururoa Atoll in the Tuamotu Islands (Ruzie and Gachon, 1985), which have discovered a series of small gorges cut into reef formations and subsequently filled by sand. Similar landforms are reported from frameworks other than atolls, such as those situated in the northern region of the Great Barrier Reef, where Orme (1978) and Orme *et. al.* (1977, 1978a,b) have used high resolution profiles to demonstrate the existence of submerged valleys, some infilled, while others remain open, without recent sedimentation, gives examples of both types, the open one in the inner approaches of Cook Passage between two outer reefs, where currents have kept the bottom free of sediment, while the infilled one is more sheltered, and has retained deposited material.

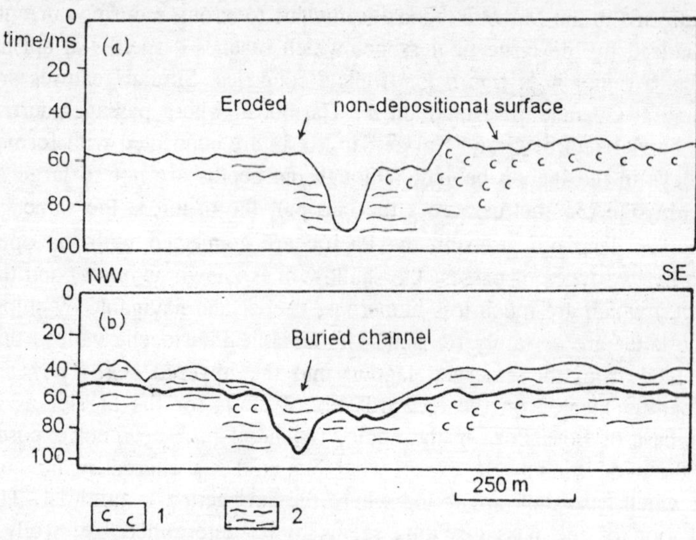

Fig. 2.4. Two NW-SE high-resolution seismic profiles across the Great Barrier Reef lagoon off Lizard Island.

Orme (1978.) noted that these submerged channels can be of various ages, but the fact stressed here is that they always bear testimony of former low sea-level phases. Sometimes the submerged channels have been overfilled by sediments, so that they stand as positive relief above the general bottom topography. This overfilling has happened in the central sector of the Great Barrier Reef, where these features have been attributed to thick backfilling by deltas during abandonment causing deposition in the lower reaches of the river. 'Hence modern shelf morphology may not be an accurate guide to the positions of Pleistocene channels across the shelf'. Another result of sea-level changes is the occurrence of *sharms* or *marsas*, quite frequent on the

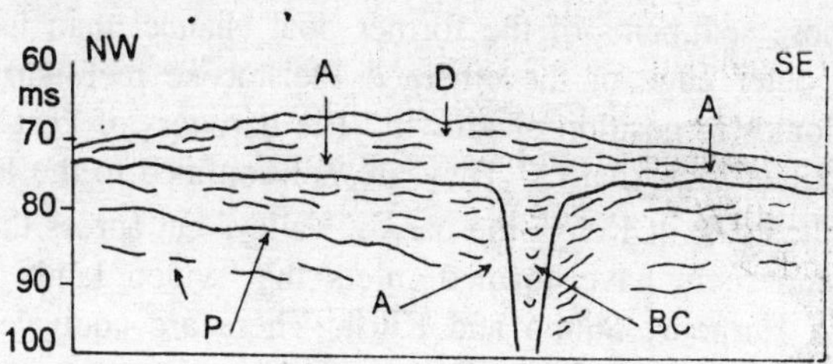

Fig. 2.5. Continuous seismic profile across the Great Barrier Reef lagoon off Townsville, showing a Holocene delta (D) built over a Pleistocene buried channel (BC). A: Pre-Holocene erosional surface; P: Older Pleistocene erosional surfaces.

Red Sea coasts of Saudia Arabia, Sudan, and the Sinai Peninsula in Egypt. A sharm is a long, narrow, deep inlet, sometimes meandering, with corals growing on its shores, and sediments washed in from the wadi(s) at the inner end: Sharm Abhor, close to Jeddah, is a fine example. Sharms are valleys that were incised by rivers during Pleistocene low sea-level phases, and may be considered as a special kind of ria formed in an arid coral environment. Various stages of infilling are found, and sometimes the only indication of the former sharm is a re-entrant in the fringing reef on the coast, as on parts of the Gulf of Aqaba or Elat. On the east coast of the Sinai Peninsula. In Sudan, where they have been cut into old, Peistocene fringing reefs, sharms often display inner widenings behind a narrow entrance, because the valley cut during the low sea-level phase was more easily carved

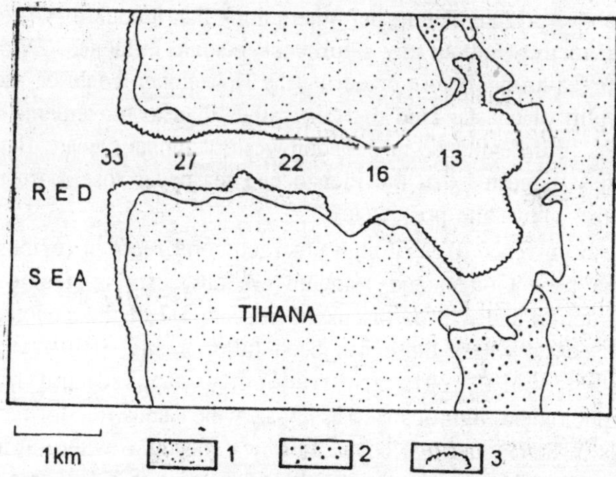

Fig. 2.6. Sharm Jubbah, on the Red Sea coast of Saudi Arabia at 27°30´N, drowned valley with Holocene coral growth on its sides. 1, Tihama; 2, alluvial deposits in wadis; 3, fringing coral reefs.

in the loose sediments of the former boat channel than in the more resistant outer edge of the emerged Pleistocene reef, structured by coral colonies in position of growth. The harbours of Port-Sudan and Suakin utilize such sharms. They are not confined to the Red Sea.

On the coast of Kenya the narrow valleys cut across the emerged Pleistocene reefs have formed inlets that widen landward, as at Mombasa Harbour, Shimo and Kilifi. These are equivalents of the Sudanese marsas, with a new fringing reef and boat channel added as a sequel to the Holocene marine transgression. In the Pacific, Erromango Island in Vanuatu has several sharms, particularly at Ipota drowned valley in Cook Bay on the east coast, which has been cut into an emerged coral reef lying at 3 to 4 metres above present sea-level. The oscillations of high and low Quaternary sea-levels, which had such important impacts on reef building and erosion, have been much studied in recent decades. Unfortunately, the facts appear more complicated in 1987 than it was though around 1940-1950. Concerning former high sea-levels, the penultimate (Holsteinian-Hoxnian-Yarmouth: the terminology differing according to countries), and the last one are the best known, but several earlier ones have occurred.

There were secondary, minor oscillations of sea-level during the Eemian, which has lasted at least 50,000 years, and the maximum level attained by the se in the Eemian is subject of much discussion. According to some authors, it would not have risen more than a few metres above present datum, but others think that it could have reached 15 or 20 metres without any significant tectonic influence. Evidences of higher sea-levels, before or during the Holsteinian, could be ascribed to the uplift of coastal land margins rather than to movements of the sea itself. From evidence in the southwestern Indian Ocean. Battistini (1972) has suggested that the Pleistocene sea never rose more than a few metres above the present level.

The occurrence of very high emerged coral reefs in regions such as Papua New Guinea and Vanuatu are obviously related to large tectonic uplifts and will be discussed in section 7 of this chapter. But even on coasts where there has been little, if any, uplift, emerged Pleistocene coral reefs are frequently exposed and have a geomorphological significance, as those which constrict former boat channels in Sudan and in Kenya. In New Caledonia, where a drilling has been made down to the pre-coralline basement at 226 metres below the present sea-level in the barrier reef at Grand Tenia, stratigraphic evidence of three successive reefs backed by sandy lagoonal deposits

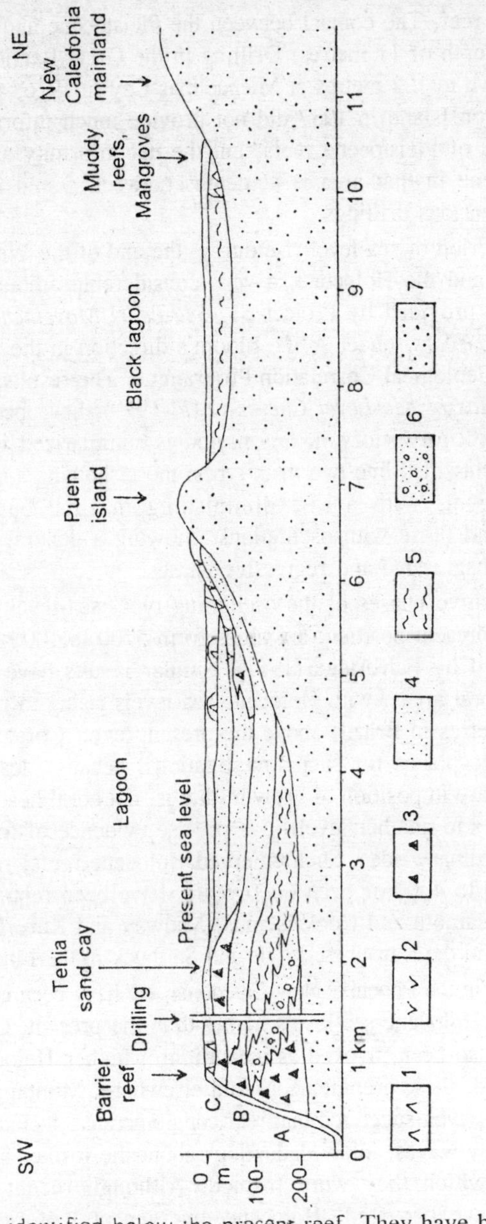

Fig. 2.7. Drilling through the New Caledonia barrier reef off Saint Vincent Bay necr Noumea, and the interpretative cross section. 1, Holocene reefs; 2, Late Pleisocene aeolianites; 3, Pleistocene reefs 4, Lagoon sands; 5, Lagoon muds; 6, Pleistocene back reef deposits; 7, Pre-Oligocene basement.

has been identified below the present reef. They have been tentatively related to Sangamon, Yarmouth, and Aftonian. They lie well below present sea-level because of the subsidence that accompanied the building of the barrier reef. The contact between the Pleistocene and Holocene reefs is at a depth of 11 metres. Drilling in the Great Barrier Reef of Australia, down to 182 metres at Michaelmas Cay in 1926, and to 222 metres at Heron Island in 1937 did not provide much information on the succession of Pleistocene reefs, but the unconformity at the base of the Holocene in that area is at depths between 5 and 20 metres according to various drillings.

For the period of sea-level rise during the end of the Weichselian-Wisconsinian and the Holocene, a very considerable amount of new data has been provided by Project 61, *Sea-Level Movements During the Last 15000 Years,* under A. L. Bloom's direction in the UNESCO International Geological Correlation Programme. The results, included in Bloom's *Atlas of Sea-Level Curves* (1977-1979), have been widely discussed. Sea-level history in this period is summarized in curves, which can be classified into two main types: those showing a continuous rise in sea-level, with a rate diminishing in time but without oscillations: and those with oscillations, showing a general rise with alternating transgressive and regressive phases.

In the positive phases of the oscillating type, sea-level surpasses frequently its present position: six times from 5700 to 1000 BP in the curve presented by Fairbridge (1961). Similar results have been put forward for coral areas, with Holocene sea-levels rising to more than 3 metres (5 metres in Brazil) above the present level. Coral areas are very favourable places for such investigations, because dead corals, provided they are in position of growth (that is, not coral heads thrown by storm waves to another level) give precise evidence of former sea level at low ordinary tides. Such emerged Holocene reefs, ranging in age from 1200 to 4000 or even 6000 years, have been reported from the Society, Tuamotu and Cook Islands, Midway and Kure Islands in Hawaii, Ifaluk in the Carolines. Jarvis and Starbuck in the Line Islands.

Enderbury in the Phoenix. New Caledonia, etc have been considered as evidence of Holocene sea-levels higher than the present. Corrosion notches have also been invoked as indication of higher Holocene sea-levels in parts of these archipelagoes and elsewhere. Montaggioni and Pirazzoli (1984) even suggested that reef conglomerates, including coral heads thrown by waves, allow deductions about the former heights of the zones in which they were formed. Although recent tectonic disturbances have occurred. Holocene emergence(s) of reefs by a lowering of sea level has certainly occurred. But an essential fact resulting from many recent works is that these conclusions cannot be extrapolated outside the areas in which they are established; in other words, that 'the search for a universal eustatic curve must be regarded as over; regional differences in response to change in the geoid mean that eustatic sea level curves can have only regional validity', and, on the other hand, that 'no part of the earth's crust can be regarded as wholly stable.'

Glacial isostasy, consisting of the plastic depression of the earth's crust under the loading of the ice sheets, with a slight complementary elevation near the edge of the ice, and rebounds rafter the melting of the glaciers, does not concern coral reef areas. More recently, however, there has been a growing appreciation of hydro-isotasy, which is a crustal response, particularly in shelf areas, to the differences in loading by the sea water alternately kept and released by the glaciers. Hydro-isostasy applies not only in glaciated areas, but also in coral areas, and its effects are expected to vary around the globe.

The gradient of the sea-level curve during the Holocene may thus be an indicator of mantle viscosity, and such considerations could provide a solution to the discussions on Holocene sea-level changes in the Great Barrier Reef area. More widely, a 'geodial eustasy' has been proposed by Morner resulting from changes in the geoid, and thus different from the eustasy of glacial origin. Satellite surveys have confirmed that bumps and depressions exist in the geoid surface, probably related to density variations within the earth, and it is likely that these highs and lows migrate and vary in dimensions. A multidisciplinary approach in sea level research has been initiated by IGCP Project 200, *Sea Level Correlations and Applications,* under the chairmanship of P.A. Pirazzoli, 1983-1987. These trends in research throw a new light in the interpretation of Holocene sea-level curves, and hence of the growth of the coral reefs since the end of the last glaciation. Clark *et al.* (1978) used a numerical model to define six contrasting sea-level zones in the world, of which four are in areas of coral reef growth. An example may be quoted from the Brazilian coast, which has coral reefs, where Holocene sea-levels 2.3 to 5.0 metres higher than the present have been recorded, and tentatively related to geoid surface deformations.

Reef Growth in Response to the Holocene Transgression

The growth of reefs during the Holocene marine transgression has been studied in several areas during recent years, in relation to local sea-level curves, and deserves special attention as a consequence of

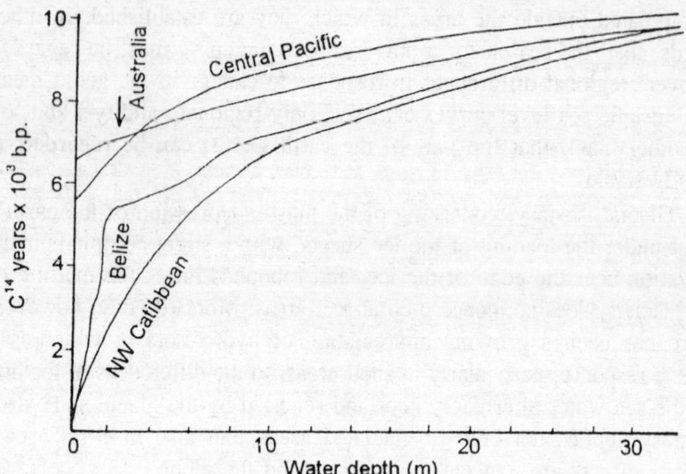

Fig. 2.8. Holocene sea-level curves for four selected coral reef areas.

shifts in sea-level. Numerous American studies on reefs in the Caribbean and the nearby tropical Atlantic, using diver-operated drilling to investigate the internal facies, have provided information on the various ways in which these reefs responded to the sea-level rise. The basement consists of Pleistocene fossil reefs, sometimes accompanied by lithified aeolianties, as on the southwest coast of Jamaica.

After detailed studies such as Adey's one on St Croix reefs (1975), and preliminary syntheses as by Adey *et al.* (1977), Neumann and Maclntyre (1985) have defined the three ways in which the Caribbean reefs have grown. A first type consists of *keep-up reefs*, i.e. reefs which were able to maintain shallow, frame-building communities throughout the sea-level rise. 'Growth rate tracks rise rate.' These reefs, which occur on windward, narrow shelves, are voluminous, wedge-shaped, with layered isochrons. At the rate of sea-level decreased, they have spread laterally. Galeta Reef, off the Caribbean coast of Panama, is a good example. It began to grow at a depth of about 1.5 metres, and maintained an *Acropora palmata* facies until 2000 BP.

The reef off Non such Bay in Antigua is also a keep-up reef, with *A. palmata* progading seaward during the last 3000 years. *A. palmata* framework is a very reliable reference for reconstructing the history of Late Quaternary sea-levels, owing to its restricted depth range (5 metres), the ease of obtaining uncontaminated samples, and its minimal compaction. A second type includes the *catch-up reefs*, which begin as shallow reefs and become deeper as rise rate exceeds accretion rate;

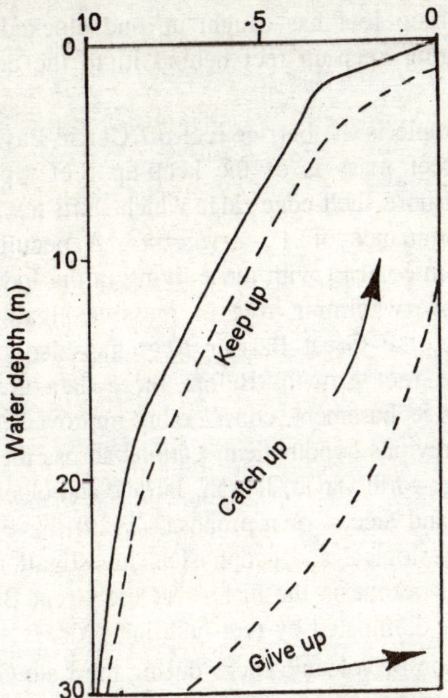

Fig. 2.9. Three reef responses to Holocene sea-level rise in the Caribbean.

then they grow upward and catch up with sea-level. They are found on protected shelves or back reefs. Their internal structure shows coalescing coral heads and irregular isochronous surfaces. *A. cervicornis*, not so shallow as *A. palmata* is generally the main component throughout the growth. Alacran Reef off Yucatan is an example, with a growth rate of 12 metres/1000 years relative to a sea-level rise rate of 3 to 4 metres/1000 years. Catch-up reefs occur also at Eleuthera, North Andros, in the Northern Bahamas.

The third type is the *give-up reefs*. 'relicts that founder as reef growth lags rise rate until growth declines or ceases. They deepen upward, are capped by a deep water facies, are usually small, scattered and of limited vertical extent'. These submerged relict reefs are common on insular shelves of the eastern Caribbean, e.g. the west coast of Barbados. An increase in turbidity or a cooling of shallow shelf water may have contributed to their demise. Shallow lagoons often undergo extreme conditions intemperature, salinity and turbidity, which have a deleterious influence on the reefs in front of them, so that such reefs have been said to be 'shot in the back by their own lagoons'. Compound reefs are also found, as the reef off the south coast of St Croix 'where an offshore catch-up reef has caught up and blocked off the ocean access of a fringing keep-up reef behind it, to the detriment of the keep-p reef'.

Another example is the barrier reef off Carrie Bay Cay in belize, where the mainreef mass is of the keep-up reef type, but is now eclipsed by an offshore shelf-edge ridge which shifts towards that catch-up type with dominance of *A. cervicornis*. A peculiarity of many Caribbean reefs, in contrast with those living in the Indian and Pacific Oceans, is the overwhelming role of massive algal ridges. In the southwest Pacific, the Great Barrier Reef has also provided many data on Holocene reef growth. Before these data are summarized, however, those on its basement, considerably improved in recent years, must be given. Previous hypothetical sections across the Great Barrier reef by Jukes, Edgeworth David, Jenson, Jardine and others, summarized in Steers (1929), and Steers' own proposal (1929), have been replaced by the result of geological and geophysical investigations. In contrast to Polynesia and Micronesia the history of the Great Barier Reef has not been generally dominated by reef-building processes.

After deposition of volcanic rocks during the Late Cretaceous, the facies were alternatively marine and continental, the latter, of progradational type, being dominant during Upper Oligocene, Upper Miocene and Plio-Pleistocene: they consisted of large deltaic systems. The shelf was thus widened by 10 kilometres off Cairns, and by 50 kilometres off Townsville. Before the Holocene marine transgression, however, Pleistocene reef formations existed over this shelf. They developed during brief periods of high Pleistocene sea-level, and were dissected during the Wisconsin regression by fluvial channels, either buried or still open, which have been raveled by seismic proofiling. Such is the irregular basement upon which the Holocene reefs grew. Their growth has been documented from 70 drill holes on 25 reefs in the central and northern Great Barrier Reef. It occurred during a rapid sea-level rise, followed by a relative stability since 6500 BP.

In a first phase, the transgression over the colonizable substrate was dominant, and the water depth over the bottom increased. In the second phase, which began about 8000 BP, reef growth was dominant, at first with a growth rate compensating the rate of sea level change, and late with a growth rate faster than the transgression, around 6500 BP, with further leeward progradative phases. A good example is found at One Tree Reef. If we refer to the terminology used for the Caribbean reefs, this strategy, common in the Great Barrier Reef, falls into the catch-up type, in which three subtypes have been distinguished according to relationships with the sea-level rise.

The give-up type has not been identified, and the keep-up type is rare, although present at Fitzroy Reef in the southern province of the Great Barrier Reef. In French Polynesian atolls, the basement of the Holocene corals is completely different from the Great Barrier Reef, consisting exclusively of reef formations which are reviewed in section 4 of this chapter. Holocene reef growth, known for Mururoa and Mataiva atolls and the Moorea barrier reef for a sea-level history similar to that of the Great Barrier Reef, appears to have begun between 7000 and 8000 BP. Vertical accretion rates have varied between 1.5 and 2.5 metres/1000 years for the reef rim at Mururoa, and 0.2 and 2.5 metres/1000 years for the lagoon at Mataiva, with the possibility of much higher rates in places subject to hurricane sedimentation. In some places, reef growth was delayed some time after the transgression, while in others it was synchronous.

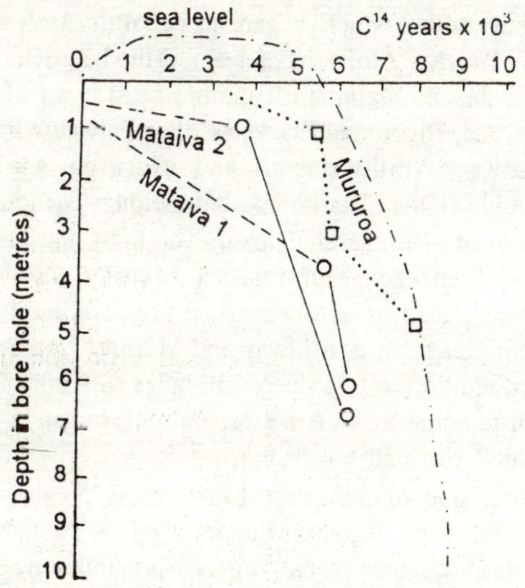

Fig. 2.10. Growth curves in relation to sea-level rise in Polynesia, at Mataiva Atoll and at Mururoa Atoll.

The reefs of the Tuamotu and Society Islands seem thus to have followed the catch-up strategy, dependent on or triggered by the rate of sea-level rise, as in the great majority of the East Australian ones, in spie of the environmental differences. At Tarawa Atoll, Kiribati, the Micronesian environment is not very different from that of French Polynesia, except for the lack of hurricanes. Cores have shown a reef growth beginning about 8000 BP, with fairly uniform vertical accretion at a rate of 5.0-5.4 metres/1000 years in two holes, and 8.2 metres in a third one (less accurate), with a rate of sea-level rise thought to have been 5-8 metres/1000 years since at least 12000 BP. Tarawa thus illustrates the catch-up model too, with a growth rate higher than that of French Polynesia, and higher than at Enewetak and Bikini. Marshall Islands in Micronesia, which seem to be of similar type. These results indicate that the catch-up strategy has been extremely pre-dominant in the Pacific, while in the Caribbean the mode of growth has been much more diversified. More data are still needed for reefs in the Indian Ocean.

Result of Deep Drilling and Geophysical work on Atolls in the Pacific and in Indonesia

Beside the information on shelf and reef evolution during the Quaternary in Eastern Australia and southwestern New Caledonia, other essential data have come from deep drilling carried out in Pacific and Indonesian atolls, which began at Funafuti Atoll in the Ellice Islands, now Tuvalu. Atolls have been drilled at Kita Daito Zima near Okinawa, Japan, Maratua off the northeast coast of Kalimanatan in the Celebes Sea, Bikini and Enewetak *alias* Eniwetok in the Marshall Island. Midway Atoll, Hawaii and Mururoa Atoll, which is administratively in the Tuamotus, but belongs structurally to the Pitcairn—Gambier—Hereheretue chain. The most detailed information has come from Enewetak, Mururoa, and Midway, all of them having been drilled more than once down to the pre-coralline basement, and deeply into this basement at Midway and Mururoa. At Enewetak there have been, in addition to two deep drill-holes, a number of shallower borings down to some tens of metres, but unfortunately they have all been in the atoll rim and not in the lagoon.

At Midway one of the deep drill-holes is on the rim, and the other on an island in the lagoon. Mururoa is by far the most investigated atoll in this respect since it has thirteen boreholes deeper than 175 metres, six in the lagoon and seven in different parts of the rim, and geophysical profiles across the atoll. All these investigations have shown the great thickness of shallow water alcareous formations in these atolls scattered around the Pacific. Two of them, Kita Daito Zima and Maratua, are raised atolls. At Kita Daito Zima the drilling stopped at 431 metres in a formation dated Upper Chattian (Oligocene). At Maratua, where two holes were drilled under the former lagoon, the deepest reached 550 metres below the crest of the rim, 429 metres below the present sea-level, also without reaching the pre-coralline basement. The age of the rocks is not certain. At Funafuti, Pleistocne and Cenozoic rocks were encountered, but the Oligocene had not been reached at a depth of 339 metres.

At Bikini, drilling to 777 metres went through Pleistocene and Cenozoic rocks without reaching the Eocene. In the two boreholes on the rim of Enewetak, the basaltic basement was reached at 1283 and 1408 metres, the oldest reef sediments, immediately above the basement, being of Eocene age, and containing corals 'that do not flourish below 150 feet'. At Midway, the basement, also basaltic, was at 173 metres in the lagoon borehole, at 503 metres in the reef rim borehole, the deepest coralline formations being earliest Miocene. At Mururoa, the two deepest boreholes found the contact between detrital coral formations and basalt at 415 and 438 metres; there the beginning of the sedimentary sequence, including the erosion of the emerged volcano, has been dated at the end of the Miocene (10 million years). These results demonstrate, with a certainty rarely obtained in geomorphology, that the seven atolls were built by corals and associated organisms during a long-continued subsidence, over a differing period in each case. In other words, the subsidence theory, proposed by Darwin and developed by Dana. Davis and others, is thus fully supported.

It remains to decide whether the atoll comes from the building, by corals and associated organisms, of a rim up to a higher level than the floor of the central lagoon, or it this saucer shape derives from some other gemorphological process; also. Whether several steps other than the classic fringing-barrier-atoll succession can be deduced from the drillings. Before these problems are examined, evidence from the top of the basement at Mururoa and Midway will be mentioned, and also from Rurutu, a raised structure in the Austral Islands, since they indicate how the sedimentation began. At Mururoa, five of the boreholes show a succession upwards of a submarine and subaerial volcanism now between 269 and 455 metres below sea-level, followed in four of the five by a transition level of variable thickness (up to 82 metres at Viviane), before dolomitized coralline strata are reached.

The transition level consists of detrital basaltic blocks, with montmorillonitic and kaolinitic clays, overlain by calcareous rocks with variable amounts of coral. Beach sands are found, deposited in places that were shallow,while coral organisms colonied in deeper water, the pioneer species varying around the emerged volcano. At Rurutu, a foundation of submarine basalt is overlain by a transition zone consisting of submarien red clays and pockets of manganiferous ores, passing upwards into submarine conglomerates including a mixture of coral reef and volcanic blocks.

At Midway the basement under the reef rim consists of pure tholiitic basalt which includes both aa and pahoehoe lavas, which have been weathered, truncated by wave action, and capped by 38 metres of volcanic clays and conglomerates with volcanic pebbles and four limestone layers. All these data show that when subsidence began, the volcanoes were surrounded by beaches made of sediments derived from marine organisms and from the volcano, which was still very active at Mururoa, and in a late stage at Midway, judging by the degree of weathering and the nature of certain pebbles. At Mururoa there were successive incipient fringing reef deposits buried by lava flows during a slow subsidence, which also fit Darwin's model. Above the basal sedimentary deposits, the various boreholes show a succession, or mixture, or massive coral framestones, *in situ* algal frame-stones, biodetrital transported facies which are sometimes sandy, sometimes silty or muddy, and include foraminifera, molluscs, and large or small pieces of coral. The degree of cementation is highly variable, even at great depths. For example, at a depth of 1310 metres in the main borehole in Enewetak.

The Eocene sediment is a 'limestone, friable to moderately well cemented, porous, white, foraminiferal—algal'. In all these drillings the internal structure of the atolls includes an enormous number of voids and caves. Several holes at Enkebi Island, Enewetak showed in the upper 80 metres an intricate mixture of rocks described as 'boundstones', 'packstones', 'grainstones,' 'wackestones', all of them either uncemented, moderately cemented, or well-cemented. French contractors in Polynesia call this kind of sediment a coral soup. The voids result in part from later solution, but some are original, deriving from cavernous reef building where the rock is not completely compacted at depth. One of the results is a freshwater lens called the Ghyben-Herzberg lens, at shallow level in the islands occurring on atoll rims, and surrounded by brackish water subject to tidal oscillations.

The hydrogeological problems raised by this lens, sometimes very large as at Tarawa and extremely important in atoll life, are more complicated than it had been previously thought. In this sedimentary framework, diagenesis has occurred, leading to dolomitization which has been recognized in all the atoll boreholes, including Funafuti and Kita Daito Zima. At Mururoa, the dolomites reach a thickness of 280

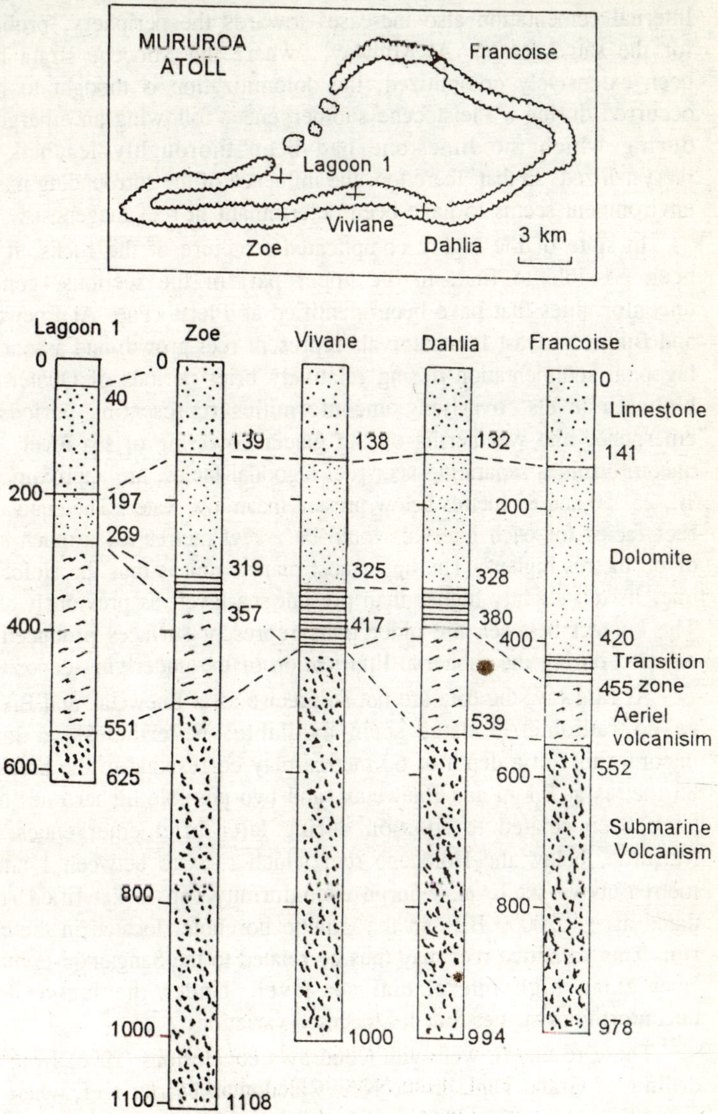

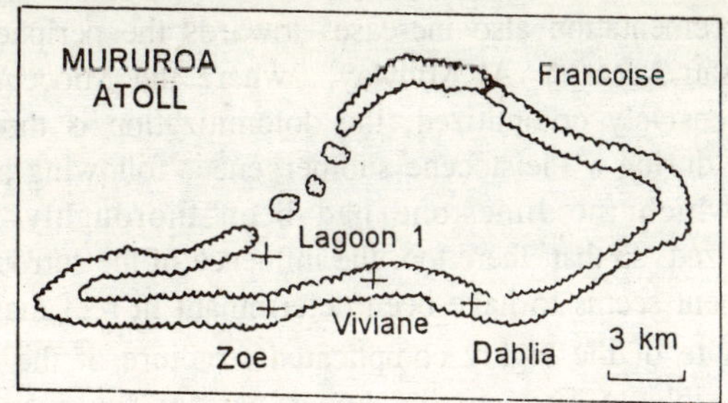

Fig. 2.11. Five drilling in Muroroa Atoll, showing succession of volcanism and sedimentation in different parts of the atoll.

metres below the peripheral islands, while they are absent from certain boreholes beneath the lagoon; they are considered to be contemporaneous with, or subsequent to, aragonite dissolution, and their distribution suggests that marine influence has been important in their formation.

Internal cementation also increases towards the periphery, probably for tho oamo roaoon. At Midway, whoro tho Miooono otrata havc been extensively dolomitized, the dolomitization is thought to have occurred during a Pleistocene submergence, following an emergence during which the limestone had been thoroughly leached and recrystallized: so that, there too, the influence of the surrounding marine environment seems to have been determinant in reef diagenesis.

In spite of the highly complicated structure of the rocks, it has been possible to find, in the upper part of the sections, general unconformities that have been identified as Pleistocene. At Enewetak and Bikini 'at least four intervals represent reef growth and associated lagoonal sedimentation during relatively brief periods of Quaternary high sea levels, overlying unconformities representing periods of emergence and weathering during glacial lowering of sea level. The uncomformities separating near-reef lagoonal facies, are approximately 10, 25, 50 and 85 metres below present mean low water. The equivalent reef facies for each interval would be a few metres to as much s 50 or 60 metres higher'. The uppermost marine unit is thus the Holocene one, its top slightly higher than present sea-level, as previously said. The contact between the older units represent surfaces produced by solution during the subaerial lithification of the underlying deposits.

At Midway, the data are not so precise as at Enewetak and Bikini, as no radiometric dates seem available; nevertheless, a large unconformity at a depth of 60 metres may correspond to the break at 85 metres at Bikini and Enewetak, and two possible higher ones may have been related to solution during later brief emergences. At Mururoa, below the Holocene reef which rises to between 1 and 4 metres above sea-level, a large unconformity caps a karstified reef, dated at ± 120000 B.P. in the Colette borehole, located in the reef rim. This karstified reef may thus be related to the Sangamon-Eemian-Ipswichian high interglacial sea-level. Under the lagoon,four unconformities suggestfour leistocensea-variations.

These results fit well with Coudray's conclusions (1976) from the drilling of Grand Tenia in the New Caledonian barrier reef, where he found three successive Pleistocene reefs capped by unconformities, plus the Holocene reef. In addition,the Sangamon-Eemian reef has been identified and radiometrically dated in places as widely scattered aa u n Hawaii; Anaa, Niau and Makatea in the Tuamotus; Mangaia in the Cook Islands; Rottnest Island in Western Australia; Mauritius; the Seychelles; and Guadeloupe Island in the Caribbean.

It may be asked, however, why there are not more than three or four reefs referable to Pleistocene high sea-levels, when various studies, including Emiliani's palaeotemperature curves (1966, 1970), have indicated a much larger number of positive thermal oscillations, as well as minor fluctuations during the Wisconsin, In Alaska, Hopkins (1967) found six marine tansgressions before the Holocene, including one at 2000000 BP, and one between 1900000 and 700000 B.P. Other studies based on deep sea cores in the northeast Pacific point to glaciations (outside the Antarctic) and interglacials older than the Nebraskian/Gunz and the Aftonian/Cromerian. It is not yet known how these variations in sea-level affected reef building, and further research is required.

On the other hand, the numerous drillings in the Mururoa Atoll have provided a so far unique picture of atoll growth, and of variations in shape in the course of subsidence. After the initial colonization by a fringing reef, subsidence of about 200 metres, occurring in several stages with interruptions, led to the widening of the fringing ref. It seems that a barrier reef stage, with a peripheralrim standing around a lagoon, as reached just before the top of the volcano was submerged: in other words, the barrier reef stage, which remains a little hypothetical, appears to have been short, whereas the fringing reef stage was comparatively long. Thereafter 'subsidence seems to have decreased and sea level variations prevailed' (as a result of early glaciations?), and the reef took the shape of 'a wide shallow lagoon extending over the whole atoll; possibly a continuous barrier reef' (actually, a rim) 'isolated this wide lagoon from the ocean'. Subsequently, the rim continued to grow vertically, while detrital, fine facies settled in the lagoon. The general difference between the rim and lagoon formations, where borings indicate the muddy facies common in the lagoon, with intervening rigid frameworks representing lagoon pinnacles or ridges; while on either side the autochthonous framework or coarser biodetrital facies is predominant or exclusive, with coarse slope deposits on the northeastern outer side. The modern rim formed progressively between 90 and 40-30 metres depth.

Summarizing this history, it may be said that, in the only present-time atoll in the world in which the stages of growth shape are known with some accuracy, the fringing reef lasted longer than expected from Darwin's model; the barrier reef stage was shorter; and the atoll stage, characterized by a peripheral reef and a central lagoon with fine sediments and scattered pinnacles or ridges, much like the present lagoon, appeared when the general structure reached a level of 90 metres below the present. Apart from these slight differences, Darwin's theory is supported, especially concerning the origin of the rim surrounding the lagoon depression which lies in the atoll centre. In the Indian Ocean, no deep drilling has been made in the open ocean reefs, except perhaps in the Maldives. Geophysial data reported by Stoddart (1973) point to moderate thicknesses of calcareous sediments in these structures, which seem to have generally been built more recently than many of their Pacific counterparts.

Karstic Saucer Theory of Atoll Origin

Another theory has been put forward to explain the shape of atolls by Yabe (1942), Asano (1942), Hoffmeister and Ladd (1945), Ladd and Hoffmeister (1945), Mcneil (1954), Purdy (1974), Bourrouilh (1977a). The karstic saucer theory is more or less a new version of Hoffmeister and Ladd's antecedent platform theory (1935-1944). It is assumed that it was introduced by Yabe and Asano, whose papers in Japanese were translated into English by Burke (1951) in a report of limited diffusion. In summary, this theory states that the shape of atolls, and also of barrier reefs, derive from an antecedent, more or less horozontal, calcareous plat-form, which emerged as the result of uplift or sea-level fall. Having emerged, it became karstified by rainfall and percolating water, which produced the widely accepted karstic landforms such as drowned dolines and blue holes, but also resulted in a surface dissolution which gave it the shape of a saucer, with a central depression and a peripheral rim interrupted by gullies through which a part of the water escaped (the other part percolating through the karst).

Subsequent submergence of such a structure resulted in revival of coral growth, which buried the karstified limestone; and the annular rampart resulting from the karstification became a barrier reef or an atoll rim. The general shape would thus result from subaerial processes and not from peripheral reef building. The explanation is not in contradiction to subsidence: it can be applied to structures as Maratua, Enewetak. Bikini and Midway, where drilling has demonstrated a great thickness of reef limestone accumulated during the slow lowering of the basement: it agrees partly with Darwin's theory since the subsidence is accepted as the cause of limestone deposition, but it disagrees on the origin of the rim. The karstic saucer theory has found support in experiments and in geomorphological observations of emerged, atoll-shaped, reefs. Experiments have been carried out in the laboratory on limestone blocks by Hoffmeister and Ladd (1945) and by Purdy (1947).

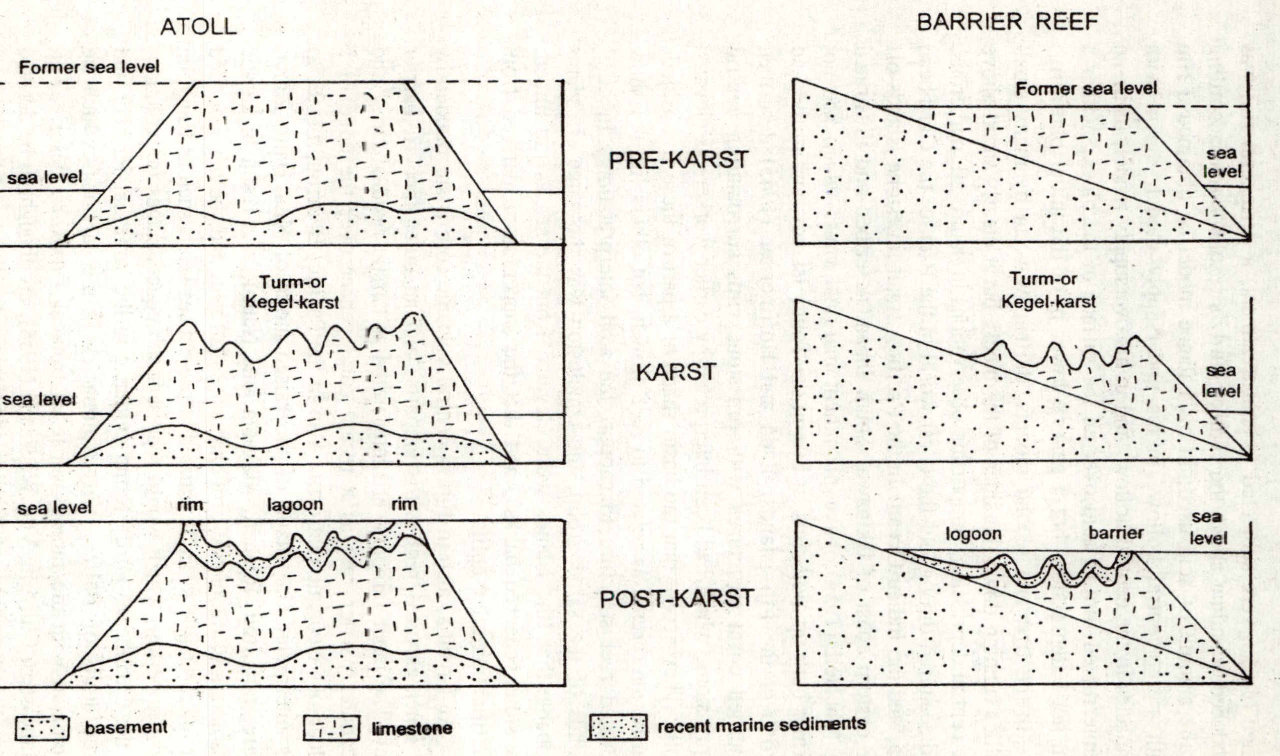

Fig. 2.12. Formation of barriers and atolls according to Purdy (1947)

When blocks were subjected to acid treatment, a peripheral rim was produced, standing above the centre. Purdy's results varied according to the amount of acid used: in his balance model he obtained a rim with a flat central hollow; in his undersupply model, a rim with pinnacles in the central hollow, and in his oversupply model, using an uninterrupted film of acid over all the surface of the block, neither a rim not a depression, but a general lower in of the initial platform.

In the first and second models, the rim was cut by many small gullies running down the sides of the block, but it clearly stood above the central depression. Supporting observations in nature are numerous, and scattered throughout the coral world in the Pacific, the Caribbean Sea, and the Indian Ocean. In the Pacific, small limestone islands off the eastern coast of Okinawa, South Japan, described and illustrated by MacNeil (1954), have prominent marginal rims, which are not former built-up atoll rims In Eua in the Tonga Islands, and in the Lau Group of the Fiji Islands, Ladd and Hoffmeister (1945) described emerged coral structures with limestone rims surrounding central depressions, which had been interpreted by earlier workers as elevated atolls. They accept that the basin shape of a part of these five islands can be interpreted as related to reef growth; but others 'do not show elevated reef structure to support the atoll interpretation'. In other islands of the same group, the conclusion was the same. It seems probable that 'the islands were elevated from the sea as unrimmed, limestone-veneered banks', and that the saucer shape derived from karstification after uplift.

At Tuvutha, also in Fiji, a central volcanic core is surrounded by an inner elevated reef with remparts, and an outer modern barrier reef. The inner ramparts are interpreted as karstic residual rims, and the outer barrier reef as a recent coral veneer over the edge of a calcareous Neogene platform. Also in the Pacific, Bourrouilh (1977a) has interpreted the shape of the raised atolls of Kita Daito Zima, Mare and Lifou, Rennell, Makatea, and Nauru, as a result of karstic dissolution of the centre of a carbonate platform, and she has extended her interpretation to atolls situated at present sea-level, such as Rangiroa and Clipperton, which would represent subsequent stages in evolution. In the Caribbean, Purdy found evidence in the large barrier reef and three atolls of Belize. Large pinnacles in the lagoons, which he compared with the famous subaerial tropical *Turmkarst* described in southeastern China by Von Wissmann (1954), are thought to be derived from the drowning of similar landforms. Sparker profiles show a correspondence between the pre-Holocene topography and present bathymetry. Also on the Belize shelf, calcareous ridges on eitherside of the English Cay Channel, a drowned pre-Holocene valley ending across the barrier reef, have been compared with the subaerial limestone narrow walls along rivers at Okinawa; and similar ridges have been obtained experimentally by applying acid to a limestone block. Counterparts of the English Cay Channel ridges exist in the Great Barrier Reef and along the meandering Longogori assage across the Mayotte barrier; but there is so far no evidence of a karstic origin at Mayotte site: it is only a possibility.

There is certainly a tendency for karstification, especially in the tropics, to result in long, narrow, steep-sided ridges, not only on platform sides, but also along river valleys; and such features have been reported from areas which have reefs in the vicinity. A coral structure which could be more easily explained by the karstic saucer theory than by Darwin's theory is the Chagos Bank in the Indian Ocean. This coral bank is bounded by the usual outer rim, has many pinnacles in its lagoon, and depths not exceeding 100 metres; but its quite anomalous size (over 170 × 105 kilometres), completel different from ordinary atolls, especially the nearby atolls of Diego Garcia and Peros Banhos, raises certainly a problem.

Unfortunately the Chagos Bank is still little known, except for its bathymetry, but an explanation by the antecedent karst theory, with a reef-building starting from a calcareous platform, is a hypothesis to be checked by further research. At the moment, which could be the conclusion about that theory in the present state of research? In 1987 we have only one set of numerous deep boreholes over the different parts of an atoll, including the lagoon: the drillings carried out at Mururoa Atoll. According to Buigues (1985), there is a pronounced difference between the framework of the rim, made of built-up reef and coarse slope debris, and the lagoon, made of fine debris except for the pinnacles and ridges; and this difference has persisted since well before the Holocene. Such a structure does not favour the karstic saucer theory, but fits the classic Darwinian idea of rim gradually built around a central lagoon which has been the site of a fine sedimentation during a long period.

Scott and a Rotondo did not consider the Mururoa data, which were not then available, when they wrote (1983,) that 'it is clear that the present atoll morphology is fundamentally karst induced, rather than growth induced.' Yet, this does not mean that the karstic saucer theory has no value. The evidence put forward is very extensive, especially for structures standing now well above sea-level, but even for some of those lying at and below sea-level. A worthwhile remark by MacNeil (1954) is that 'it is necessary to interpret the history of each atoll separately instead of trying to explain all atolls by a single theory'. Beside the general fact of subsidence combined with variations in sea-level, and tectonic uplift in a number of cases, it appears wise to accept the possibility of more than one explanation of reef rim formation until many more borings have been made. It is quite necessary to drill into many pinnacles in atoll lagoons, to establish the extent to which they are Holocene structures, or Holocene coatings over leistocene structures. The 'Turmkarst lagoon theory', a companion to the karstic saucer theory, must be tested in a large number of sites before a general valid conclusion is possible.

Tentative Synthesis of General Reef Evolution on the Pacific Plate

The data resulting from the deep drillings carried out in seven atolls are indispensale in any further interpretation of atoll genesis; but these data are scattered over the Pacific, while the islands in this ocean are grouped into sets or archipelagoes, generally running in a southeast and northwest direction, as in the Hawaiian, the Caroline, the a Tuamotu, the Society Islands. In these groups the coral reefs and their basements vary in type from one end to the other, and belong generally to younger types in the southeast and older in the northwest. The Society Islands are typical in this respect, with a succession starting at Mehetia, on a still active volcano, passing through extinct volcanaoes with barrier reefs from Tahiti to Maupiti, and ending in atolls at Mopelia, Scilly and Bellingshausen. Such a pattern has been recognized for a long time, for example by Davis who wrote (1928): 'the eruptive construction migrated eastward with the passage of time and (the islands) have, since their eruptive construction, suffered erosion and subsidence proportional to their age'.

Later Chubb (1957) commented in the same way on the other island chains of the Central Pacific. But it is only since the late nineteen sixties and seventies that development of the knowledge of deep-sea floor evolution has allowed a general explanation, ending in a genetic classification much more sophisticated than had been previously possible. Two general models have been proposed independently by Scott and Rotondo (1983a,b) and by Brousse (1985), the second one resting upon a large recent bibliography. The sea-floor

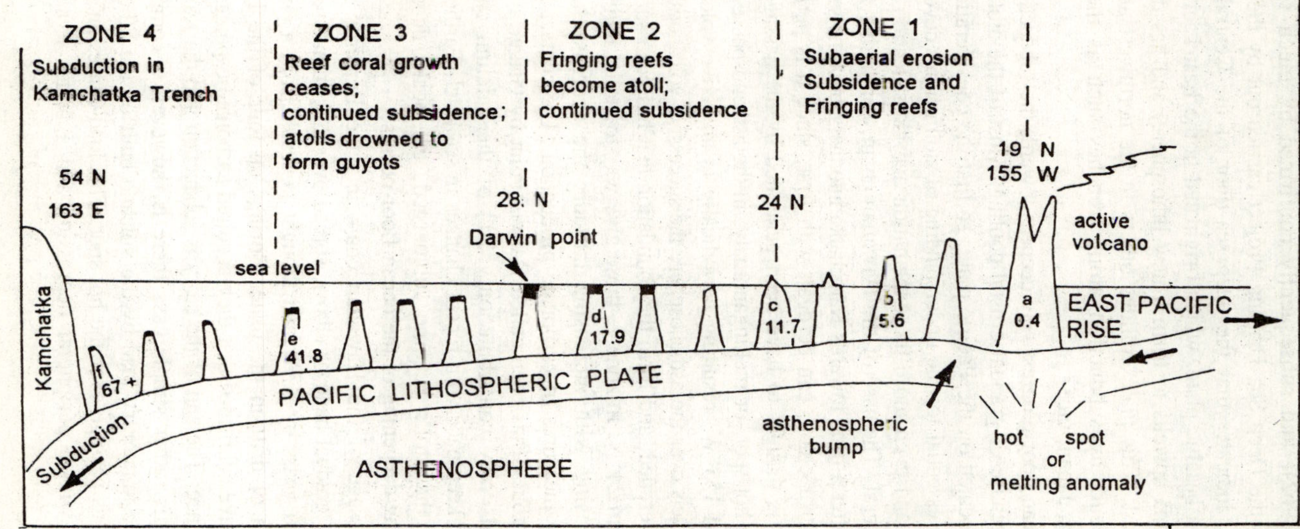

Fig. 2.13. Evolution of the Hawaiian-Emperor seamount chain, combined. Ages of six geologic features in million years (a) Hawaii; (b) Kauai; (c) Gardner Pinnacles; (d) Midway Atoll; (e) Suiko seamount; (1) Seamount at 53°N, 165°E.

spreading theory, having been proposed separately by Dietz (1961) and by Hess (1962), and subsequently confirmed by much research, particularly the Deep Sea Drilling Project carried out by the *Glomar Challenger*, indicates that the deep sea-floor of the Central Pacific consists of a basaltic plate, which originated in the East Pacific Rise, where magma upwells to form a new lithosphere, and has drifted to the northwest. This plate collides others in the northwest and west, and disappears by subduction along trenches such as the Kuril-Kamachatka, Japan, etc.

The ocean floor progressively deepens towards the subduction zone, probably because of gradual crustal cooling. Most of the atolls of the world are located on the Pacific plate, so that the consideration of its dynamics is essential for the understanding of atoll origin, development and decay, and even more the explanation and classification of coral reefs in general. The case of the Hawaiian chain, continuing for 6340 kilometres after a bend towards Kamchatka into the submarine seamounts of the Emperor chain, can be dealt with first, since it is perhaps the simplest. It is thought to have been generated by the passage of the plate over a hot spot, or meltinganomaly, in the asthenosphere, located at 19°N and 155°W, resulting periodically in an active volcano. As the plate travels over the asthenosphere the successive volcanoes become extinct, are eroded, and form the foundation that has been surrounded by reefs from 24°N. In zone 2, these give way to atolls, their growth rates decreasing with increasing latitude: the decrease has been indicated by measures of carbonate production. The contact between zone 2 and zone 3 has been called Darwin Point by Grigg, a location where corals may contribute only 20° of the calcium carbonate necessary to keep pace with recent changes in sea-level. Farther north, in zone 3, the atolls become submerged, the peaks deepen, an accompanying deepening of the ocean floor occurs.

The passage of the Hawaiian chain into the Emperor chain is marked by a gradual bend or elbow, spread over 260 kilometres, dated at 42 million years by K/Ar and thought to result from a change in the direction of drift in the plate. Borings on four of the Emperor seamounts have confirmed that they acquired carbonate cappings before their subsidence. The end of the story is subduction into the Kamchatka trench. In this rather simple case there is, however, a complication related to the temporary load by the volcano built on the lithospheric plate during its passage over the hot spot. This load causes an isostatic adjustment, with a peripheral depression and a bump at a greater

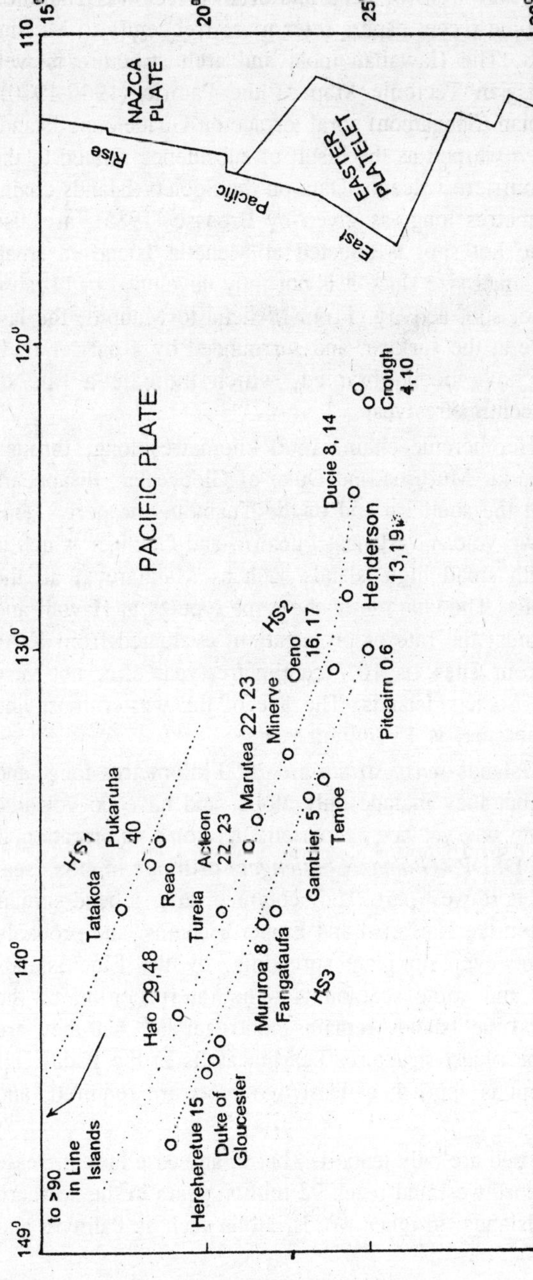

Fig. 2.14. Assumed formation of atolls in the Southeast Pacific by drifting of the Pacific plate over three hot spots (HS1, 2, and 3). Solid lines: plate boundaries. Lines of dots: tracks after passage over hot spots. Circles: atolls, high islands and seamounts. Number indicate the age of the structures in Ma. Eastern part of HS2 surficial expression is trapped by a fracture zone, hence the deviation of the chain, for HS1 and HS2 chains: for HS3 chain.

distance. It seems that the first author to have clearly described and analysed this moat and arch (or deep and arch) effect was Hamilton (1957), although some recent papers refer more frequently to McNutt and Menard, 1978. The Hawaiian moat and arch structure is well shown on the Russian Tectonic Map of the Pacific (1966-1970). Similarly, the Eemian (Sangamon) coral terrace on Guadeloupe Island, Carribean, has been warped as the result of subsidence related to the formation of La Soufriere volcano. Data on the Society Islands chain, which is 800 kilometres long, as given by Brousse (1985), are also very accurate. The hot spot is situated at Mehetia Island, a small active volcano, of small size since it is not fully developed or because of a decrease in hot spot activity. From Mehetia to Maupiti, the last volcano still visible at the surface, and surrounded by a barrier reef, eight K/Ar ages have been obtained, which indicate a rate of propagation of 12 centimetres/year.

The Pitcairn-Hereheretue chain. 1800 kilometres long, through Gambiere, Fangataufa. Mururoa and Duke of Gloucester, disappears in the northwest at the southern end of the Tuamotus properly. This chain consists of two volcanic islands, Pitcairn, and Gambier which is an almost-atoll with small high islands such as Mangareva; all the others are true atolls. The chain is of the same type as at Hawaii and in the Society Islands; the rate of propagation, evaluated from K/Ar measurements at four sites, is 10.7 centimetres/year, i.e. not very different from the Society Islands. The age of the westernmost and oldest atoll, Hereheretue, is 15 million years.

The Tuamotu Islands *sensu stricto* are 2000 kilometres long, and also run SE-NW, but they include only atolls, and have no volcanic outcrop. No drilling has yet been reported, but some information is available from the DSDP (*Glomar Challenger*) drilling in the sea-floor around their northwest part. They continue, after a bend similar to the elbow between the Hawaiian and Emperor chains, and probably related to the same event in plate spreading, by the Line Islands chain, also atolls and some seamounts. The interpretation of the Tuamotus and the Line Islands remains controversial, but they are certainly among the oldest structures bearing atolls in the Pacific. If the hot spot concept is applied, at least two spots are required, and perhaps more.

The ages proposed are only tentative, but in any case they increase from southeast to northwest and reach 92 million years in the northern atolls of the Line Islands: in other words, atolls such as Palmyra and

Washington began to grow during the Turonian. Near Rangiroa in the northwestern Tuamotus, a *Glomar Challenger* boring has found volcanogenic siltstones containing redeposited reef fossils dated for a part of them at 51 million years (Lower Eocene). At the other end of the Tuamaotus, the eastern part of the Hao—Crough chain shows a bend which is considered to be the expression of a fracture zone deviating the signature of the hot spot.

Henderson Atoll has been uplifted at 30 metres, an uplift related, according to McNutt and Menard (1978) and Okal and Cazenave (1985), to the moat and arch effect of the recent island of Pitcairn in the HS3 chain. Similarly, in their northwestern section, the Tuamotu Islands are not far from Tahiti, the last but one volcanic island of the Society chain. According to Pirazzoli and Montaggioni (1985), this is another example of the moat and arch effect. The weight of Tahiti seems to have produced a rise of the southern parts of the atolls of Tikehau, Rangiroa, Kaukura in the Tuamotus resulting in an anomalous height of Pleistocene reefs of these islands: the jagged *feos*, which are unknown as such elsewhere in the Tuamotus. The raised atoll of Makatea would have also been influenced by the rise of the arch, but its altitude (113 metres) cannot be fully explained in this way. The westernmost atoll of the Tuamotu chain, Mataiva, is a very unusual structure.

The Austral-Cook Islands chain is 2500 kilometres long. It begins in the southeast at the MacDonald active submarine volcano, and the propagationrate has been calculated from K/Ar dating at 11.4 ± 2.3 centimetres/year. Mangaia in the west is 19 million years old. But, after a normal sequence from fringing reef to barrier reef, a volcanic reactivation has occurred in several islands so that the reefs of these islands have been raised 70-100 metres. One explanation, put forward elsewhere by Scott and Rotondo (1983), would be that these volcanoes, after their initial formation over a hot spot (where MacDonald is now), have passed over an asthenosphere bump. Although this explanation is not certain, it should pr[illegible]ably be taken into account for several isolated atolls in the Pacific.

The Caroline Islands chain is a rather simple case, like the Hawaiian-Emperor chain, with volcanic islands in the south-west, the almost-atoll of Truk in the middle, atolls in the west, and final engulfment into the Mariana trench. There are only some complications at the western end; and, as this chain does not reach higher latitudes when it enters deeper water, the problems of coral growth found on the Hawaii-Emperor chain do not occur.

The Kiribati-Marshall group is more difficult to classify. It lacks high volcanic islands in its southeastern end. The many, fine atolls of the Marshall Islands are scattered on two main lines, and are associated with numerous seamounts; some of them are flat-topped, having reached the surface (the so-called guyots), others not. In the northwest thing group shows the usual deepening of the sea-floor, and includes only submarine seamounts. It is divided into the Dutton and Michelson ridges, and is subducted into the Mariana and Izu-Bonin trenches. Seamounts smaller than 40 kilometres in diameter have been broken up by the fracturing of the plate surface as it bends prior to subduction, whereas the larger ones seem to be subducted intact.

The Kiribati-Marshall group represents probably, like the Tuamotus, a senile chain produced by a long-inoperative hot spot. The Eoceneage of the basement in the Enewetak drilling, which is still far from the western end, is evidence for such senility. The Samoa chain is exceptional in that it includes active volcanoes with fringing reefs in the west and only one atoll, Rose and seamounts in the east: in other words, the trend is reversed here. This anomaly might be related to the location of Samoa just at the border of the Pacific plate, i.e. close to Melanesia alogn the Tonga trench; in Melanesia, the polarity plate, i.e. close to Melanesia along the Tonga trench; in Melanesia, the polarity of the trenches has been sometimes reversed in the course of time, as in the Vanuatu arc system in the Late Teritiary.

The exception could be that this chain, owing to its marginal position, has not followed the general rules of the Pacific plate, but this suggestion needs to be checked. A general model is proposed for the Pacific plate, the Samoa chain being left out, after Scott and Rotondo (1983a) with some simplifications. The moat and arch effect modifies initially the subsidence which continues subsequently. The passage over an asthenospheric bump results in a raised atoll, which will sink later, leading to submerged atolls and seamounts (guyots). It may be that the moat and arch effect acts also at the passage over a bump, giving a subtype of raised atoll, supposed by Scott and Rotondo but not shown here. Some atolls (principally the Phoenidx and the Tuvalu) do not fit this model: they will require further explanation, but most of the coral structures found on the Pacific plate are satisfactorily explained in this way.

The reefs of the Western Pacific and Southeast Asian seas, outside the Pacific plate, where tectonic effects are essential, are examined in section 7 of this chapter. A problem is whether the Pacific plate

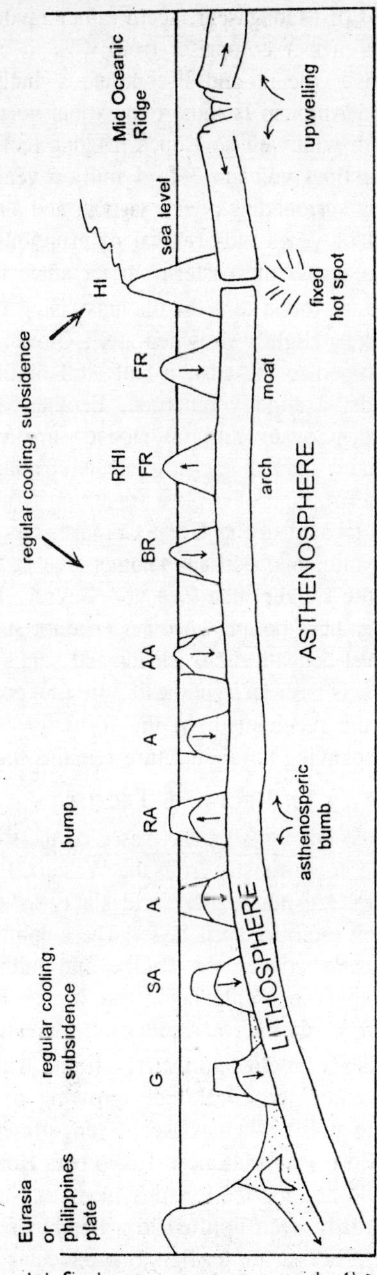

Fig. 2.15. Proposed model for island development on the Pacific plate. HI–high island; RHI-FR–raisd high island with fringing reef; FR–fringing reef; BR–barrier reef; AA–almost-atoll; A–atoll; RA–raised atoll; SA–submerged atoll; G–guyot. Arrow indicate subsidence or emergence.

coral reef model finds counterparts outside that area, particularly in the Indian Ocean. The Comores Islands and their eastern surroundings, lying at the northwest of Madagascar, seem rather favourable at first. The Comores Islands proper comprise, from west to east, the Great Comore, with an active volcano and discontinuous, incipient, fringing reefs; and Moheli and Anjouan Islands, two extinct volcanoes, 40,000 to 1,500,000 years old, with well-developed fringing reefs. Then comes Mayotte Island, an extinct volcano 3 to 4 million years old, with a fine double barrier reef surrounding a wide lagoon, and a recent (perhaps only a few thousand of years old) revival of eruptions, which have built a small volcano across the eastern barrier since its formation.

Still farther east are found three seamounts rising from the deep-sea floor: Zelee Bank, a slightly drowned atoll; Geyser Bank, topped by a crescentic reef open to the east, a half atoll or tilted atoll; and the Glorieuses Islands, a slightly emerged, Eemian/Sangamon reef. The cores of Zelee, Geyser and Glorieuses are volcanic. The resemblance with chains on the Pacific plate is striking, the general pattern suggesting the passage of a plate travelling from west to east over a hot spot now located at the Great Comore, the influence of a further bump at Mayotte, and perhaps another one at the Glorieuses since they tend a little higher than Zee and Geyser. Unfortunately, recent data on that area have not provided evidence for such a sequence. The nature of the crust beneath these islands and banks is assumed to be oceanic, but this is still a hypothesis. In the present state of knowledge (1987), the mechanism of the formation of this set of volcanoes and accompanying coral structure remains uncertain.

Impact of Regional Tectonics

Major objections to Darwin's theory, based on the study of uplifted coral structures, have come mostly from the Western Pacific, outside the Pacific plate (they are not reallly valid since, as we have seen, Darwin himself had left room for such cases.) These uplifted frameworks occur also in Indonesia between the Pacific and Indian Oceans, in Melanesia, in the Japanese archipelago, and in the Red Sea: they deserve attention here as they have resulted in interesting landforms.

The Japanese Islands lie close to the trench into which the Pacific plate is engulfed, so that the coral reefs growing on their eastern margin are subject to uplift when phases of engulfment occur. The southern part of the Boso Peninsula near Tokyo bers Holocene fringing reefs dated at 6500 BP, and located in embayments at Numa, Tateyama City, which have already been uplifted to a height of 12-18 metres. This is still a simple process, not leadign to spectacular morphological features: but a more imposing result of a strong and long-continued

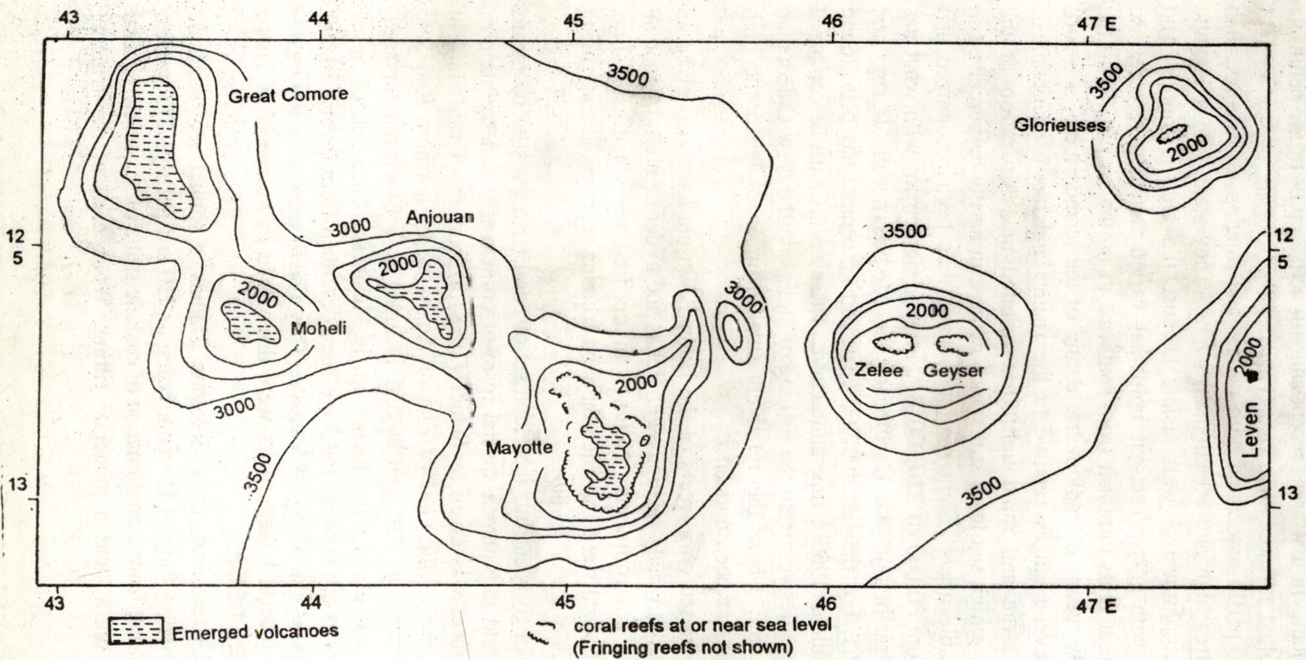

Fig. 2.16. The Comores chain, with the 3500, 2500, 2000 isobaths. Leven Bank is attached to the Madagascar shelf.

uplift is a *set of theatre-like terraces*, the most impressive being on the north coast of the Huon Peninsula, Papua New Guinea, facing New Britain. In that area, recent and still active uplift of the peninsula has resulted in more than twenty successive fringing or barrier reef terraces, partly associated with deltas, and ranging up to more than 600 metres high and 80 kilometres long. They stand like a gigantic Greek amphitheatre, well visible because the area is treeless and covered with grass. Tectonic uplift has varied along the terraces, so that their heights diminish from southeast to northwest parallel to the coast. By means of radiometric dating of the terrace reefs, a record of sea-level changes relative to the rising land has been obtained. Such a large and rapid uplift has resulted in a separation of the successive high sea-level still stands which can hardly be perceived in more stable areas. Eight sea-levels have thus been recognized and dated from 30,000 to 220,000 BP, in good agreement with what has been described in the Caribbean, the Mediterranean, Japan, and California. For the last 400,000 years, an agreement has been found with Emiliani (1966) temperature curve. These results must be compared with those derived from the drilling of atolls and the New Caledonian barrier reef, where the relatively small number of leistocene high sea-levels has been noticed.

The raised coral reefs of the Huon Peninsula have also given an opportunity to study relations with deltas and to define littoral, lagoonal and reef environments. Notches in uplifted reefs indicate short stillstands, and were cut when the next fringing reef was already being formed at a lower level.

Although beautiful, the Huon set of coral terraces is by no means unique, and falls into a special form of coral reefs. A description of such terraces was given by Schick (1958a,b) from the Tiran Island at the entrance of the Gulf of Elat or Aqaba in the northern Red Sea. This set rises to 320 metres and includes at least eleven terraces, with more doubtful relics higher up. In Melanesia, other sets of raised reef terraces are known, and have been described from several islands (Efate, Malekula, Santo) of Vanuatu. They rise to more than 300 metres in some places. Counterparts were discovered in Indonesia at Timor and Atauro Islands.

At Guadalcanal in the Solomon Islalnds, a similar reef staircase exists in the vicinity of Honiara, rising to 210 and perhaps 270 metres, and resulting from a combination of eustatic shifts of sea level, general tectonic uplift, and a moderate tilting. When each of the fringing reefs emerged, another one was initiated at a lower level, with a shallow boat channel in between. The terraces are treeless, as on the Huon Peninsula, and thus quite visible. Sections show that the boat channels periodically received, during floods, a terrigenous sedimentation. Alternating with coral growth, such successions are similar to the deltaic deposits reported at Huon by Chappell, and testify to the west climatic environment in which these two staircases were formed.

Another very fine set of coral terraces exists at Hispaniola Island, in the northwestern part of Haiti, but it has not yet been investigated in detail in the field.

Perhaps the most fantastic example of tectonically-influenced reefs is the elevated double barrier reef of Marovo, at the eastern end of the New Georgia group in the Solomon Islands, described by Stoddart (1969c,d). This double barrier reef, facing east, consists of two lines of coral islands surrounding the volcanic (mainly andesitic) island of Vangunu, and, to the south, the other volcanic island of Gatukai. The inner barrier is separated from the outer by water more than 180 metres deep. The outer barrier rises from 13 to 25 metres above the datum, the inner one being from 0 to 7.5 metres high. This double barrier, which passes into a single one in the north and in the southwest, surrounds a lagoon very irregular in depth, with many basins and steep-sided shoals, depths of more than 55 metres having been charted. The outer barrier islands have generally vertical or overhanging cliffs on their seaward side, and are deeply incised by elevated intertidal corrosion notches. The largest and most extensive notch, which was measured is at +4 metres, is floored by a layer of algal limestone, and has a maximum depth of 15 metres; its visor or upper rim is a complex feature with two or more concave sections; slatactites and stalagmites occur in the notch, just as in a subterranean cave. A smaller notch exists at a lower level, and submerged terraces have been reported at depths of 10, 15 and 45 metres below sea-level. From an analysis of the site and projection of underwater and subreef contours of the Vangunu and Gatukai volcanoes, Stoddart (1969c) estimated that the Marovo reef prism was 'not less than 300m and possibly more than 600m thick at its outer edge.' The most probable sequence of events is that it was initiated in the Late Tertiary on the subsiding volcanoes 'in the Darwinian manner' the barrier reef then resulted from further subsidence. This occurred in stages where the barrier is double, and was more uniform where the barrier is a single formation. This barrier was subsequently raised tectonically in several phases; emergence of the 4 metres notch has been ascribed to tectonic rather than eustatic changes. The irregular topography of the lagoon seems to be of karstic origin, developed during leistocene low sea-level(s). Modern coral growth is very moderate. What remains to explain is the very large depth—more than 180 metres—of the channel between the two barriers; it may be due to local downfaulting. The Marovo reef is thus a fine example of the combined effects of subsidence, tectonic deformation and eustatic shifts in sea-level.

Variation in tectonic influences from place to place in the New Georgia group is obvious, for example, on the north coast of Vella Lavella Island where a fringing reef passes laterally on both sides into a barrier reef, with a progressive development of lagoons between the reef and the mainland. This is probably a result of local subsidence of the mainland, and is a striking illustration of Darwin's theory. Similar features exist at the eastern end of Guadalcanal.

In all the tectonic influences so far analysed, no evidence against Darwin's theory has thus been found to explain thic reefs. Only thin reefs, as those forming theatre-like coral terraces, escape this theory. On the other hand, the karstic saucer theory apparently does not find a support in the Marovo barrier reef and lagoon, since the supposed karstic features in this lagoon are not an essential fact. However, raised reefs such as those described in the Fiji Islands seem to exemplify combinations between the shaping of large internal depressions in reefs by subaerial solution and the effects of tectonic uplift.

As a general conclusion on reef genesis. Steers and Stoddart's statement at the end of their 1977 review may be reminded as being always quite acceptable: Darwin's subsidence theory provides generally adequate explanation of the development of the main reef types; Daly's theory in itself is inadequate, but quaternary changes of sea-level have been of crucial importance, particularly in permitting the karstification of emerged reefs. Regional tectonic upheavals explain certain reef features in marginal regions such as Melanesia and the Red Sea, as Darwin noted. But if extensive drilling has greatly clarified the evolution of Central Pacific reefs, similar boreholes are needed in the Indian Ocean, the Red Sea and the Caribbean.

3

METAZOAN ANCESTORY

ORIGIN AND NATURE OF THE METAZOA

Although protozoans have attained great success within the limits of a plasma membrane, the very nature of such an organization imposes certain restrictions, especially that of size. If too large, an undivided mass of protoplasm lacks that mechanical and integrative organization which makes for successful occupation of the many kinds of habitats present in the biosphere. A much greater degree of diversification can be found in a group of different types of cells in which they can follow different specializations.

As a result of such limitations, many protozoans have become colonial to overcome such restictions. At first these colonial forms consisted of irregularly arranged groups of cells that were chiefly independent of each other in a gelatinous matrix. Later development included protoplasmic strands linking the cells with each other. The cells of a colony may have been alike at first, but by division of labor and differentiation, some of the cells were somatic and some were germ cells. Examples of these sequences are found in some present forms. In the life cycle of many protozoans there are stages that consist of many cells.

Of the different theories that try to explain the origin of metazoans or multicellular animals, a logical and much favoured one considers the origin of the Metazoa as coming from the Protozoa through stages that resemble the colonial protozoans of the present time. The requirements of life are very much the same in both protozoans and metazoans. Life activities are essentially alike, and similar principles must be followed in life's existence. All animals, small or large, poorly integrated or highly organized, must have food, oxygen, osmoregulation, waste disposal, reproduction, and other functional features. The Protozoa are able to carry out all these functions within one cell; the Metazoa distribute these functions among groups of cells by division of labor. It is understandable that a hard and fast line between some colonial forms in which there has been division of labor and differentiation and simple metazoans cannot be made. Some of the colonial protozoans may be considered multicellular animals, especially if they have attained an antero-posterior axis and have cells differentiated with respect to that axis.

No one route of metazoan origin is known with certainty, and all theories are highly speculative, with only suggestions of evidences for those proposed. The ingenious way in which zoologists have applied biologic principles in support of their ideas for the origin of metazoans is worthy of some admiration, although flaws can be found in all of them.

Origin of the Metazoa

Although the evolution of the isolated cell has been very creative, as seen in the many types of protozoans, it was not until the transformation of the unicellular to the multicellular state that animal life could realize most of its potentiality. As multicellular forms, animals could increase in size and become adapted to many environmental niches denied to the protistan form of life. As groups of cells, metazoans by specialization could form tissues, organs, and systems as functional assemblages. Many new architectural patterns were now available. At some stage in their evolutionary history the protistans gave origin to the higher plants and animals (Metaphyta and Metazoa). Both groups have reproductive tissues and usually organs such that gametes (sperm and eggs) are differentiated. Metazoans possess larval stages, whereas metaphytes do not. Metazoans have heterotrophic nutrition and are mostly motile, whereas metaphytes are photosynthetic and sessile.

Pressures Exerted on Unicellular Animals

The basic structural pattern of the protistans and the way their patterns can be modified indicate that they must have had enormous evolutionary potentialities. But they encountered certain disadvantages in their adaptive evolution. All their differentiation had to occur within the limits of a plasma membrane. The nucleus can control the chemical activities of nonly a small amount of cytoplasm. Although many protistans overcame this restriction somewhat by having many muclei

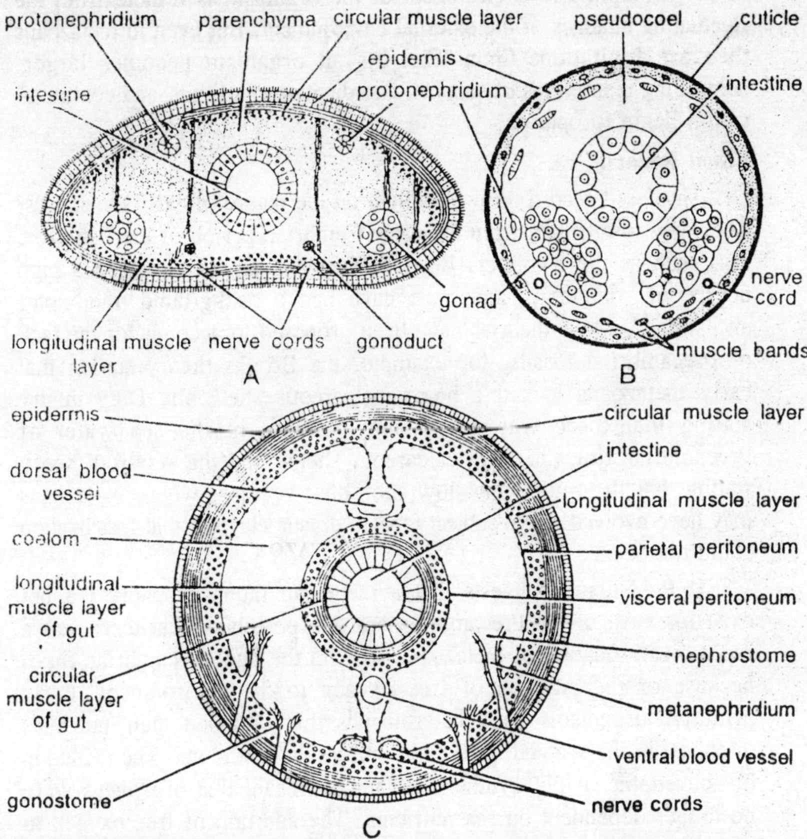

Fig. 3.1. Diagrammatic cross sections of grades of structure. A—Acoelomate grade; B—Pseudocoelomate grade; C—Eucoelomate grade.

and by having specialized organelles, they still lacked adequate circulatory and conducting systems for specific integration. As the body increased in size, diffusion alone was insufficient to supply needed materials, and more efficient transportation was required, which may the chief advantage they obtained when some of them formed colonies.

Single-celled forms were thus confronted by two selective pressures. They had to be larger to exploit successfully all the potential environments in which to live, and they needed that complexity of structure necessary to adjust to these possible environments. Size has been very important in the evolution of animals, and size is achieved by cell aggregations that have some selective advantages. Larger size enables organisms to be less dependent on environment because much of the protoplasmic environment of the organism is remote from the fluctuating changes of the external environment. But even in metazoans thee are limitations upon size. As an organism becomes larger, controlling restrictions of support, coordination, nutrition, and communication begin to operate.

Fossil Record

The fossil record throws on light on the nature of early metazoans. The early fossils date from the early Cambrian period of the Paleozoic era, such as the trilobites. But trilobites had already attained a high degree of complexity and must have had a background of simpler ancestors. Various theories have been proposed to account for the lack of Precambrian fossils, for example, the Brooks theory stating that early metazoans lacked a heavy calcareous shell; the Daly theory stating that there was insufficient calcium in the sea water of Precambrian times to form calcareous shells; and the Axelrod theory stating that the littoral (shallow shoreline) regions, where metazoans may have evolved, were subject to such drastic changes that fossilization could not occur.

J. R. Nursall suggests that a metazoan fauna probably did not exist for most of the Precambrian era. He postulates that there was a rapid diversification of metazoan forms in the late Precambrian times because of the addition of free oxygen to the environment. Early primitive organisms were heterotrophs that obtained their nutrients dissolved in the sea and lived under anaerobic conditions. The evolution of autotrophic or phtosynthetic organisms meant that organisms were no longer dependent on sea nutrients. The addition of free oxygen to the atmosphere by photosynthesis made oxidative metabolism possible. This more efficient method of extracting energy by food breakdown made possible a greater range of organization such as the tissues and organs of the multicellular organism. The fossil record then would be a natural outcome of the greater diversity of life that resulted.

Theories of Metazoan Origin

It is thought that all many-celled animals arose from flagellate protistans and that the sponges (Parazoa) may have arisen from a different group of flagellates than did the other metazoans or else may have diverged very early from the metazoan stem. Early protists experienced many selective pressures. Some protists apparently changed little, others became phtosynthetic and plantlike, and still others became heterotrophic, such aks rhizopods and ciliates. Somewhere in their

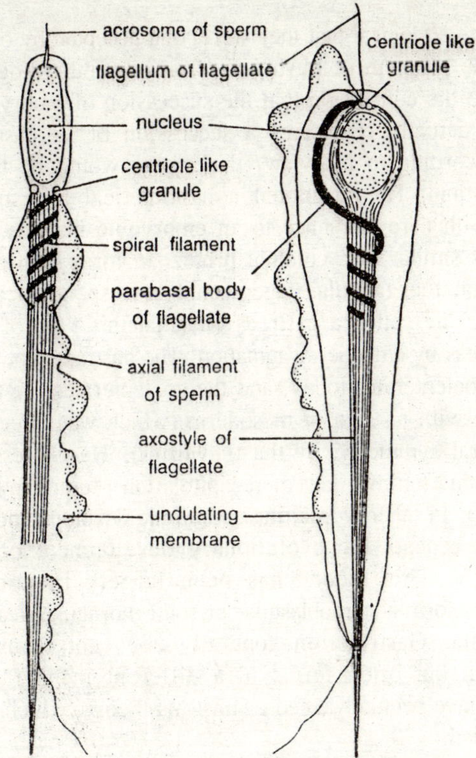

Fig. 3.2. Comparison of a spermatozoan (A) and a polymastigote flagellate (B).

evolution, protists gave rise to terrestrial green plants (Metaphyta) and the multicellular animals (Metazoa) as we now know them both.

There are many theories regarding the exact origin of the metazoans. The polyphyletic origin of different groups has made it difficult to arrive at definite conclusions. This Problem is all the more difficult because of the lack of fossil and biochemical evidences. Whreas there is general agreement that metazoans came from the protists, there is no agreement regarding which groups of protists may be the actual ancestor. At present speculation is based largely on evidence presented by the study of the conditions of living forms, as in embryology and comparative anatomy. So many phases have to be accounted for in any theory that great ingenuity has to be exercised in explaining how this or that condition arose.

The first serious attempt at explaining metazoan origin was that proposed by Haeckel in the nineteenth century. Many aspects of his theory werre so logical that they were long accepted by others. Haeckel attacked the problem of phylogeny from an embryologic standpoint He came to the conclusion that the succession of embryonic stages of an animal actually represents a succession of its past evolutionary stages. According to his view, the zygote would be the unicellular protistan stage. He postulated a hypothetical ancestral animal, a blastaea, with a resemblance to an embryonic blastula. This type of animal was similar to a colonial protozoan form such as *Volvox*. He believed that the gastrula stage would correspond to a hypothetical adult animal he called a gastraea and that this type of animal arose from blastaeas by embolic invagination. The gastraea type is represented by living coelenterates today, and the coelenterates, he thought, were diploblastic with no sign of mesoderm (which was later added, along with bilateral symmetry, by the flatworms). He expressed his views in brief by his familiar law of recapitulation ("ontogeny recapitulates phylogeny"). In other words, the embryonic development of the zygote (ontogeny) repeats the evolutionary development of the phylum (phylogeny). This theory has been largely invalidated. Many coelenterates form a stereoblastula, or solid morula, instead of a hollow coeloblastula. Gastrulation generally does not form by embolic invagination but often forms in a different manner. Many other objections have been advanced against his theory, which is considered oversimplified.

Two other theories may be briefly described. One of these is the *colonial theory*, which states that the metazoans evolved from colonial flagellates and may have been in the form of hollow, spherelike bodies similar to *Volvox* or other colonial forms. A common series of colonial

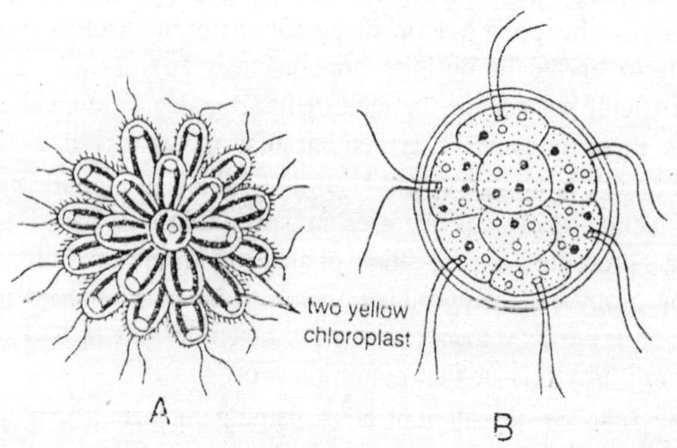

Fig. 3.3. A—Synura vella, a chrysomonad; B—Pandorina, a phytomonad.

organisms occurs in the holotrophytic order Volvocida, which by asexual reproduction forms colonies of individuals, such as *Gonium*, with 4 to 16 zooids; *Eudorina*, with 32 zooids; and *Pleodorina*, with up to 128 zooids. Such a progenitor may exhibit polarity, functional differentiation of cells (mostly reproductive), and sexual reproduction. In some the differentiation and division of labor may have extended further; individuality in cells finally became lost, and the whole colony became a single organism. The colonial theory considers that blastula-like colonies took place in the development as Haeckel believed, but that the inner layer was solid instead of hollow as Haeckel maintained.

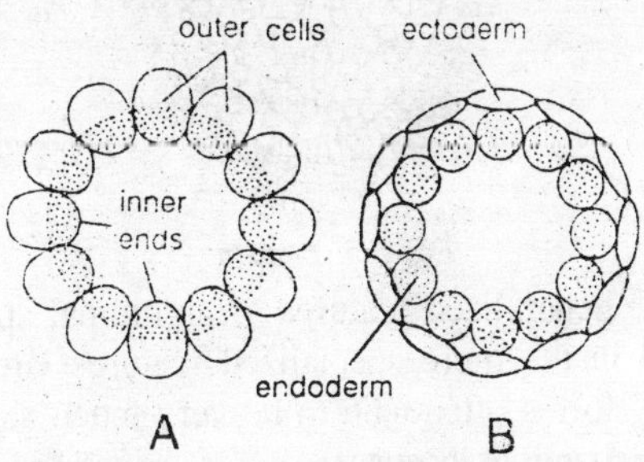

Fig. 3.4. A and B, Endoderm formation by primary delamination, according to Lankester's theory.

The *syncytial theory* of metazoan origin states that the progenitor of the metazoans is to be found in a multinuclear ciliate form that already had organelles. Simple division of the protist by developing membrances divided the single cell into multiple cells, each with one nucleus, thus forming a flatworm-like metazoan. According to this compartmentalization theory coelenterates arose from ancestral proflatworms and not the reverse. Of the two theories, the colonial theory is preferred by most biologists.

Organization of the Metazoan

Most animals have two major traits in common, a structural one and a functional one. With the metazoans, the structural organization is usually manifested by organs and organ systems. The functional trait revolves around the type of nutrition, which is usually heterotrophic, that is, its food cannot be made within its body (photosynthesis) but must be obtained from preexisting organisms. These two traits set the pace for the basic nature of the animal.

The structure of each animal has a specific architectural plan or design; all its constituent parts are arranged in some pattern. Such a structural arrangement is fitted for adjustments to its particular habitat.

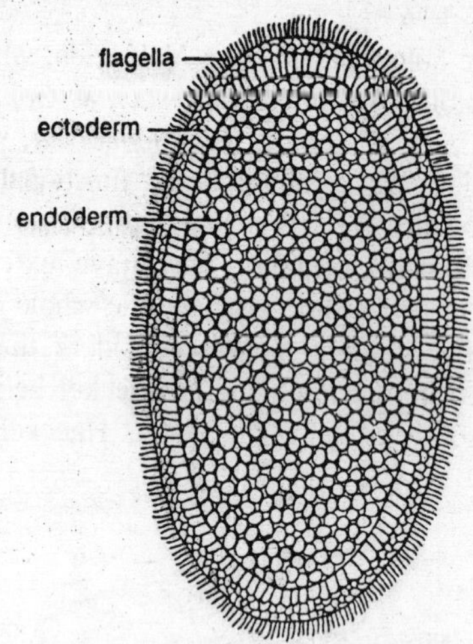

Fig. 3.5. Planula.

Usually the more generalized features of an animal appear first in its organization and in the more specialized features later. This is why embryos of many forms often tend to be very much alike, but become different as specializations occur.

Basic organization features depend on the organization level of the animal. However, there is a principle of uniformity thoughout the animal organization. The life processes are about the same in all animals. They all maintain a continuous exchange of energy and materials with their environment. There is a basic unity in structural and functional patterns no matter how different animals may seem to be.

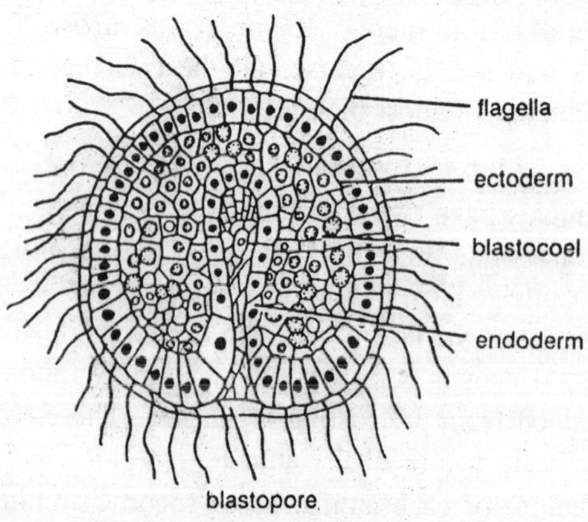

Fig. 3.6. A typical gastrula showing endoderm formation by embolic invagination.

Differentiation and Divison of Labor

All animals, even the simplest, undergo division of labor that leads to specialization of body parts and increased effectiveness of the living process. A unicellular organism carries on the entire life process of many functions. In a multicellular form these functions may be distributed among several or many cells. No cell is limited to a single function, for every cell must have the materials for metabolic and self-perpetuative functions to survive. The fewer additional functions a cell must perform, the more specialized it is.

Whatever the design of an animnal, that partern may be sufficient for its own particular relation to its environment. Environments vary in difficulty, and it may require a greater complexity of design to adjust to some particular niches. The organisms that fit into a more difficult environment usually have a higher division of labor and specialization and reach a complexity of structure not found in lower forms.

Symmetry and Polarity

Symmetry has adaptive significance and various types are found in animals. Symmetry refers to the regularity of body parts and the geometric relations that exist among their structures. An animal may have primary symmetry as an embryo and a different (secondary) symmetry as an adult, Symmetry may be modified to meet the conditions of life, and most animals tend to be asymmetric in regard to distribution of organs.

Polarity refers to a diffefrential gradation of materials along an axis. There may be also a gradient of activity in polarity. Motility of an animal has a significant symmetric effect on the architecture of an animal.

In *spherical symmetry* there is a symmetric center and every plane passed through the center divides the body into similar halves. Some colonial protozoans approach this type of symmetry, which is especially fitted for floating or rolling.

In *radial symmetry* and animal can be divided into similar halves by any plane passing through the longitudinal (oral-aboral) axis. Such symmetry is particularly suitable for forms that are adapted to a sessile existence and is characteristic of the coelenterates and, to some extent, the ctenophores. However, in the ctenophores the symmetry is modified and is called *biradial symmetry* because of the presence of some part that is single or paired so that usually only one or two of the longitudinal planes divide the form into truly similar halves.

Bilateral symmetry, in which two halves of the animal, one on each side of a dorsoventral plane, are more or less mirror images of each other, is by far the commonest type. Such symmetry is usually associated with cephalization and motility, and the animals are usually elongated in the direction of movement.

Asymmetry refers to a lack of symmetry and is demonstrated by most sponges and by the coiled shells of adult snails.

Animals are commonly grouped according to their symmetry. The radiate animals, or those with radial or biradial symmetry, are called the Radiata; the bilateral animals, or those with bilateral symmetry, are called the Bilateria.

Segmentation and Tagmatization

Segmentation refers to a series of linearly arranged somites, or metameres, each of which is similar to those in front of and behind it. Such a condition must have appeared at least twice in evolutionary history—in annelids and in chordate animals. Because of the close relationship of the annelids and arthropods, the arthropods are also segmented.

Metamerism is a mesodermal phenomenon and is restricted to the body wall musculature and body cavity. A corresponding metamerism is found in certain other organs such as blood vessels, nerves, and excretory organs. The only organs that are not involved in the metamerism are those derived from endoderm—in other words, the digestive organs. Some other wormlike forms show conditions of pseudometamerism, especially among the flatworms. Several theories have been advanced to explain the origin of metamerism. Some state

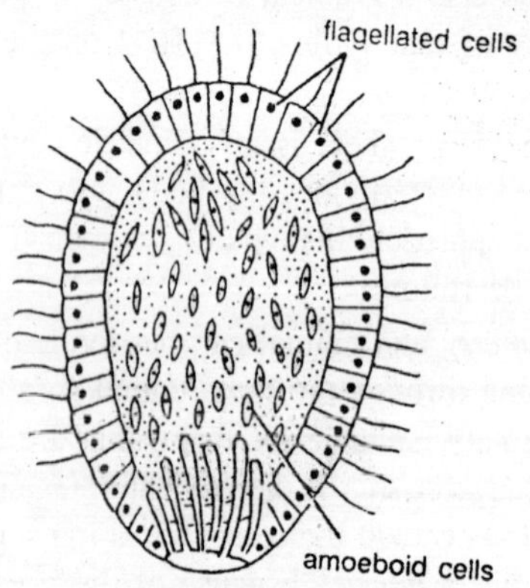

Fig. 3.7. Parenchymula larva.

that somites arose as incompletely divided chains of zooids, and others maintain that metamerism first arose in the muscular tissue as an aid to locomotion and that this influenced other mesodermal organs to follow suit. In general the problem of its origin is obscure.

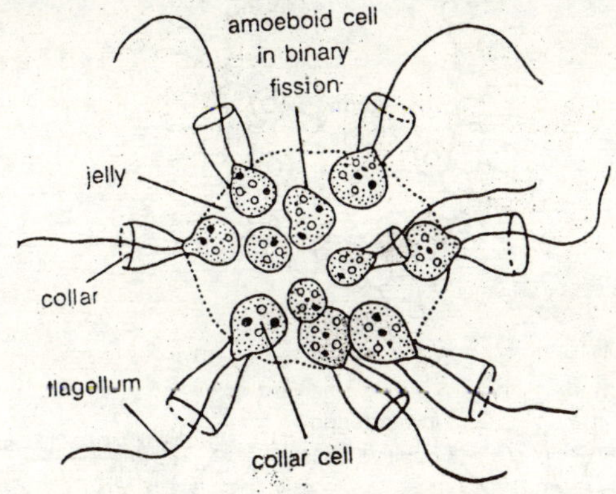

Fig. 3.8. Proterospongia haeckeli.

Whatever its origin, there is no doubt about the significance of metamerism in the locomotion of animals. With the aid of septa the coelomic cavity is divided into a series of repeated units. The pressure of the coelomic fluid within these units creates a type of hydrostatic skeleton that gives form to soft bodied organisms. Varying the amount of turgor within indivudial segments increases the capacity for local changes of shape and becomes a valuable aid to locomotion. Even in animals possessing a skeleton the serial arrangement of the body wall musculature ensures more effective and efficient body movements.

Metamerism is modified in various ways. Segments may be united into functional groups, each group being structurally separated from other groups and specialized to perform a certain function for the animal. These groups are called *tagmata*, and tagmosis is especially common in the arthropods. In most cases, metamerism is best shown in the embryonic condition but is frequently masked in the adult.

Embryology of the Metazoans

Types of Cleavage

In the embryology of the metazoans there is an early period of cleavage that differs somewhat from ordinary cell division (mitosis). In ordinary mitosis in the adult body the individual cells undergo a period of growth between successive divisions, but in cleavage on such interval of growth occurs. and each succeeding division produces smaller

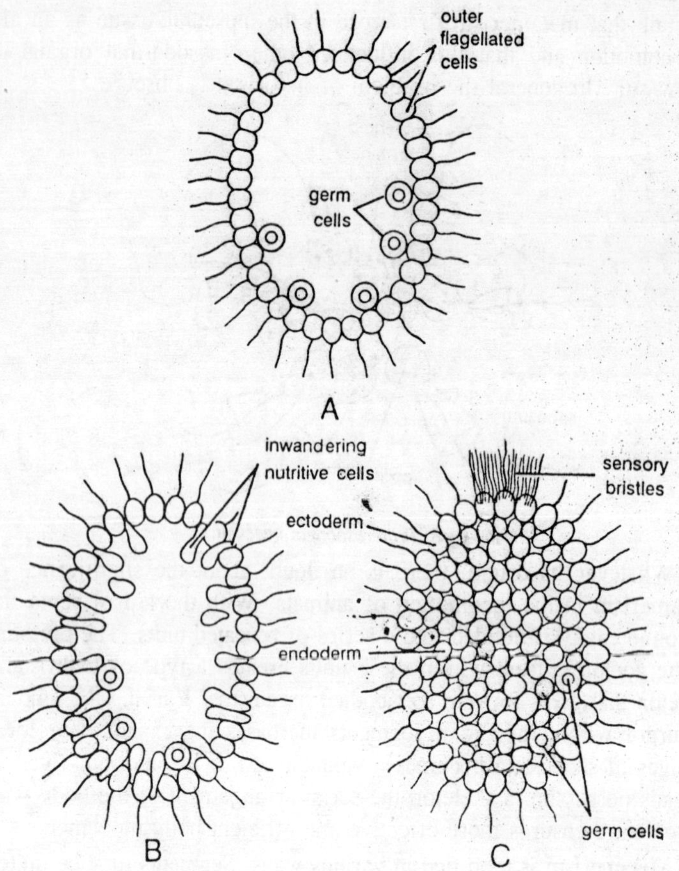

Fig. 3.9. Formation of endoderm according to Metschnikoff and Hyman. A—Hypothetical ancestral blastaea. B—Multipolar ingression. C—Hypothetical stereogastrula or a planucloid ancestor of Eumetazoa.

blastomeres, or cells. Cleavage is considered as terminating upon the establishment of the blastula.

The pattern of cleavage depends to a great extent on the amount of yolk in the egg. Yolk tends ot retard cleavage by presenting a barrier to the orderly rearrangement of nuclear material and the formation of the spindle. The locations of cleavage planes depend largely on the positions of the mitotic spindles. The locations of cleavage planes depend largely on the positions of the mitotic spindles, which are controlled to a certain extent by the organization of the cytoplasm and local differences in the egg cortex. Invertebrate eggs differ enormously in size from the tiny microscopic eggs of the bryozoans to the very large ones of the squid.

There are two major types of cleavage—spiral and radial. In radial cleavage the early cleavage planes are either parallel with or at right angles to the polar axis, thus producing blastomeres in tiers or layers of cells. Radial cleavage is *indeterminate* (regulative), and the early blastomeres are equally potent. If the first 2 or 4 cells are separated from each other, each cell is capable of continuing cleavage and forming a small blastula or even a diminutive larva. It is thus possible for twinning to occur. In this type of cleavage the localization of specific formative substance is delayed and decided by interactions with adjacent cells (embryonic induction).

In spiral cleavage, on the other hand, the cleavage planes are diagonal to the polar axis and produce alternate clockwise and counter-clockwise quartets of unequal cells around the axis of polarity. Unlike radial cleavage, spiral cleavage is *determinate* (mosaic), and the fate of each cell is fixed early in development. When a 2 celled embryo is separated into 2 blastomeres, each cell is capable of producing only half a larva. The formative substance of the cytoplasm is early localized, and all blastomeres are necessary for the formation of a complete embryo.

Cell lineage is a type of embryologic study that involves tracing blastomeres, or the cells formed during differentiation of the zygote, to their ultimate fate in tissues and organs. To trace the lineage of cells, a method of designating each blastomere and its descendents is employed. This system can be used only in those forms that follow a rigid, stereotyped pattern, as in those eggs that have determinate cleavage.

As will be pointed out later, spiral cleavage occurs chiefly, in the Protostomia and radial cleavage in the Deuterostomia. However, there may be specialized modifications of cleavage patterns in both these groups.

In spiral cleavage the egg divides into four equal blastomeres in the first two meridional cleavages. These blastomeres are called A, B, C, and D. The third division is unequal and at right angles to the first two and results in four large macromeres and four small cells forming the first quartet of micromeres. These micromeres are designated 1a, 1b, 1c, and 1d, and the macromeres are 1A, 1B, 1C, and 1D. In successive cleavages, second, third, etc. quartets of micromeres are given off by the macromeres. Micromeres of the second quartet are numbered 2a, 2b, 2c, and 2d, whreas the macromeres become 2A, 2B, 2C, and 2D. When a third quartet is given off (3a, 3b, 3c, and 3d), the macromeres become 3A, 3B, 3C, and 3D, etc. In this scheme lower case letters represent micromeres; capital letters represent macromeres.

As the original quarters divide, their subsequent divisions are reparesented by exponents; thus the two daughter cells of 1b become $1b^1$ and $1b^2$. When these two daughter cells divide, more exponents are added. In this way the daughter cells of $1b^1$ become $1b^{11}$ and $1b^{12}$, and those of $1b^2$ will be $1b^{21}$ and $1b^{22}$. The division of $1b^{21}$ produces $1b^{211}$ and $1b^{212}$. The exponent 1 is given to the cell nearer the animal pole. The macromeres are located at the vegetal pole. It will be noted that all the blastomeres named "a" are derived from blastomere A of the original four blastomeres (A, B, C, D), those named "b" come from blastomere B, etc., and each set occupies about one quadrant of the embryo. The number before the letters designates the quartet from which the cell is derived, and the exponent indicates the number of subsequent divisions after the formation of a particular quartet of cells. The lower the number the nearer the micromere is to the animal pole; the macromeres are found at the vegetal pole. Cleavages are oblique and alternate right and left. It is thus possible to follow the origin of any particular cell.

The first three quartets and the cells derived from them form the ectoderm and the ectomesoderm. The fourth quartet procues the endoderm of the enteron. Micromere 4d of the fourth quartet gives rise to the endomesoderm, but there are some variations of this pattern in some species.

Blastula and Gastrula Formation

The blastula may be hollow and have a blastocoel (coeloblastula), or it may be solid and have no blastocoel (stereoblastula). Or in meroblastic cleavage, which is common in birds and certain invertebrates, cleavage mayu be restricted to the upper part of the zygote so that the blastula is only a plate of cells on top of an undivided cell mass. Cleavage in centrolecithal ova, found in some invertebrates, produces a continuous layer of cells surrounding a central undivided yolk mass (a periblastula).

Gastrulation occurs in many ways, giving rise to the germ layers and to the archenteron, or parimitive gut, which opens to the outside by the blastopore. The blastocoel may persist, or it may be obliterated completely. The outer layer of the germ layers is the ectoderm and the inner layer is the endoderm. Gastrulation and germ layer formation are chiefly processes of morphogenesis rather than of differentiation, which comes later in the development of organs.

Mesoderm and Coelom

Mesoderm is formed in a variety of ways. It may be derived from ectoderm or endoderm, since all cells of a gastrula belong to one or the other of these two primary layers. Mesoderm may be considered the secondary germ layer. In the radiate phyla, mesoderm is derived from ectoderm (ectomesoderm); in other metazoans it comes chiefly from the endoderm (endomesoderm). Most Bilateria derive mesoderm from both ectoderm and endoderm. From the germ layers the various tissues and organs are derived.

The formation of mesoderm is best understood in relation to the formation of the body cavities. In the radiate and lower bilateral animals the digestive tube is the only body cavity of the adult. Such forms are called *acoelomate* because there is no true body cavity, or coelom. In radiate forms a few cells from the ectoderm form between the ectoderm and endoderm a mesenchymal layer that secretes a jellylike matrix, the mesoglea, in which there may be some cellular ingression. In acoelomate Bilateria (flatworms and others) the mesoderm completely fills the region between the ectodermal integument and the endodermal digestive tube. This middle layer, the parenchyma, may contain jelly-secreting mesenchyme and specialized cells of organ systems.

In the *pseudocoelomates* the body cavity is a remnant of the blastocoel; it is lined on the outside by the body wall tissues and on the inside by the tissues of the digestive tube, with some mesoderm aggregated in places.

The *eucoelomates* (animals with a body cavity that is a true coelom) include the remainder of the Bilateria. The coelom may be formed in either of two ways—the schizocoelic or the enterocoelic method.

In the schizocoelic method the adult mesoderm arises in the embryo from two endoderm derived cells (primordial mesoderm cells), one on each side of the archenteron. At first it is a mass filling the space between ectoderm and endoderm, but by delamination it splits into an outer somatic layer and an inner splanchnic layer on each side of the archenteron. The somatic layer adheres to the body wall and the splanchnic layer to the digestive tube. This results in a coelomic space completely surrounded by mesodermal tissue.

In the enterocoelic method the mesoderm arises as paired lateral pouches from the endoderm. When the pouches lose continuity with the endoderm, their inner splanchnic portions remain against the developing digestive tube while their outer somatic layer is applied against the body wall. Between the two layers is the coelom as in the other method.

Other patterns of coelom formation are also found. For example, in some lophophorates loose mesenchyme may form and become arranged in a regular manner to form coelomic spaces.

The value of the coelom cannot be overestimated. It makes the activity of the digestive system independent of the body wall, Filled with fluid, the coelom can act as a hydrostatic skeleton. It may also aid in transporting food, waste, and other materials. It may form tubular organs connecting the coelom to the outside (coelomoducts) for the transport of gametes. Also, it forms a space for the location of organ systems that are connected to the body wall by mesenteries.

Protostomia and Deuterostomia

The terms Protostomia and Deuterostomia were introduced arbitrarily by K. Grobben (1908) to describe the two groups into which the bilateral animals of the animal kingdom are divided. According to this concept, after the origin of a bilateral ancestor two evolutionary lines of these animals developed. One line became the phyla in which the blastopore of the gastrula forms the mouth (Protostomia); the other line became the phyla in which the blastopore forms the anus (Deuterostomia). The radially symmetric coelenterates and ctenophores are naturally excluded.

Other fundamental differences besides the method of formation of the mouth and anus are found in the two groups. All protostomes have of the arthoropods, the prostomes have spiral cleavage. And in protostomes the coelom arises as a schizocoel by a splitting of the mesoderm. Deuterostomes have radial and indeterminate cleavage, and the coelom forms as an outgrowth of the enteron (enterocoelic method). In those protostomes that have a larva, it is usually of a trochophore type; in deuterostomes the larva of indirect development is of a dipleurula type.

Accordingly, the bilateral animals may be said to have a diphyletic origin—the protostomes leading to the annelids, arthropods, mollusks, and others and the deuterostomes leading to the vertebrates and related groups. There are many exceptions to be noted among the animals that make up the two groups. The development of the mesoderm varies greatly among the various animals. The development of the coelom is known to vary; in vertebrates, for example, is arises as a schizocoel

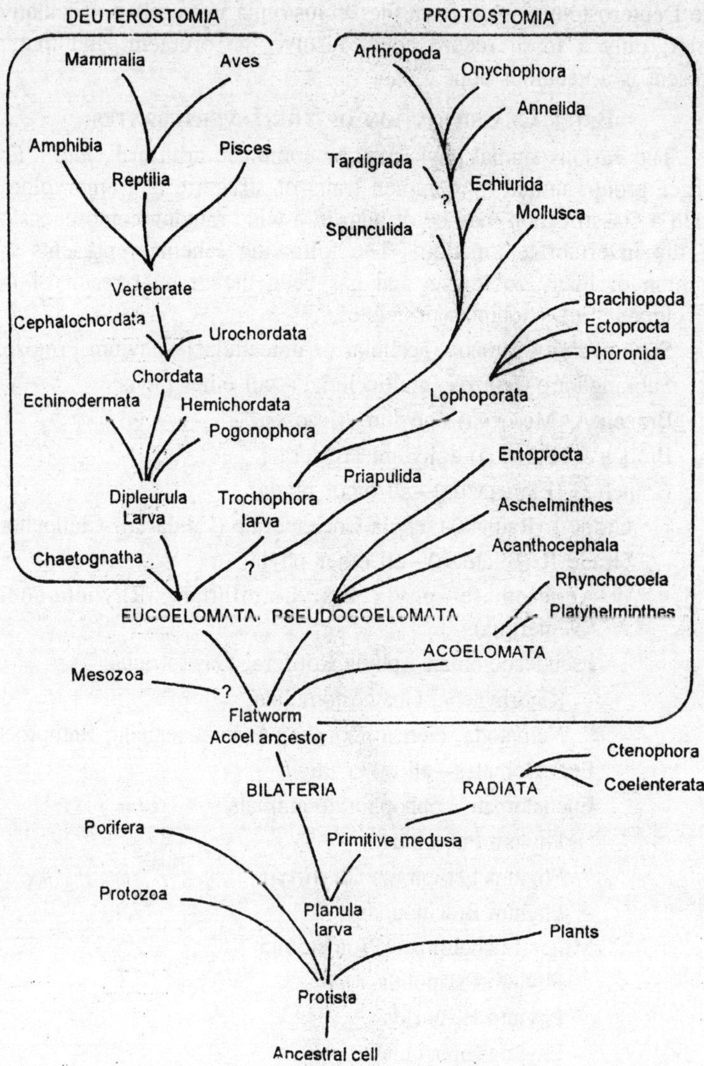

Fig. 3.10. Animal phylogeny (hypothetical). Basis for such a tree phylogeny is evolutionary relationship, or common ancestory, but such information is scanty.

instead of an enterocoel. There are many modifications in every phylum, for nature follows no set plan of development for each major group.

In general, the concept of Protostomia and Deuterostomia may be considered logical, in spite of many shortcomings, for each group has tended to follow potentialities inherited from a common ancestry. How the Deuterostomia arose from the Protostomia is largely a speculative point; only a fossil record could resolve the problem, and that at present is a nebulous hope.

Brief Classification of the Invertebrates

The various animal phyla may be combined arbitrarily into a few larger groups united by common traits of structure and embryology. Such a classification may be of help in a wide ranging comprehension of the invertebrate kingdom. The following scheme represents the opinion of many zoologists and has been the gradual result of the development of taxonomic principles.

Subkingdom Protozoa (acellular or unicellular)—phylum Protozoa
Subkingdom Metazoa (multicellular)—all other phyla
Branch A (Mesozoa)—phylum Mesozoa
Branch B (Parazoa)—phylum Porifera
Branch C (Eumetazoa)—all other phyla
- Grade I (Radiata)—Phyla Coelenterata (Cnidaria), Ctenophora
- Grade II (Bilateria)—all other phyla
  - Acoelomate—phyla Platyhelminthes, Rhynchocoela (Nemertina)
  - Pseudocoelomate—phyla Rotifera, Gastrotricha, Kinorhyncha, Gnathostomulida, Nematoda, Nematomorpha, Acanthocephala, Entoprocta
  - Eucoelomates—all other phyla
  - Eucoelomate Lophophorate animals
    - Phylum Phoronida
    - Phylum Ectoprocta (Bryozoa)
    - Phylum Brachiopoda
  - Minor Eucoelomate Protostomia
    - Phylum Priapulida
    - Phylum Echiurida
    - Phylum Sipunculida
    - Phylum Tardigrada
    - Phylum Pentastomida (Linguatulida)
    - Phylum Onychophora
  - Major Eucoelomate Protostomia
    - Phylum Mollusca
    - Phylum Annelida
    - Phylum Arthropoda

Eucoelomate Deuterostomia

Phylum Echinodermata

Phylum Chaetognatha

Phylum Hemichordata

Phylum Pogonophora (Brachiata)

Phylum Chordata

Subphyla Urochordata (Tunicata), Cephalochordata

Table 3.1: Outline Resume of the Invertebrate Phyla

Grade of Organization	*Phyla*
I. Unicellular (or acellular), or colonial with undifferentiated somatic cells	Protozoa
II. Multicellular (Metazoa) with differentiated cells and typically larval stages	
A. Stereoblastula grade	Mesozoa
B. Cellular grade (Parazoa)	Porifera
C. Tissue grade (Eumetazoa, in part)	Coelenterata,
D. Organ system grade (Eumetazoa, in part)	Ctenophora
1. Acoelomate	Platyhelminthes, Rhynchocoela (Nemertinea)
2. Pseudocoelomate	Rotifera, Gastrotrichia, Kinorhyncha, Nematoda, Nematomorpha, Acanthocephala, Entoprocta
3. Eucoelomate	
a. Protostomes (blastopore forms mouth; spiral, determinate cleavage)	
1. Lophophorate protostomes	Ectoprocta, Brachiopoda, Phoronida
2. Schizocoelous protostomes	Priapulida, Sipunculida, Mollusca, Annelida, Arthropoda
b. Deuterostomes (blastopore forms anus; radial, indeterminate cleavage)	Chaetognatha, Pogonophora, Hemichordata, Echinodermata, Chordata

4

General Form

Main Axis and its Poles in Multicellular Animals

Remembering what we have said about symmetry in Protozoa, we may say that the general direction of the development of symmetry and, on the whole, the structural plan of individual protozoans, beginning with homaxonic forms, amounts to a constantly-increasing definiteness of location of part and consequently to a decline in symmetry: we see a transition from forms possessing a centre of symmetry to forms lacking such a centre, from multiaxial forms to uniaxial, from homopolar to heteropolar, from forms with an axis of rotation of a infinitely large order to forms with an axis of rotation of a definite order with a definite number of antimeres, and ultimately to bilaterally-symmetrical and dissymmetrical forms. If we turn to examine the symmetry of multicellular animals we find something analogous there. Comparing the structural plans of all Metazoa from radially-symmetrical Medusae to disssymmetrical snails we also find gradual diminution of the number of elements of symmetry, a decrease in repetition of parts of account of the constantly-increasing differentiation of members of the body, on account of decrease in the number of identical parts due to increasing heterogeneity of such parts.

Thus in broad outline the laws of development of the general structural plan are the same in Protozoa and in Metazoa. In Protozoa the decline in symmetry is associated, more or less, with general increase in complexity or organisation; that is, animals standing higher in their general level of organisation possess a impoverished type of symmetry. In Metazoa a similar correlation is much more strongly expressed, so that the grades of development of symmetry in them coincide with the grades of their general level of organisation, as far as establishment of bilateral symmetry and heteronomous metamerism. There are important differences, however, between Metazoa and Protozoa with regard to symmetry. In the first place, the symmetry of Metazoa is much more uniform than that of Protozoa. Among multicellular individuals we do not find at any stage anaxonic forms without symmetry or traces of it, nor homaxonic or stauraxonic-heteropolar forms possessing a centre of symmetry. It is true that several multicellular animals have a spherical form at the fertilised-ovum stage. Several other stages of development also have an approximately-spherical external form, e.g. the blastula of sea-urchins, etc. But on closer examination that blastula proves to be not at all homaxonic: it has a single main axis of symmetry and two different poles. At one pole the cells are smaller and contain less yolk, and at the other they are larger and contain more yolk. In precisely the same way fertilised ova and other stages of multicellular animals that at first glance appear to be polyaxonic prove to the actually at least monaxonic-heteropolar, and sometimes even bilaterally symmetrical. The anaxonic or homaxonic form is found only in unfertilised ova of

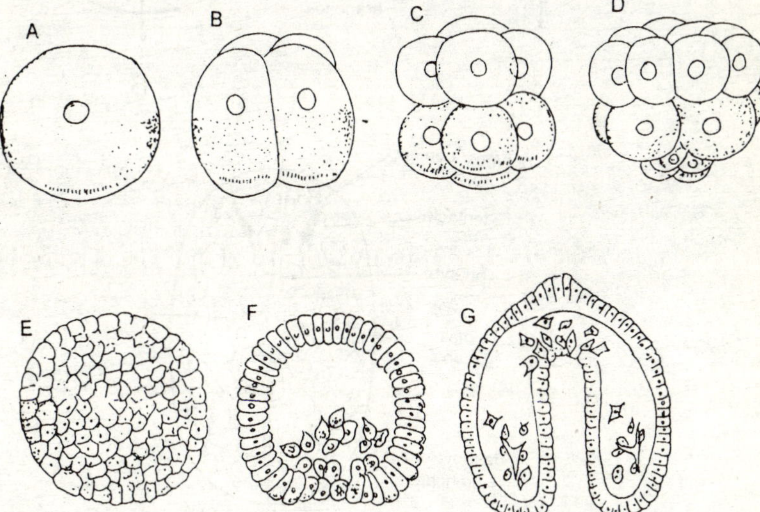

Fig. 4.1. Monaxonic-heteropolar structure of different stages of development of Metazoa, taking the sea-urchin Strongylocentrotus lividus as an example. A—ovum before cleavage. B—4-blastomere stage. C—8-blastomere stage. D—16-blastomere stage. E—blastula. F—immigration of mesenchyma. G—gastrula.

several lower forms (sponges, acoelous turbellarians, hydroids), but we must not regard an unfertilised ovum as a multicellular individual.

Therefore the monaxonic-heteropolar form with an axis of symmetry of an infinitely large order is the original form of symmetry for all multicellular animals. Moreover, it is a highly important circumstance that the main axis of the body and the two poles (designated animal and vegetative) are homologous in all groups of lower Metazoa and in those higher groups where they can still be distinguished. There is no possibility of tracing such homology of axis and poles within the phylum

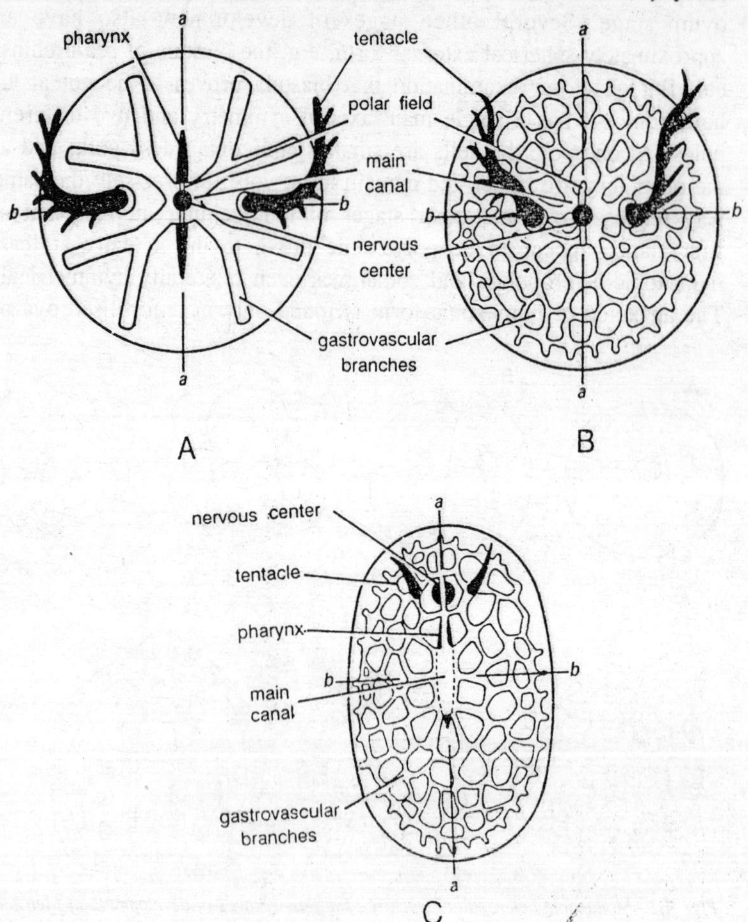

Fig. 4.2. Gradual change from a ctenophore to a polyclad, according to Ctenophore-polyclad theory. A—Ctenophore; B—Hypothetical intermediate stage; C—Polyclad.

of Protozoa. At most we can sometimes speak of homologies within separate classes, but sometimes we have to restrict ourselves to smaller groups. The organisation of Protozoa is too diversified and at the same time too limited in characteristics to permit far-reaching comparisons of structural plans. Moreover, in those cases where we can form an opinion on the subject, the monaxonic state (like bilateral symmetry) arose independently in different branches of Protozoa, by different paths, and on very different morphological foundations, so that it is impossible to speak of homologies.

All multicellular animals, then, possess a single structural plan, at least in their lowest representatives and in early stages of development; therein they differ greatly from Protozoa, which do not possess even that unity of plan. What are the characteristics of the animal and vegetative poles in the lower Metazoa? Such characteristics are fairly numerous, and although each of them may be lacking on some occasions, taken all together they always permit correct orientation of the animal. The *vegetative* (sometimes called *oral*) pole is characterised primarily by the fact that the mouth is formed there. In the lower forms it is actually a mouth, but in higher forms it is a primitive embryonic mouth (blastopore). An exception is presented by sponges, which have no mouth at and, all in which a blastopore occurs only rarely (Calcarea). Secondly, the vegetative pole is distinguished by the fact that in the early stages of development the formation of the inner embryonic layer, or *phagocytoblast* (by that term I.I. Mechnikov designates the entire combination of endoderm and mesoderm) is concentrated there. It is true that in the most primitive types of formation of phagocytoblast, found in some sponges and hydroids, it has a multipolar origin, i.e. it arises over the whole periphery of the embryo. But in most sponges and hydroids, not to mention all other classes, formation of the phagocytoblast or at least of the greater part of it is concentrated at the vegetative pole. Thirdly, when the animal is swimming by means of cilia the oral pole is almost always directed backwards and is physiologically posterior; ctenophores, which usually swim with the oral end of the body in front, constitute an exception. Giving that feature exaggerated significance, E. Reisinger denies the homology of the oral end of Ctenophora with that of other lower Metazoa.

The *animal* pole, sometimes called *aboral*, has the following characteristics: in the first place, in cases where the lower Metazoa make a transition to a sessile mode of life, attachment of the free-swimming larva to the substrate almost always takes place at the animal pole. The only exception is the aberrant ctenophore *Tjalfiella*, whose larva attaches itself to the substrate by its oral pole. The creeping ctenophores Platyctenida also apply their oral end to the substrate. Secondly, abundance of sensory elements and of neural elements in general is characteristic of the animal pole. Such abundance is a feature of coelenterates and worms, and also of ctenophores, which alone among coelenterates fully retain the free-moving larval mode of life. At the same time the majority of Cnidaria make a transition with age to a sessile mode of life, attaching themselves to the substrate by the aboral end, as a result of which the latter loses most of its neural elements, being converted into a simple foot. Still poorer in neural elements is the aboral end of the body (exumbrella) of metagenetic medusae (i.e. medusae with alternation of sexual and asexual generations—Leptolida), in which the aboral pole corresponds to the place where the medusa breaks away from the maternal colony. Among hypogenetic (i.e. not having alternation of generations) medusae, on the other hand, some Narcomedusae retain a sensory organ at the aboral pole, but only up to and including the actinula stage. The aboral sensory organ is retained, in the form of an apical plate, in the actinula of the hydroid *Tubularia* (Leptolida Athecata).

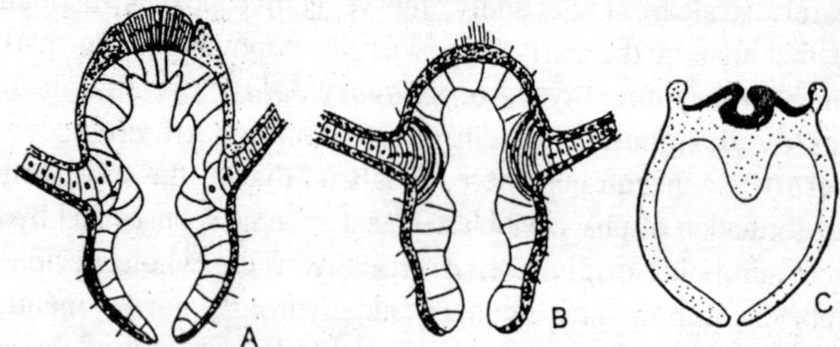

Fig. 4.3. Representatives of Hydrozoa with well-developed aboral sense organ and its homologues. A—larva (actinula) of Tubularia. B—larva of the narcomedusa Solmundella. C—larva (conarium) of the siphonophore Velella (Disconantae).

Such cases of retention of its original wealth of neural elements of the aboral pole in Cnidaria in stages later than the planula are of great interest. They confirm the view that the aboral concentration of neural elements is a very ancient characteristic of the lower Enterozoa (R. Woltereck).

Returning to the features of the animal pole we must remark that besides the above it is distinguished by characteristics opposite to those of the vegetative pole: with progressive ciliate movement it is usually directed forward; the animal pole lies in the *kinetoblast* (ectoderm) region. On the whole we see that each pole of the body of primitive Metazoa possesses numerous characteristics, although every one of these proves to be inapplicable in some particular cases. The large (and important in the field of comparative anatomy) group of sponges (Porifera) has neither mouth nor nervous system, and there is no doubt that the absence of both in that group is primary. Nevertheless the direction of movement and method of attachment of the larvae, as well as other indications, enable us to determine the homology of the poles of the body even in sponges. In a word, we can always recognise the oral and aboral poles by an assemblage of features. Therefore when we speak of the heteropolarity of multicellular animals we refer to it not in general terms but as something definite, determinable by a single definite method.

The line joining the aboral and oral poles of the body is usually called the main axis or primary axis of the body. In the lower Metazoa it is an axis of symmetry; in higher bilateral forms the primary axis ceases to be an axis of symmetry. The name 'primary axis' is retained even there for the line joining points homologous with the oral and aboral poles. Only in those animals where it is impossible to discover the location of the poles of the body does it also become impossible to determine the primary axis.

The fertilised ova of most Metazoa possess monaxonic-heteropolar symmetry, and their poles also are called animal and vegetative. The ovum and the multicellular animal, however, are units of different structural order, and the similarly-named poles of the ovum and of the larva have qualitatively-different groups of characteristics. The characteristics of the animal pole of the ovum are: (*i*) maximum physiological activity; (*ii*) relative poverty in yolk, unless the latter is distributed with absolute uniformity; (*iii*) proximity of the centrosome, and sometimes also of the nucleus, to the animal pole; (*iv*) appearance of cleavage grooves at it, when they do not appear simultaneously over the whole surface of the ovum; etc. The characteristics of the vegetative pole are the reverse of these. In the lower Metazoa, however, the characteristics of the two poles of the ovum are far from being all, and always, displayed, and their homology has to be determined—as in the larva—by an assemblage of characteristics.

The succession of similarly-named poles in the ovum and in the embryo is fairly constant in higher animals (beginning with Scyphozoa and sea-urchins), but in lower Metazoa it has not yet been quite set;

in sponges and metagenetic hydromedusae the animal pole of the ovum becomes the posterior pole of the flagellate larva. In the ontogenesis of Leptolida we at first observe the ovum with its cellular, 'protozoan' entirety and polarisation, which breaks down in the early stages of fission, and the polarisation of the blastula arises as the expression of the entirety of a newly-developing multicellular individual. This circumstance is of the greatest importance in understanding the early stages of evolution of the ontogenesis of Metazoa.

Symmetry in the Structure of Sponges and Archaeocyatha

Those Metazoa in which the main axis of the body continues to be an axis of symmetry throughout the lifetime of the animal are called radial. They include three phyla: sponges (Porifera) and coelenterates (Coelenterata) among modern Metazoa, and the Cambrian Archaeocyatha. Radial animals in this sense are not in any way a taxonomic unit: in the natural system of animals there is no such group, since in the total of their characteristics coelenterates stand much farther from sponges than from other Metazoa. Along with the latter, coelenterates are included in the superdivision Enterozoa, as contrasted with sponges. Sponges occupy an extremely individualised position in the system (forming the superdivision Enantiozoa). Therefore the group of radial animals as defined above would be a combination of three phyla with only one feature in common—the nature of their symmetry. We prefer to apply the name Radialia only to coelenterates, to distinguish them from the rest of the Enterozoa, which are to be grouped together as Bilateria. We may remark that there are many bilaterally-symmetrical forms among coelenterates, but all these forms retain incomplete radial symmetry (see below) along with bilateral symmetry.

The majority of Archaeocyatha, sponges, and coelenterates are, however, constructed on a radial pattern. In what forms do we find radial structure among sponges? The simplest structural type, stauraxonic-heteropolar with an axis of symmetry of infinitely large order, we find principally in isolated sponges of the orders Calcarea, Triaxonia, and Tetraxonia. Among sponges, however, with their often incomplete individuality, radial symmetry is rarely so geometrically-perfect as in many of the higher coelenterates. Sometimes the body axis is curved; sometimes the body walls are deformed by irregular growth; in extreme cases radial symmetry is scarcely discernible. All of these deviations from true radial symmetry represent nothing else

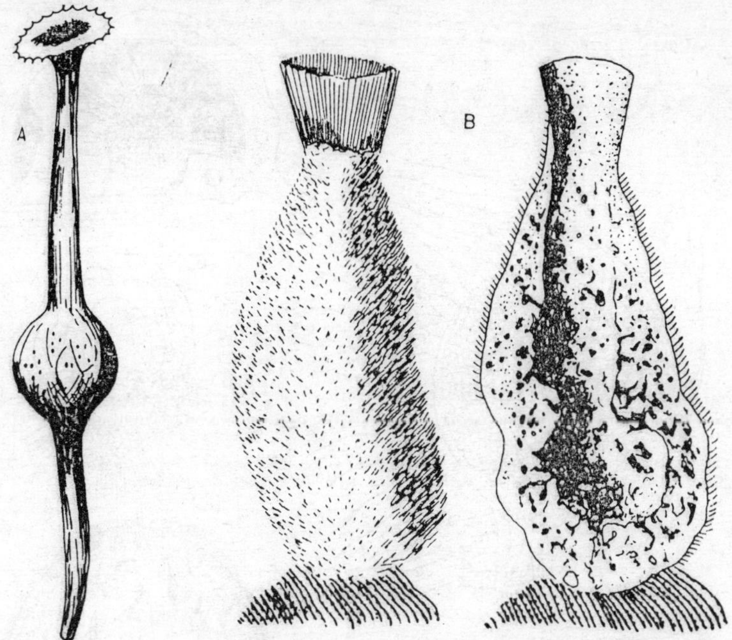

Fig. 4.4. Radial symmetry in solitary sponges. A—Disyringa dissimilis (Tetraxonia); the axis of symmetry is slightly curved. B—Leucandra aspera (Calcarea), external aspect and longitudinal section, incomplete radial symmetry.

than traces of anaxonism that has not been completely overcome. The radial symmetry of many sponges has not yet fully dominated their bodies. It is not that sponges have passed beyond the limits of perfect radial symmetry—they have not yet developed into it. In sponges, which of the poles in animal and which vegetative?

The orifice, as is well-known, is not a mouth, as it serves for expulsion and not for ingestion of water; water enters the body through numerous pores. Therefore the position of the orifice cannot be used to locate the oral pole. We can determine the homology of the poles of sponges only on the basis of their larval structure and development. There are two principal types of sponge larvae: amphiblastula and parenchymula. The *amphiblastula* occurs in the development of calcareous sponges (Calcarea), mainly in members of the suborder Heterocoela (*Sycandra*, *Grantia*, etc.). In this case the developing blastula consists of two kinds of cells: half of it is formed of tall narrow flagellate cells and the other half of large cells rich in yolk. The amphiblastula swims with the flagellate cells in front, and after

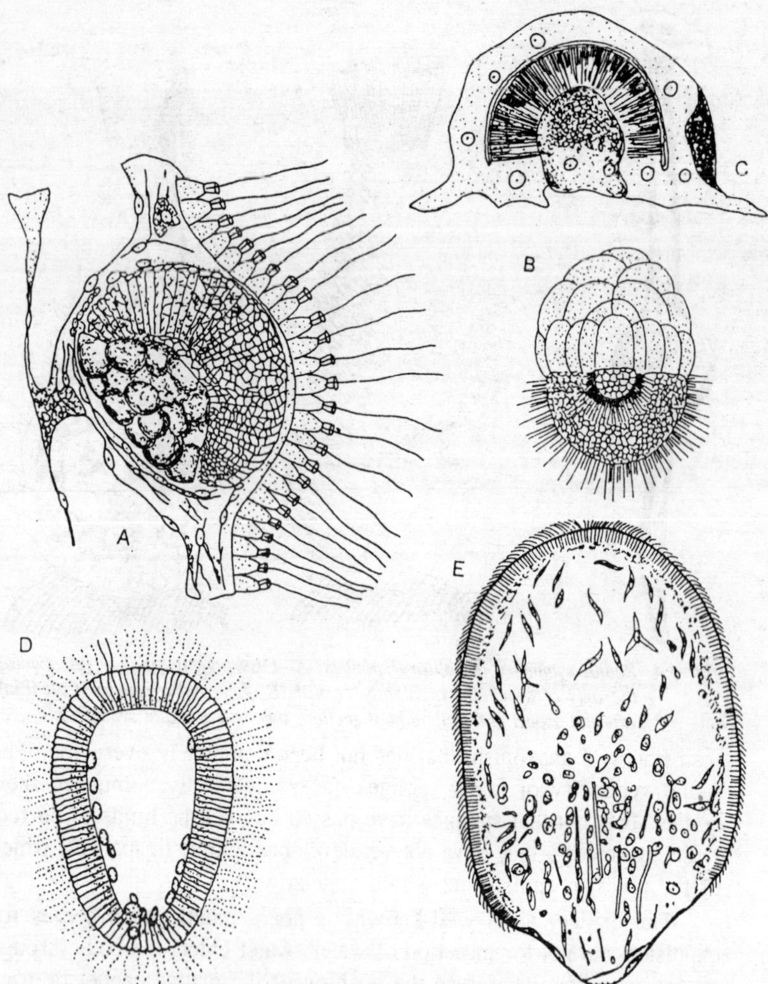

Fig. 4.5. Sponge larvae. Amphiblastula, Sycandra (Calcarea): A—in the maternal body. B—free-swimming. C—attaching itself ('inversion of embryonic layers'). D—coeloblastula of Clathrina reticulum (Calcarea). E—parenchymula of Myxilla rosacea.

some time it attaches itself to the substrate by the end. In course of time the orifice is formed at the opposite end of the larva. To judge by the swimming larva's direction of movement and method of attachment, the end of it bearing the flagellate cells must be considered to be the animal pole, and the opposite end covered with large granular cells the vegetative pole. This interpretation is supported, as we shall see later, by study of the homology of the two kinds of cells of the amphiblastula. The *parenchymula* larva is found in all non-calcareous sponges whose development has been studied, and also in some Calcarea (e.g. *Clathrina reticulum*).

In many siliceous sponges (Triaxonia, Tetraxonia) we find irregular cleavage, due to the presence of a large amount of yolk in the ovum. On that account the cells are smaller at the pole with less yolk and larger at the other pole; the smaller cells (micromeres) usually indicate the animal pole, as is well-known. Later on the micromeres overgrow the mass of the larger cells or macromeres. Afterwards the micromeres take the form of flagellate cell. In swimming the larva keeps in front that end this is called the animal pole during fission, and later it attaches itself to the substrate by the same end. Thus all the features characterising the animal pole coincide, and the homology of that pole is unquestionable. On the whole, examination of both larval types leads to the same conclusion: the end of the body by which a solitary larva is attached to the substrate corresponds to the animal pole, and the orifice is at the vegetative pole. In spite of the fact that the latter therefore corresponds to the mouth in location, there are no grounds for considering the two formations homologous. In any case we find even in sponges, in spite of their profound difference in body structure from all other Metazoa, the primary poles and the main axis homologous with the poles and the axis of Metazoa. In solitary sponges, however, the main axis of the body is an axis of symmetry of an infinitely large order.

The systematic position of Archaeocyatha is debatable. The presence of pores and the permeation of the whole body by a system of canals appears to bring them close to sponges. In the microstructure of their skeletons, on the other hand, Archaeocyatha have nothing in common with either sponges or corals. They evidently form a completely independent group of lower Metazoa. In their degree of regularity of symmetry they do not surpass sponges, but the presence of radial partitions between the two walls of the calyx gives their symmetry some similarity to that of the multiradiate Rugosa.

Among coelenterates, monaxonic symmetry with an axis of symmetry of an infinitely large order is no longer the dominant type of symmetry. We find it mostly in larvae. Flagellate blastulae, typical of metagenetic hydromedusae, possess rotary symmetry of an infinitely large order, expressed in the mosaic arrangement of the blastoderm cells. Planula-larvae also possess an axis of symmetry of an infinitely

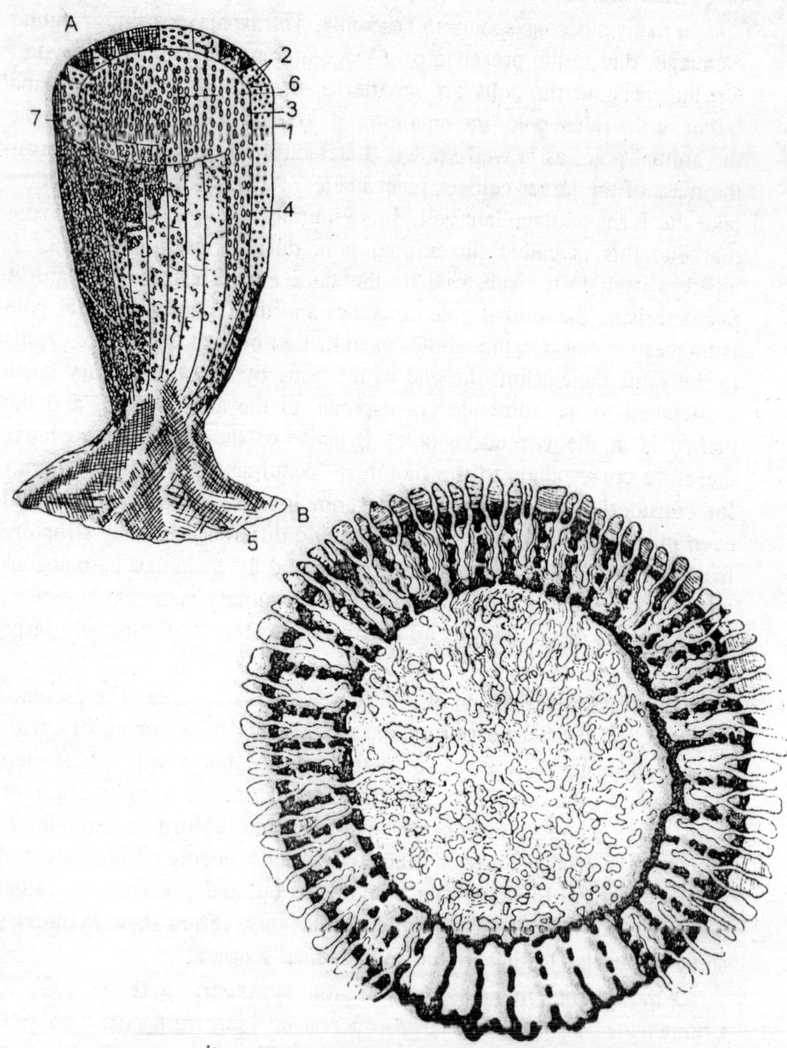

Fig. 4.6. Structure of a typical representative of Archaeocyatha, Ajacicyathus. A—diagram of structure of a solitary individual. B—cross-section of calyx, showing cavity of cup filled with slightly-calcified tissues, and also canal system permeating them. 1, outer wall of cup with small pores; 2, intervallum; 3, inner wall of cup with large pores; 4, radial partitions; 5, foot; 6, central cavity; 7, interseptal cavities.

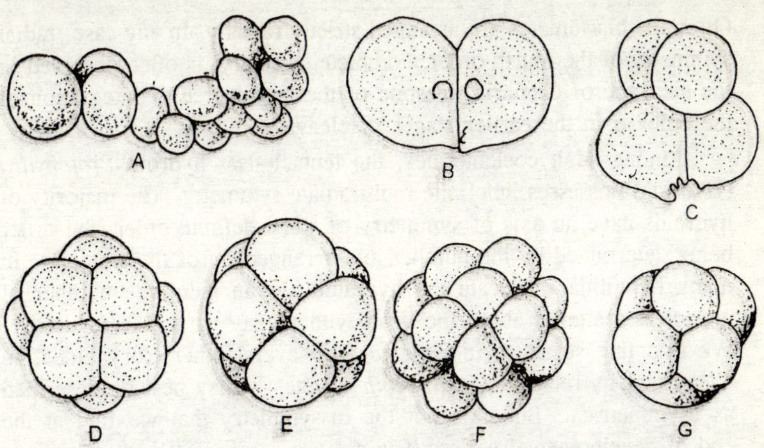

Fig. 4.7. Types of symmetry of ova of Cnidaria during cleavage.

large order. The *planula* occurs in the development of most hydroids, in Octocorallia among Anthozoa, and in a very few other forms. The planula has an oblong-oval, elongated, and sometimes worm-like form. Its anterior end is usually broadened and the posterior end often sharpened. It is circular is cross-section. The main axis passes through both ends of the body, and every plane drawn through it is a plane of symmetry.

The cleavage of Cnidaria takes place in very different ways in different forms and is usually distinguished by wide individual variation. In some hydromedusae, e.g. *Oceania*, an irregular arrangement of blastomeres is found; such an 'anarchical' type of cleavage is considered primary by L. N. Zhinkin (1951). At the same time cleavage in many other hydromedusae is of a primitive spiral type. Stage 4 in many cases consists of two pairs of blastomeres lying across each other (*Rathkea fasciculata*, *Tiara pileata*). That is a biradiate arrangement of blastomeres, which we see also in the same cleavage stage in Turbellaria Acoela. In stage 8 of both these medusae the blastomeres are arranged as in the quadriradiate spiral cleavage of annelids and molluscs. O.M. Ivanova-Kazas (1959), referring to the analogy with the development of Volvocidae, believes it possible that some primitive form of spiral cleavage was the form from which all cleavage in Metazoa is derived

Among other Cnidaria, spiral cleavage has been described in *Lucernaria* by W. Wietrzykowski (1912), in sea-pens by S. E. Berg (1941), and in Actiniaria and Ceriantharia by K. C. Nyholm (1944).

Often the blastomeres are arranged strictly radially. In any case, radial symmetry of the fourth order, complete or rotary, is often observed in the cleavage of Cnidaria, in spite of the low degree of integration of the embryo in the earlier stages of cleavage.

Among adult coelenterates, the tentacle-less hydroid *Protohydra* (Hydrida) possesses indefinite multiradiate symmetry. The majority of hydroids have an axis of symmetry of some definite order, the order being determined by the number and arrangement of the tentacles. In many Leptolida, Athecata the hydranth has an indefinite number of tentacles, scattered about the head without any particular regularity. We find that structure in *Zanclea* and several other Corynoidea, in *Clava* and its relatives, in *Myriothele*, etc. It may best be described as asymmetrical. But it is not the dissymmetry that we find at the end of developmental series. It is not a case of 'lost' but of 'not yet attained' symmetry. It is an indeterminateness of arrangement that precedes, and stands lower than, true radial symmetry. Here the tentacles merely destroy the animal's symmetry, and do not determine its order as they do in higher-ranking hydroids. If we disregard the tentacles, we have in *Clava* the same monaxonic symmetry with an

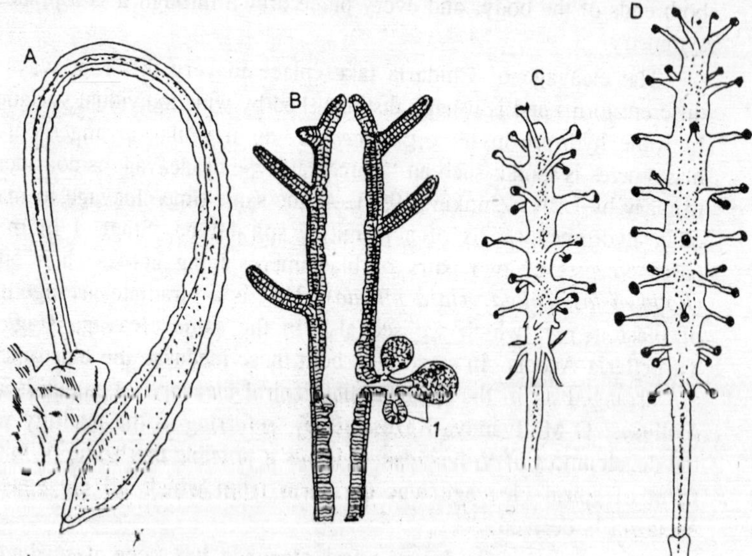

Fig. 4.8. Lower types of radial symmetry in hydroids. A—*Protohydra leuckrti*; B—hydranth of *Clava multicornis*, C—*Staurocoryne filiformis*; D—the same, one of the later polyps.

axis of rotation of indefinitely high order that we have seen in *Protohydra*; and if we take the tentacles into account we find an even less definite position, representing vestiges of anaxonism.

In most hydroid polyps, unlike *Clava*, the tentacles have a more-or-less-definite number and arrangement. One of the first steps in that direction is displayed by *Pennaria*, in which a basal corona of filiform tentacles is formed; the latter thereby acquire a radial arrangement, although they continue to be of an indefinitely high number. The remaining claviform tentacles retain the indeterminate arrangement typical of *Clava*, and are scattered over the whole surface of the hydranth above the corona of filiform tentacles. In *Tiarella* the remaining tentacles are also grouped in two coronae, so that altogether three coronae of tentacles are formed; in *Tubularia* or *Corymorpha* there are two coronae of tentacles; but in most hydroids (such as *Obelia*) only a single corona of tentacles remains. In all these forms all the tentacles without exception are arranged in coronae, i.e. they assume a true radial arrangement, and consequently the radial symmetry of the hydranth becomes complete. But that tells us nothing of the order of symmetry; the order of symmetry in such hydranths is determined by the number of tentacles in the separate coronae. In *Corymorpha nutans* the number of tentacles in each of its two coronae is indeterminate and very large: here we still find an axis of symmetry of an indefinitely high order. In many hydroids, e.g. *Clavatella*, the hydranths have eight tentacles arranged in a single corona, so that octoradiate symmetry develops. In the hydranths of *Cladonema* there are also eight tentacles, but they are arranged in two coronae; the two oral ones are arranged crosswise, and the four basal ones in the intervals between them; as a result, in spite of the eight tentacles, we have not octoradiate but quadriradiate symmetry, since the order of symmetry is determined not by the total number of tentacles but by the number in each separate corona.

Symmetry of uneven orders—triradiate, quinqueradiate, etc.—occurs rather exceptionally in hydroids and in coelenterates generally. The hydranths of *Eutyma mira* have quinqueradiate symmetry, each possessing five long tentacles alternating with five short ones. In *Staurocoryne filiformis* the stem-hydranth of a colony presents a curious combination of triradiate and quinqueradiate symmetry, as its head possesses a basal ring of five filiform tentacles and six or seven distal rings of cephaloid tentacles, three tentacles in each ring. In other hydranths of the colony the cephaloid tentacles are arranged in seven to nine rings, four in each ring. At the same time *Staurocoryne* is an interesting example of orderliness in the arrangement of a large number of cephaloid tentacles.

We do not know of any hydropolyps possessing only two tentacles located opposite each other in biradiate symmetry (although a biradiate form of theca sometimes occurs, e.g. in *Campanularia platycarpa*). Nevertheless a few cases of true bilateral symmetry are known among them. Such symmetry occurs in distinctive forms in different groups and manifests itself variously. Therefore these cases do not represent a general direction of development of hydroids, but appear as a result of several essentially divergent, and only formally parallel, lines of development.

One of the most unusual forms of these is the gigantic hydroid *Branchiocerianthus*. The hydranth is placed obliquely on a long cylindrical stalk, and the hypostome lies eccentrically on the oval oral disc; around the mouth the points of attachment of the gonophores are arranged in a horse-shoe pattern, and nearer the periphery of the disc a similar horse-shoe-shaped zone, interrupted above the point of attachment of the stalk, is occupied by numerous small tentacles. Bilateral symmetry therefore does not arise in this case from decrease in number and appropriate positioning of a few tentacles; it is created by the curving of the hydranth itself, by the eccentric location of the hypostome and the mouth, and by the horse-shoe form of the peristome, the latter being due to the formation of a permanent zone of growth and regeneration of tentacles and gonophores where the horse-shoe is interrupted. Herein there is a definite analogy with Ceriantharia (see below), whose bilateral symmetry is also based on the presence of a zone of tentacle growth and regeneration interrupting the corona of tentacles. In the hydroids *Lar sabellarum* (family Williidae) and *Monobrachium*, on the contrary, the development of bilateral symmetry is closely related to a great decrease in the number of tentacles. In the hydranths of *Lar* the oral orifice has the form of a transverse fissure, on one side of which the peristome is thickened, while two long tentacles are situated on the other side. The only plane of symmetry passes between the two tentacles and divides in half both the fissure-like mouth and the thickening on the other side. In *Monobrachium* (Limnomedusae), described by C. Mereschkowsky (1877), the hydranths each retain a single tentacle, which is necessarily placed eccentrically, since the oral pole is occupied by the mouth. On that account only a single plane of symmetry may be drawn through the body of the hydranth there also, dividing in half both the unpaired organs—the mouth and the tentacle. *Monobrachium* represents an extreme stage in *oligomerisation* of the tentacles of hydroids, i.e. it is the final member of the series of successive reductions in the number of tentacles.

The plane of symmetry of *Monobrachium* is obviously not homologous with that of *Lar*, since in the former it passes radially through the tentacle and in the latter interradially between two tentacles. There are no data on which to establish the homology of either of these planes of symmetry with that of *Branchiocerianthus*, the basic form of which is that of Tubulariidae with their indefinitely large number of tentacles. The bilateral symmetry of *Lar* or of *Monobrachium* is produced, as we have seen, by decrease in the number of tentacles, which in turn is related to the small size of the individual hydranths. The bilateral symmetry of *Branchiocerianthus* is due to the presence of a permanent zone of growth of the hydranth and of regeneration of tentacles, that is, to increased complexity in the organisation of the hydranth resulting from its extremely large size. Thus the origins of bilateral symmetry in these cases are as different as the way in which it is manifested.

In the group of families Sertulariina (Aglaopheniidae and some others) the bilateral forms of the hydranths has arisen as a result of their method of colonial attachment: the hydranths are attached to a common stalk obliquely and without a long stem. That position produces bilaterally in the hydrothecae and an oblique arrangement of the diaphragms separating the cavities of the hydranths from that of the coenosarc; on that account the heads of the hydranths also assume a bilateral form, in spite of the fact that the number of tentacles is large and there is no break in their arrangement. In this case the bilateral symmetry is intensified by the presence in many forms of a blind saccular appendix on the side of the hydranth, formed abaxially, i.e. away from the stalk of the colony, or rather from the branch on which the hydranth is located. Thus bilateral symmetry arises independently in different groups of hydropolyps, as a result of entirely different developmental series, and is manifested in entirely different ways; and the similarity between the bilaterally-symmetrical forms in this class is purely coincidental.

The great majority of hydroid medusae possess radial symmetry, the order of the axis of symmetry and the number of antimeres varying greatly among them; but the number is always quite definite, being determined by that system of organs of the medusa that is represented by the smallest number of antimeric units. The radial canals usually constitute such a system; in Leptolida the gonads also serve that purpose, and in some forms the tentacles or the manubria. The highest number of antimeres is found in Narcomedusae, since among them forms occur with a great number of tentacles and at the same time without symmetry of a lower order in the gastrovascular system, gonads, or manubrium. *Solmaris*, for instance, possesses neither manubrium nor radial canals,

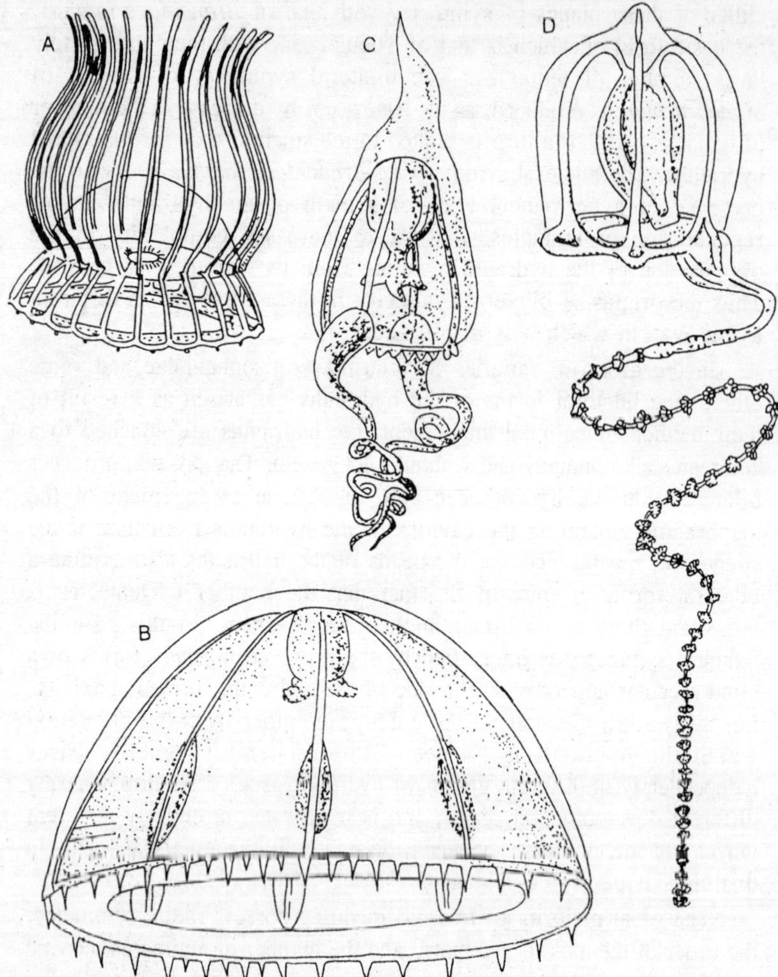

Fig. 4.9. Radial symmetry in Hydromedusae. A—Solmaris coronantha; B—Homoeonema typicum; C—Amphinema titania; D—Corymorpha nutans.

but is has a crown of 24 tentacles. Thus *Solmaris* is a 24-rayed form. Among Narcomedusae there are also 64-, 8-, 6-, and 2-rayed (biradiate) forms. Among all other hydromedusae quadriradiate symmetry is usually predominant. They usually have four radial canals, very frequently (Leptomedusae—Leptolida Thecaphora) also four gonads, and sometimes four marginal bodies (rhopalia) and four tentacles. Sometimes the number of tentacles may be much higher: in such cases the order of symmetry is determined by the radial canals and other organs, and the numerous tentacles create an order of symmetry of some higher order corresponding to their number, which is usually the multiple of four or six. Biradiate forms often occur among hydromedusae, arising through reduction of the number of tentacles to two, located opposite each other. They usually retain four radial canals, a quadriradiate manubrium, and four rhopalia (*Dinema*, *Amphinema*—Leptolida Athecata and others), so that we may speak of incomplete quadriradiate symmetry; if the vestiges of two vanished tentacles remained there, we could speak of heteronomous antimerism. Finally, among Anthomedusae we find bilateral forms that retain only a single, but well-developed, tentacle, and consequently a single plane of symmetry passing through the manubrium, the single tentacle, and the radial canal laying opposite it; an example of such a bilaterally-symmetrical Anthomedusa is the medusa *Corymorpha nutans*.

In the most primitive forms of hydroids, therefore, there is an axis of rotation of an order equal to infinity, and an infinitely large number of planes of symmetry intersecting each other along that axis; their symmetry is indeterminately multiradiate. The vast majority of hydroids possess symmetry of some definite, almost always even-numbered order, the number usually being a multiple of four. In a few forms, diverse in shape and scattered through the whole class, we see bilateral symmetry appearing.

Symmetry of Scyphozoa

Scyphozoa, or higher medusae, present much greater conformity in regard to symmetry than do hydroids. The majority of Scyphozoa possess the most perfect and complete radial symmetry to be found among animals. At the same time predominance of the quadriradiate type of symmetry is very clearly seen among them. In the order Charybdeida (Cubomedusae) quadriradiate symmetry is found in a pure form: an umbrella of cubical shape with four rhopalia, four tentacles or bundles, and four pairs of gonads. In most other Scyphozoa (orders of Peromedusae, Discomedusae) quadriradiate symmetry is combined with incomplete octoradiate symmetry or symmetry of some order that is a multiple of eight, and which does not embrace the whole organism but only separate systems of organs: marginal lobes, marginal tentacles, and rhopalia. An example of such composite radial symmetry is the familiar *Aurelia aurita*. In it, as in all Discomedusae, we see a four-armed manubrium and consequently a quadriradiate type of symmetry. In precisely the same way the gonads and subgenital fossae are arranged in the form of a cross, being subject to the same quadriradiate symmetry. There are also, however, eight branched (radial) canals, alternating with eight unbranched (interradial) canals, and eight marginal rhopalia: these give *Aurelia* incomplete octoradiate symmetry. Finally, the margin of the umbrella bears an indefinitely large number of marginal tentacles and marginal lobes.

In other Discomedusae the number of rhopalia, tentacles, canals, etc. may be various multiples of eight: 8´3, 8´5, 8´7, 8´8 and so on. In the medusa *Kuragea* (family Pelagiidae) there are 8´1 rhopalia, 8´7 tentacles, 8´8 marginal lobes, and 8´2 radial canals. It is difficult to find better examples of composite symmetry, expressed in possession of incomplete symmetry of the eight order, 8´*n,* together with complete symmetry of the same type but of a lower (fourth) order.

Some exceptions to the law of multiples of four, and even to true radial symmetry, are found only in some specialised forms, mainly in the suborder of Discomedusae Rhizostomeae.

Not the slightest tendency towards development of biradiate or bilateral symmetry is observed in adult Scyphomedusae. The scyphistoma (sessile stage of Scyphomedusae), however, originally possesses only two tentacles opposite each other, and consequently biradiate symmetry. In some scyphistomta traces of biradiate symmetry persist even after all the tentacles are formed, being displayed in a slight flattening of the body.

Symmetry of Anthozoa

Anthozoa, or higher polyps, possess much less perfect radial symmetry than do Scyphozoa. Their radial symmetry is usually incomplete and does not embrace the entire organisation. Externally all Anthozoa appear more or less radial, from the form of the body and the arrangement of the tentacles; but in their internal structure we see a strong tendency to bilateral symmetry. Anthozoa are divided into several orders, partly modern and partly extinct. The skeletal structure of the latter does not always permit us to construct an adequately-clear picture of the anatomical structure of the animal as a whole, but usually enables us to determine the nature of its symmetry. The most perfect radial symmetry among modern Anthozoa is found in the subclass Hexacorallia. The bodies of solitary sexradiate polyps (Actiniaria) are approximately cylindrical in form, with a flat basal disc at one end and a circumoral disc at the other. The circumoral disc carries a crown of tentacles, and in its centre is the fissure-like mouth.

The general form of the body and the arrangement of the tentacles give Actiniaria the appearance of multiradiate symmetry, which the fissure-like mouth converts into biradiate symmetry. A flattened pharyngeal tube descends into the body from the mouth, usually provided at both narrow edges with ciliated grooves—*siphonoglyphs*; they are anatomically identical, so that the gullet also is biradiate; but physiologically one siphonoglyph serves for admission, the other for expulsion, of water, on account of a difference in the direction of movement of the cilia in each of them. In some forms the water currents serve for feeding and respiration, in others for respiration only. Septa project from the walls of the body into the intestinal cavity, in the upper part of the body adhering throughout to the gullet, and in the lower part ending in free thickened margins (mesenterial filaments). The order of formation of the septa is strictly determinate for each family of Actiniaria and in most cases is that partly. The first twelve septa are called primary septa or *protomesenteries*. They

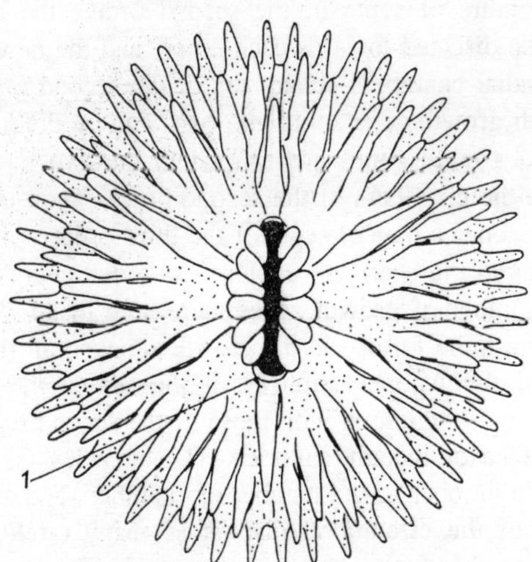

Fig. 4.10. Diagram of arrangement of tentacles in a true sexradiate sea-anemone: tentacles of five orders. View from oral pole: 1, siphonoglyph.

are arranged in six pairs, with muscular crests running along each septum and arranged in perfectly regular order. In the two pairs of septa that adjoin the narrow edges of the gullet with the siphonoglyphs, the crests turn away from each other; these septa are called directive mesenteries. In the other four pairs of septa the muscular crests are turned towards each other. The area enclosed between the two septa of a pair is called a radial chamber, and the area enclosed between two septa belonging to two different adjacent pairs is called an intercameral area. At the level of each radial chamber a tentacle arises from the surface of the circumoral disc, so that there are six primary tentacles. Because of its possession of six radial chambers and six tentacles the young sea-anemone at that stage would have sexradiate symmetry, but the flattened gullet with the two siphonoglyphs and directive mesenteries with their muscular crests directed away from each other, and the uneven development of tentacles, give the animal a definite biradiate symmetry, beside which its sexradiate structure in the arrangement of chambers and tentacles proves to be merely incomplete symmetry.

In the majority of solitary Hexacorallia there is constant individual growth not only in length but also in diameter. The growth in diameter takes place mainly in the meridional growth zones, which correspond to the intercameral areas. Consequently there are originally six growth zones. Later a pair of new septa arise in each of these, making a total of six pairs of septa of the second order; the crests of these septa are also directed towards each other, and the new pairs of septa form new radial chambers—chambers of the second order. From that moment each growth zone splits into two, and on the location of each of them two growth zones are formed in the two new intercameral areas beside the chambers of the second order. Then in each of these new growth zones a pair of septa of the third order is formed—a total of twelve pairs, and each pair encloses its radial chamber of the third order, the number of growth zones increasing to 24. Still later, new septa continue to be formed in the growth zones in all the intercameral areas, round the whole periphery of the sea-anemone. All of the successive septa, beginning with those of the second order, are called secondary septa or metamesenteries. The tentacles, as we have seen correspond in number to the number of chambers and are arranged on the surface of the circumoral disc in as many circles as there are orders of chambers, the number of tentacles in successive circles increasing regularly (Golard's law). The first ring usually consists of six primary tentacles, the second also of six, the third of twelve, the fourth of 24, and so on in the series $6 + 6' 2^0 + 6' 2^1 + 6' 2^2 + 6' 2^3 + ... + 6' 2^{n-2}$, where n is the number of the circle, counting as the first circle of tentacles the one corresponding to the first six chambers.

Adult sea-anemones, therefore, possess biradiate symmetry due to the structure of the gullet and the arrangement of the muscular crests of the directive messenteries, then incomplete sexradiate symmetry due to the arrangement of the primary chambers and the primary tentacles, and finally incomplete symmetry of the order $6'2^{n-2}$ due to the arrangement of the secondary chambers, the growth zones, and the secondary tentacles. In other words, here—as in medusae—we find composite radial symmetry.

The presence of bilateral symmetry in the early stages of formation of septa does not prove the primitive nature of that type of symmetry in Hexacorallia. It is explained rather by the fact that in young individuals, because of their small size, the number of septa is still very small, but their structure and the arrangement of their musculature are the same as in adults; for that reason the septa cannot be arranged to correspond with adult symmetry. In their architectonic plan the

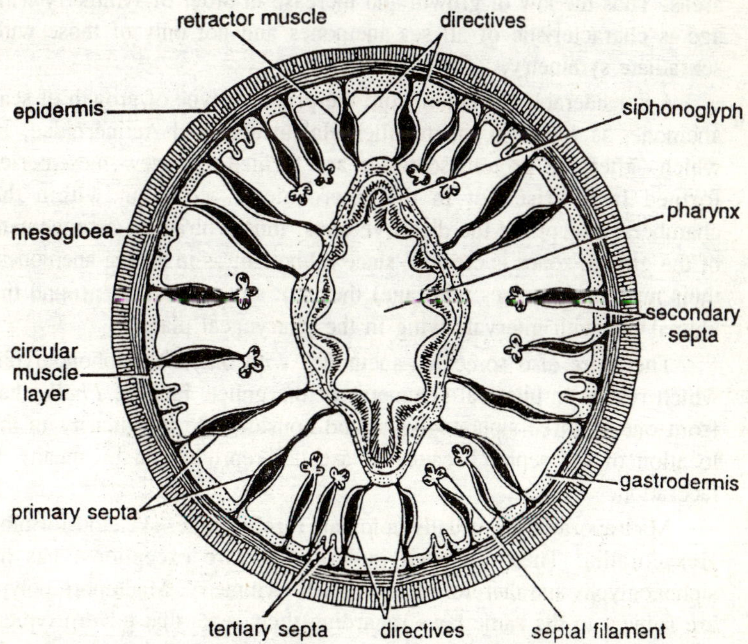

Fig. 4.11. Metridium. Cross section through the pharynx.

young stages of sea-anemones are less primitive than the adults, having an incomplete number of septa as compared with the adults and therefore a reduction of the order of symmetry. Similar oligomerisation—reduction in the number of homotypic organs—is fairly common as a difference between juveniles of various animal species and adults of the same species.

As is easily seen from the above facts as a whole, adult sea-anemones display complex combinations of biradiate and multiradiate symmetry, with the order of symmetry continually increasing with age in some of them, whereas in the early stages of development of the same animals there is usually a certain degree of bilateral symmetry due to incomplete development of the basis complement of protomesenteries.

It should be noted that we find among sea-anemones, besides sexradiate forms, also forms of fundamentally octoradiate or decemradiate type, i.e. with eight or ten pairs of protomesenteries; there also the formation of metamesenteries proceeds according to the above laws, that is, they are formed in pairs in the intercameral areas. Thus the law of growth and increase in order of symmetry with age is characteristic of all sea-anemones and not only of those with sexradiate symmetry.

A considerable deviation from the principal type of growth of sea-anemones is found in the families Halcuriidae and Actinernidae, in which—after the protomesenteries are formed—the new mesenteries formed later arise not in the intercameral areas but within the chambers, except for the directive ones: thus a biradiate arrangement of the growth zones is created, since (although, as in all sea-anemones, their number increases with age) they are situated not all around the animal but with intervals lying in the pharyngeal plane.

There are also some sea-anemones with only one siphonoglyph, which results in bilateral symmetry of the gullet. Finally, *Phellia* has from one to three siphonoglyphs and considerable irregularity in the location of the septa, .because of asexual reproduction by means of laceration.

Madreporaria are usually colonial, rarely single, skeleton-forming Hexacorallia. Their flattened gullet (with rare exceptions) has no siphonoglyphs and therefore has biradiate symmetry. Madrepore polyps are subject to the same laws regarding their septa that govern typical Actiniaria. Included here, however, are small colonial polyps that do not grow continually, and in which the numbers of chambers and tentacles are limited. Thus *Madrepora* and *Porites* have only protomesenteries, and four of these are usually incomplete. Therefore these forms are arrested in their individual development at a stage intermediate between *Edwardsia* and Halcampula.

The order of Zoantharia is close to Actiniaria. It includes both solitary and colonial forms. Their gullet has one siphonoglyph, which introduces into the animal's organisation an element of bilateral symmetry instead of the biradiate symmetry of Actiniaria. The first twelve septa (protomesenteries) are precisely similar in the arrangement of their muscular crests of the first twelve septa in Actiniaria, and their order of formation is fairly similar in both groups. Only two growth zones are formed in Zoantharia, these being in the intercameral areas lying beside the pair of directive mesenteries adjoining the siphonoglyph. New septa (metamesenteries) arise in pairs, as in Actiniaria, but the zone of their formation is always restricted to the area between the latest-formed pair and a directive mesentery. Therefore the number of growth zones does not increase with age but always remains at two. All the protomesenteries except the directive ones, and all the metamesenteries already formed, are gradually moved by the newly-formed pairs towards the directive mesentery lying on the side of the gullet opposite the siphonoglyph. In this way the growth zones are bilaterally symmetrical from the time of their first appearance and retain their bilateral symmetry to the end. Meanwhile all the formed chambers and septa are arranged radially, and in this respect the animal possesses incomplete radial symmetry.

Bilateral symmetry is shown still more clearly in the subclass Ceriantharia. Their gullet also has only one siphonoglyph, as in Zoantharia. The order formation of septa in them is the same for protomesenteries and metamesenteries. There are three pairs of protomesenteries, differing from the metamesenteries only in details of structure. The first pair of mesenteries to appear is that adjoining the siphonoglyph and bounding the directive chamber; their longitudinal muscles are directed away from each other. All succeeding septa are formed one on each side of the single plane of symmetry passing through the gullet; the first pair formed after the directive mesenteries arises outside the latter, and each succeeding pair outside the preceding one, so that soon the zone of formation of septa is concentrated near the central plane of the body on the side opposite the siphonoglyph and the directive mesenteries. Consequently the growth zone in Ceriantharia is single, medial, and located on the side opposite the siphonoglyph, in

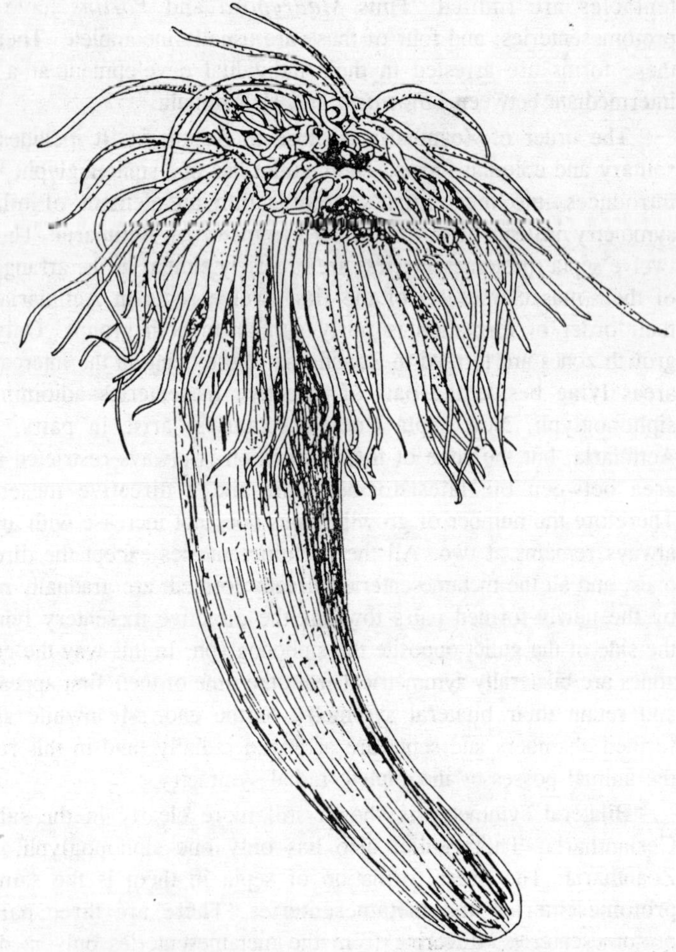

Fig. 4.12. Pachycerianthus multiplicatus. The radially-symmetrical form of the body creates incomplete radial symmetry, existing side by side with dominant bilateral symmetry.

contrast to Zoantharia, in which there are two zones lying on both sides of the directive chamber adjoining the siphonoglyph. But in both cases the position of the growth zone is consonant with bilateral symmetry.

In Ceriantharia that symmetry is particularly clearly displayed because of the location of the longitudinal musculature of the septa: on both side of the pharyngeal plane the longitudinal muscles lie on the sides of the septa turned towards the growth zone. New formation of tentacles proceeds simultaneously with the formation of septa and in the same order. If we look at a ceriantharian, as it spreads its magnificent crowns of tentacles over the sand in every direction, at first glance it appears to be an animal of complete radial symmetry. In fact, the general form of the body, circular in cross-section, and the approximately-radial arrangement of the septa and tentacles create an incomplete radial symmetry of an indefinitely multiradiate type, existing in the organisation of Ceriantharia side by side with their dominant bilateral symmetry.

While large single polyps (most Actiniaria, the madreporarian *Fungia*, Ceriantharia) grow continuously and new septa are formed in them throughout their whole lives, in colonial forms we observe—together with a decrease in the size of individuals—a tendency to elimination of constant growth and to establishment of a fixed number of septa, as we have seen in the example of colonial Madreporaria. A similar phenomenon is observed also in the purely colonial subclasses (Antipatharia and Octocorallia).

Antipatharia form arborescent colonies with quite small individuals. The number of tentacles is six or eight, so that there is incomplete sexradiate or octoradiate symmetry. There is no permanent growth zone in the bodies of antipatharians, although the final number of septa (ten or twelve) is not attained all at once. Bilateral symmetry is displayed mainly in the gastral septa, whereas the form of the body is more or less biradiate. The plane of symmetry of the polyp is its pharyngeal plane. The polyps of Antipatharia stand on the branches of the colony in such a way that the plane of symmetry of each polyp is perpendicular to the axis of the branch; here also, evidently the bilateral symmetry of the separate zooids is due to their location in the colony, as we have seen above in the case of sertulariid hydroids.

The large group of Octocorallia also consists entirely of colonial forms, often forming colonies of very complex structure. In contrast, their separate zooids are comparatively small, not constantly growing, and of simple and uniform structure. Each polyp has eight plumose branched tentacles and eight gastral septa. During the development of the polyp all eight septa are formed simultaneously. Thus Octocorallia have octoradiate symmetry. Their gullet is flattened, however, as in other Anthozoa, and bears a single siphonoglyph, which makes these polyps bilaterally symmetrical and their octoradiate symmetry incomplete. Here also the pharyngeal plane is the only plane of symmetry. The muscular crests of the septa in both halves of the body are located on the sides of the septa turned towards the siphonoglyph. Therefore their arrangement also is bilaterally symmetrical and partly reminds us of the arrangement of the longitudinal muscles of Ceriantharia; in both subclasses the longitudinal muscles of all septa in the body lie on the sides of the septa turned towards one edge of the pharyngeal plane. But in Ceriantharia they lie on the sides directed away from the siphonoglyph, and in Octocorallia on those facing the siphonoglyph; in this respect the groups are directly antithetical.

Among the extinct palaeozoic groups of Anthozoa, Tabulata (B. Sokolov, 1955), Heliolithida, and Rugosa have been well studied. They all had a purely ectodermal skeleton, and in all groups true sclerosepta are found. In these features all three groups approach Hexacorallia, although in the structure of their colonies many Heliolithida are convergently similar to structure of their colonies many Heliolithida are convergently similar to the modern *Heliopora*, a member of Octocorallia.

The most ancient Tabulata (Ordovician) often had 6-12 or 8-16 sclerosepta of two orders and in this respect they remind us of Madreporaria. Quadriradiate forms also are found among them (Tetradidae, etc.). Often the number of septa is not absolutely constant and strict radial symmetry is not displayed.

The most primitive of Tabulata, Auloporidae, generally do not have sclerosepta, if we overlook the longitudinal striations on the inner surface of the theca. An auloporoid incipient stage is known in the development of many higher Tabulata.

Heliolithida form a large group that existed from the middle Ordovicin to the middle Devonian. They are all characterised by a constant number of sclerosepta (twelve) of a single order. Neither in Tabulata nor in Heliolithida is there any hint of bilateral symmetry.

Among Rugosa multiradiate forms, with a large number of radially-arranged septa, are most common. Among Rugosa also, however, constant growth and constant formation of new septa are typical of solitary forms only. In colonial forms, even when there is a large number of septa, these are usually of not more than two orders. As well as multiradiate forms there are forms with definite quadriradiate (for which reason E. Haeckel incorrectly gave the whole order the name of Tetracorallia) and bilateral symmetry. Bilateral symmetry in the arrangement of the septa of many Rugosa arises from general bending of the body in a single plane, which becomes a plane of bilateral symmetry: W. Weissermel (1897) attributed that bending to life in places where strong currents constantly ebb and flow, and N. N. Yakovlev (1910, 1934) attributed it to lateral attachment of young individuals, with subsequent rising upwards of the growing oral end. Whatever the cause may be, Rugosa present the same picture of development that we see in hydroids; from the main mass of radially-symmetrical forms, bilaterally-symmetrical groups separate out. Many Rugosa have six primary septa, like Madreporaria and Actiniaria. The origin of these modern groups from Rugosa is almost beyond doubt.

Let us return to comparison of the various orders of Anthozoa with one another. McMurrich distinguishes three characteristic stages in formation of septa in the majority of modern corals (Octocorallia, Hexacorallia, Ceriantharia):

1. the *Protocnema* stage, with eight septa: Octocorallia remain in this stage throughout life, but is also exists in the development of the other groups, apparently including Rugosa; among Actiniaria the genus *Edwardsia* and *Halcampa duodecimcirrata* remain in it throughout life;
2. the *Halcampula* stage, with twelve septa arranged in six pairs: all modern Hexacorallia, but not Ceriantharia, pass through this stage; some Actiniaria (genus *Halcampa*) and many Madreporarian remain in this stage throughout life;
3. the *Zygocnema* stage, in which new septa are formed always in pairs; this phase also is typical only of Hexacorallia, including to some extent fossil Rugosa.

Antipatharia are entirely excluded from the above scheme. McMurrich's scheme certainly greatly simplifies comparison of the ontogenesis of different groups of coral polyps. Which of the members of the above series is the most primitive? Many authors suggest that Octocorallia are the most primitive—small forms, without constant growth, stopping their development at the Protocnema stage,which is only a transient stage in the development of the other orders. We can hardly agree with that view. Of all modern Anthozoa, Actiniaria (the majority of them) certainly have the most primitive structure with their biradiate symmetry combined with radial symmetry of orders that are multiples of 6, 8, or 10.

In Zoantharia bilateral symmetry, arising from the structure of the gullet, is also displayed in the structure of the septa, and further in the arrangement of the growth zones. The basic structural plan of Zoantharia is less primitive than that of Actiniaria, although the close relation of the two groups is quite evident. Ceriantharia stand much higher than either of the above groups, and detailed comparison of them with Actiniaria is therefore more difficult; but in them also the bilateral symmetry of the gullet and the single growth zone are certainly less primitive characteristics than the biradiate gullet and the numerous meridional growth zones of Actiniaria. The same applies to Octocorallia with their bilateral symmetry. The absence of constant growth and formation of new septa, however, gives that group a certain simplicity of structure, which many are inclined to interpret as primitiveness; but these features are rather signs of simplification due to the colonial habit of Octocorallia. We find similar features in Hexacorllia in some of the colonial Madreporaria. The separate zooids of Octocorallia do not possess the power of constant growth, because that is possessed by the colony; a certain suppression of the individuality of the zooids by the colony takes place, resulting in smaller size and simplified structure of the zooids, whose development is arrested at the Protocnema stage. The same remarks, *mutatis mutandis*, apply also to Antipatharia.

At present there is a widespread tendency to regard bilateral symmetry as the original forms of symmetry of Anthozoa. Two points are advanced in favour of this view: the bilateral structure of the early stages of many Hexacorallia, and the apparent dominance of that type of symmetry among the palaeozoic Rugosa. As I have already stated the bilateral symmetry of the early stages is probably attributable to juvenile oligomerisation, and cannot serve as an indication of the primitive nature of that type of structure. The palaeontological argument is also untenable. We have seen that, of the three large groups of palaeozoic corals, two (Tabulata and Heliolithida) generally have no traces of bilateral symmetry, whereas its occurrence among Rugosa is far from being universal; and the most ancient Rugosa, which are close to the most ancient Tabulata, are (like the latter) more or less radially-symmetrical. In general, the original form of symmetry in Anthozoa is likely to have been indeterminately multiradiate. It is most probable that the morphological prototype of Anthozoa was a small solitary polyp, close to Auloporacea (Tabulata), with a large number of ill-developed septa. Increase in the size of septa leads, by the law of balance of organs (*balancement des organes*—G. Cuvier), to decrease in their number, to establishment of a fixed small number of septa (16, 8, 12, 6, 4) and a corresponding number of sclerosepta in coral-forming types. Moreover, increase in the size of the septa leads to their appearance at different times in ontogenesis. In view of the fact that well-developed septa are attached to the gullet, and that the latter has assumed a flattened form, the first pairs of septa to appear during ontogenesis are arranged bilaterally. On that account the cessation of development at one of the stages with a less-than-complete number of septa produces bilateral adult forms, as in the case of Octocorallia. Reduction of the siphonoglyphs to one intensifies the bilaterality. We have seen other methods of origin of bilaterality among, for instance, the large, solitary Rugosa.

The general reason for the tendency to development of biradiate and bilateral symmetry in the class Anthozoa is the flattening of the gullet. Later we shall see that the flattened shape of the gullet, especially in Actiniaria, is an adaptation to its role as a valve that passively prevents expulsion of water from the gastrocoelomic cavity during contraction of the musculature of the body. Thus it is out of the question to regard the bilateral symmetry of Anthozoa as primitive, and the same applies to regarding quadriradiate symmetry as primitive. Both result from extreme specialisation.

The symmetry of modern Actiniaria is fairly close in type to the composite symmetry predominant in Scyphozoa. In Scyphozoa there is a combination of quadriradiate symmetry with incomplete symmetry of order $4n$, and in Actiniaria a combination of biradiate symmetry with incomplete symmetry of order $6n$. In all other modern Anthozoa, however, and also even in some Actiniaria, biradiate symmetry gives place to bilateral, which is decided mainly by the structure of the gullet and the arrangement of the growth zones, and partly by the details of the septa. The arrangement of the septa, the intestinal chambers, and the tentacles, and also the general shape of the body largely retain radial symmetry.

In this way the symmetry of Anthozoa present both similarities to and substantial differences from the symmetry of Scyphozoa and Hydrozoa; apart from Anthozoa bilateral symmetry occurs in Coelenterata only among hydroids, and then rather as an exception; on the other hand, among modern Anthozoa bilateral symmetry is predominant. We see, besides, that the planes of bilateral symmetry in the various groups of hydroids where that symmetry has developed are not homologous with one another. In all modern Anthozoa, in contrast, the plane of bilateral symmetry is the same pharyngeal plane; that is probably true also of the order Rugosa. Therefore in this class bilateral symmetry has appeared, not as a casual result of several different developmental series, but as the product of a single fundamental tendency dominant in the entire class. There are, on the other hand, no constant 'ventral' and 'dorsal' sides homologous in all bilateral members of the class Anthozoa. In fact, the relative positions of the siphonoglyphs, the growth zones, and the muscular crests of the septa are very diverse, and in each of the subclasses these features enter into new combinations. Therefore we have equally good grounds for taking as a criterion the homology of any one of the above features, and in all three cases we would have to base our homology only on the selected feature, since the others are not correlated with it. In these conditions we cannot speak of the possibility of distinguishing a 'ventral' or 'dorsal' side that would be homologous in all the subclasses. In this respect the bilateral symmetry of Anthozoa stands as far below the bilateral symmetry of worms, which we shall discuss later, as it is above the bilateral symmetry of hydroids.

Symmetry of Ctenophora

In their structural plan the small group of ctenophores is of outstanding comparative-anatomical interest in view of its intimate links with the main stems of Enterozoa. In particular, knowledge of ctenophore symmetry is absolutely necessary for comprehension of the origin of bilateral symmetry in Bilateria, although among ctenophores themselves we find no hint of bilateral symmetry.

Ontogenetically ctenophores are the coelenterates that develop most directly, i.e. with the least differences from their larval condition. The larvae of all coelenterates are heteropolar creatures, the aboral end of the body being especially rich in neural elements, which occur there in greater quantity than in the whole of the rest of the body. As stated above, in the vast majority of Cnidaria the aboral complex of neural elements is reduced to some extent after attachment of the planula; ctenophores retain the free-swimming mode of life, and in them the neural elements of the aboral pole form a complexly-differentiated aboral sense organ. Ctenophores also retain the ciliary method of movement characteristic of coelenterate larvae, although in them the ciliary apparatus undergoes considerable specialisation. The direct type of ontogenetic development of ctenophores may be regarded as primitive, and in that case we may consider Cnidaria a lateral, descending branch, much modified by adaptation to a sessile mode of life. But A. A. Zakhvatkin (1949) is more likely to be correct in believing the sessile mode of life of the adult stages of the lower Metazoa to be primitive. From the point of view the ctenophores, swimming by means of ciliary movement, may have arisen through elimination of the attached imaginal stage and sexual maturation of the free-swimming, planula-like larva, with perfection of the organisation of the latter. In other words, the development of ctenophores might be an example of progressive evolution, based on neoteny. From the standpoint of this hypothesis, an analogous explanation might be given for the development of Turbellaria and other groups of Bilateria, and the similarity of their method of origin would account for their similarities to ctenophores.

Examining the anatomical structure of an adult of the typical ctenophore *Pleurobrachia* or *Hormiphora* (order Cydippoidea), or of the cydippoid stage of members of other orders, we find the aboral sense organ at one pole and the mouth at the other. The aboral sense organ has the form of a deep pit, within which a complex statolith rests on four bundles of long cilia, as if on springs. Each of the four bundles of springs gives off a narrow band of ciliated epithelium. These bands diverge meridionally and divide dichotomously, and the ends of their eight branches approach the eight meridional rows of swimming-plates. At the same time the ctenophore has two tentacles located opposite each other in a single plane, and therefore it becomes possible for only two planes of symmetry to exist in its body: one passing through both poles and tentacles and called the tentacular plane, and the other, perpendicular to it, called the pharyngeal plane. In the latter plane there are two polar lamellae, adjoining the aboral pole and probably representing sense organs; some believe them to be vestiges of a second pair of tentacles that have changed their functions. The oral orifice also extends along the pharyngeal plane, its centre falling at the oral pole. The oral orifice leads into the gullet, flattened in the pharyngeal or oral plane, which derives its name therefrom. The gullet leads by a narrow oesophagus to the stomach (in the old terminology, the infundibulum), from which three sets of gastrovascular canals branch out. Two canals, *metagastric* or pharyngeal, descend to the oral pole along the sides of the gullet; they lie in the tentacular plane. Above them, in the same plane, a pair of *mesogastic* canals leave the stomach. Each of them very soon branches into one unpaired and two paired canals. The unpaired branches approach the bases of the tentacles; the paired ones again branch dichotomously, and each of their branches unites with one of the *meridional* canals. The latter run from the aboral to the oral pole beneath the rows of swimming-plates, each of them being blind at both ends and connected with the rest of the gastrovascular system only through the mesogastric canals. Finally, an axial canal called the *acrogaster* by T. Krumbach leads from the stomach to the aboral pole: it opens to the exterior through two pores beside the aboral sense organ. In *Mnemiopsis* (Lobifera) expulsion of undigested residues takes place though the pores of the acrogaster, and through them only.

Thus the form and location of the gullet and the gastrovascular canals and the location of the tentacles and the polar lamellae result in the existence of two planes of symmetry, and consequently in the biradiate symmetry of the ctenophore. Besides, the four springs of the statolith and the four ciliated bands arising from them produce incomplete quadriradiate symmetry, and the eight rows of swimming-plates and the eight meridional canals passing under them create incomplete octoradiate symmetry.

In the gastrovacular apparatus of ctenophores there is, however, another unique peculiarity, which breaks down even the biradiate symmetry without in any way converting it into bilateral symmetry. Four branches diverges upward from the acrogaster, of which two are developed and end in the two above-mentioned external pores, the so-called excretory pores; the other two branches are not fully developed and end blindly. The two existing apertures are so located between the four semi-antimeres of the animal that a straight line joining them cuts both the pharyngeal plane and the tentacular plane at an angle of 45°. In other words, they are situated asymmetrically with reference to both of these planes. If we take the pores of the acrogaster into account it becomes impossible not only to draw two planes of symmetry in the body of the ctenophore, but to draw even one. There is no mirror-image symmetry, but rotary symmetry remains. If the ctenophore is rotated around the body axis through an angle of 180° one aperture coincides with the other. In *Coeloplana gonoctena*, unlike the Cydippoidea just described, the pores of the acrogaster lie perfectly symmetrically in the pharyngeal plane on each side of the aboral pole, and therefore do not destroy the mirror-image symmetry. Nevertheless, *Coeloplana* has special aboral papillae located according to the principles of rotary symmetry (an axis of rotation of the second order, without a plane of symmetry). This shows that the tendency to purely rotary symmetry is deeply established in the nature of ctenophores, in contrast to adult Cnidaria, where it scarcely exists.

To summarise, we see in the organisation of Ctenophora co-existence of several types of symmetry, just as in Hydromedusae, Scyphozoa, and Anthozoa, If we take all parts of the organism fully into account, it will appear that the only element of symmetry in the body of a ctenophore is an axis of symmetry of the second order; but along with purely rotary symmetry of the second order we see in it incomplete biradiate symmetry, dominant through almost the entire organisation, and also incomplete quadriradiate and incomplete octoradiate symmetry. The same type of symmetry that dominates in the adult animals is also displayed in the embryo, beginning with the cleavage of the ovum.

At the 8-blastomere stage the embryo resembles a lamella curved in the tentacular plane, whose concave surface is turned towards the animal pole and the convex surface towards the vegetative pole. All eight cells are arranged strictly on the principle of biradiate symmetry, showing at the same time incomplete octoradiate symmetry. In such an embryo we can easily discern two planes of symmetry, and further development shows that one of these is pharyngeal and the other tentacular. Subsequent cleavage is uneven, and each of the first eight cells detaches several small micromeres, one after the other, in the direction of the animal pole. Each of these eight blastomeres is of future significance in the formation of one of the eight antimeres of the adult animals with one of the meridional rows of swimming-plates.

The symmetry of the dividing ovum of Ctenophora is of great interest from the point of view of comparison with worms, Bryozoa, molluscs, and lower Deuterostomia, in whose cleavage we observe related types of symmetry.

The entire class of Ctenophora is unique in respect to symmetry. All ctenophores are constructed on the same plan, and in all ctenophores we find the same combination of orders of symmetry. Nevertheless we find great diversity of external body form among them. That is caused mainly by flattening or stretching of the body in three principal directions: in the direction of the main axis of the body, in the pharyngeal plane, and in the tantacular plane. Typical forms are those of Cydippoidea, whose bodies (of circular or slightly-elongated shape) are more or less equally extended in the pharyngeal and the tentacular planes and whose internal structure corresponds to the above-described type. Beroidea, with the same body form, differ slightly in the absence of tentacles and in some features of internal structure.

The creeping ctenophores Platyctenida are much flattened in the direction of the main axis, some of them being at the same time more or less elongated in the direction of the tentacular plane (e.g. *Coeloplana gonoctena*). All Platyctenida have the shape of flat tablets, their aboral side being physiologically uppermost and the oral side being converted into a creeping sole. Lobifera (Bolinopsidea) have an oval body, but at the oral end if forms six projections: two large lobes at the ends of the mouth in the oral plane, and beside each of them a pair of small auricles. If one moves these lobes aside one obtains a considerable extension of the animal in the oral plane. In Cestidea, to which *Cestus veneris*, the largest ctenophore, belongs, the body has a ribbon-like shape. The oral pole is in the middle of one of the edges of the ribbon and the aboral pole in the middle of the other, so that one edge is oral and the other aboral. The body is extended in the oral plane and flattened in the tentacular plane; the mouth is correspondingly lengthened. Four rows of swimming-plates parallel to the oral plane are also stretched out along the full length of the animal, and four rows parallel to the tentacular plane are very short in comparison. This in Cestidea, as in Platyctenida, biradiate symmetry is sharply intensified, while quadriradiate and octoradiate symmetry are much reduced.

Ganeschidea, with a general cydippoid appearance, differ in internal structure and in displacement of the tentacles nearer to the mouth.

Comparison of the Symmetry of Ctenophora and Cnidaria

One of the most important features of the radial symmetry of Coelenterata is the wide distribution of composite forms of symmetry. Composite radial symmetry consists in the conjunction, in the body of a single animal, of two or several different orders, the various organs being arranged in symmetry of different orders around a single axis of symmetry common to the whole animal.

As a rule, all higher orders of symmetry in a coelenterate are multiples of the lower orders of symmetry in the animal's body, and all planes of symmetry of a lower order coincide with part of a plane of symmetry of a higher order. For this reason, the presence of organs disposed in accordance with the symmetry of a higher order does not disturb the symmetry of a lower order, resulting from the destitution of another group of organs. In the medusa *Amphinema*, for instance, the radial canals are governed by quadriradiate symmetry and the tentacles by biradiate symmetry, but the two planes of biradiate symmetry coincide with two planes of quadriradiate symmetry, and therefore the two types of symmetry do not conflict with each other. In *Amphinema* we find complete symmetry of the second order and incomplete symmetry of the fourth order. Such composite symmetry is, in one form or another, characteristic of almost all large coelenterates of complex structure: all Anthozoa, the higher Scyphomedusae, most Hydromedusae, and Ctenophora. In these cases the organs located nearer to the animal's axis of symmetry are fewer in number, and their form and arrangement are governed by symmetry of a lower order (usually biradiate or quadriradiate, and in many Anthozoa bilateral), whereas organs located farther from the body axis, nearer the periphery, are more numerous, and their arrangement is governed by incomplete symmetry of one or more higher orders.

In Actiniaria, for instance, the gullet, which occupies the most axial position, is constructed on lines of biradiate or bilateral symmetry; the arrangement of the septa located farther from the body axis and of the tentacles of the first order is governed by sexradiate symmetry, and that of the metamesenteries and the tentacles corresponding to them by symmetry of increasing orders, the relationship of their orders of symmetry being determined by Golard's law. In Discomedusae the arms, gonads, and subgenital fossae are arranged in quadriradiate symmetry, the radial canals in octoradiate symmetry, and the marginal lobes, tentacles, etc. in symmetry of several different orders of the general formula $8n$. The simplest and most regular case is the composite symmetry of Ctenophora, which follows the series $2^1+2^2+2^3$: the gullet, tentacles, and a number of other ctenophore organs are arranged in biradiate symmetry, the statolith supports and the ciliated cavities in quadriradiate symmetry, and the meridional canals and rows of swimming-plates in octoradiate symmetry. The simplest case of composite symmetry found in ctenophores may be called *dichotomous composite* symmetry. We shall meet this type of symmetry again as we proceed.

The frequent, but not obligatory, development of composite symmetry in radial animals is due to the following circumstance: the farther from the axis of symmetry that any zone of the body is situated, the larger it is and the more widely distributed in it homotypic organs can be (other conditions being equal), and consequently the higher will be the order of symmetry resulting from their radial arrangement. This is particularly clear in the case of the tentacles of Actiniaria, which are arranged in rings on their oral disc, the number of tentacles in successive rings increasing with the diameter of the rings and with their approach to the periphery. The branching of the gastrovascular canals of Scyphomedusae in the direction away from the body axis towards the periphery is due to the large size of the animal's body and the need for it to receive equal service from widely-distributed organs such as these canals are. The large number of marginal tentacles and marginal sense organs in large medusae is similarly due to the need to serve the considerable length of marginal of the umbrella.

We have found an exception to the rule of multiple numbers in orders of composite symmetry only in hydroid polyps. For instance, we have seen a combination of triradiate and quinqueradiate symmetry in the stem hydranth of a colony of *Staurocoryne filiformis*. If one of the quinqueradiate symmetry planes coincides with one of the triradiate symmetry planes, as it actually does, that plane becomes a plane of bilateral symmetry for the animal as a whole. Thus the stem hydranth of *Staurocoryne filiformis* possesses complete bilateral symmetry combined with a manifestation of incomplete radial symmetry of the third and fifth orders.

The development of adult symmetry from the symmetry of the embryo (in the first stages of cleavage of the ovum) is a subject of high importance. In ctenophores that development is indubitable: in their embryos the combination of biradiate and octoradiate symmetry is well shown at the 8-blastomere stage and is continually retained to the end of development; the operative elimination of one of the blastomeres at stage 8 leads to disappearance of one of the antimeres of the adult animal. In Cnidaria the absence of determinate cleavage makes developmental succession of the adult animal's symmetry from the symmetry of the cleaving ovum rather improbable. Absence of the effect of ovum-cleavage symmetry explains why many Cnidaria possess orders of symmetry that are not powers of 2 (sexradiate symmetry in most Hexacorallia, decemradiate in several Actiniaria, quinqueradiate in *Eutyma mira*, etc.); on the other hand the predominance of quadriradiate symmetry still existing among Cnidaria is, from that point of view, attributable to sheer coincidence. In favour of the simple-coincidence theory is the great variety of orders of symmetry in Hydromedusae and Narcomedusae, which are produced from ova, while there is marked predominance of quadriradiate symmetry in Leptomedusae and Anthomedusae, which always reproduce by budding. If the quadriradiate symmetry of adult Hydrozoa were derived from the quadriradiate symmetry that is sometimes observed in the cleavage of their ova, we might expect the reverse of that situation.

Ecological Significance of the Symmetry of Sponges and Coelenterates

The two ends of a sessile animal—one attached to the substrate and the other projecting freely into the water—are subjected to entirely different environmental action and fulfil different functions, and therefore are differently constructed. On the other hand the animal is subjected to uniform external action from all directions perpendicular to its main axis. These elementary considerations explain the wide distribution of radial monaxonic-heteropolar symmetry among sponges and sessile coelenterates. It is cases of bilateral symmetry among these animals that call for special explanation. There are no such cases among sponges, but among polyps bilateral symmetry is not frequent and is produced by various causes.

In the first place, as M. S. Gilyarov (1944) correctly states, one of the possible reasons for appearance of bilateral symmetry among sessile, primarily-radial animals is the existence of mechanical forces acting perpendicularly to the direction of the force of gravitation. In this case the organism acquires greater stability by curvature of the axis, leading to the appearance of bilateral symmetry, as we have seen in solitary quadriradiate corals affected by strong currents or lateral attachment. Moreover, the proximal ends of some lateral parts are attached to the vertical or oblique axis; in such cases also a distribution of forces takes place, producing bilateral symmetry in these parts, as may be seen in the leaves of higher plants or in separate antimeres of radial animals. We frequently find separate polyps in such a situation in colonies of hydroids or Anthozoa; we have seen above examples of bilateral symmetry arising in that way in the hydranths of Sertulariina. It is possible that the position of zooids in colonies may cause bilateral symmetry in some other polyps.

In some cases bilateral structure is due not to mechanical factors but to orientation in the direction of available food. Thus the hydranths of *Lar*, whose hydrorhiza covers the tube of a sessile polychaete, are symmetrically oriented in the direction of the aperture of the host's tube, from which their food comes. At the same time their bilaterality is doubtless connected with the strongly-expressed oligomerisation of the tentacular apparatus, which in turn is due to the small size of the separate hydranths.

The bilateral symmetry of Ceriantharia and several Actiniaria is basically determined by the structure of the gullet and septa, and is scarcely apparent in the external form of the animals, which remains approximately radial. In this case the appearance of bilaterality is due to the animal's growth conditions and to the functioning of its pharyngeal-septal apparatus. All free-swimming solitary coelenterates (medusaes, ctenophores) have a monaxonic-heteropolar structure, which in most medusae is multiradiate and in ctenophores is biradiate. The suitability of that structure for active swimming below the surface of water has already been mentioned in discussion of the ecological significance of forms of symmetry in Protozoa. A few cases of appearance of bilaterality in some medusae (e.g. *Corymorpha nutans*) through reduction of the number of tentacles to one are probably due to the very small size of such medusae, like the appearance of bilaterality in the hydranths of *Lar* and *Monobrachium*.

5

MATING

First Steps in Development of Genital Apparatus Coelenterates

Sponges have no genital apparatus. Genital cells arise from archaeocytes and may be located in any part of the body. A single archaeocyte gives rise to either one oogonium or a bundle of spermatozoids. The latter are often surrounded by a special nutritive cell. In some sponges a few cells form something of the nature of a follicle around each oogonium, which feeds by phagocytising these cells. The spermatozoids are formed mostly in the external layers of the parenchyma, and the oogonia near the flagellate chambers. Sponges are either hermaphroditic or unisexual.

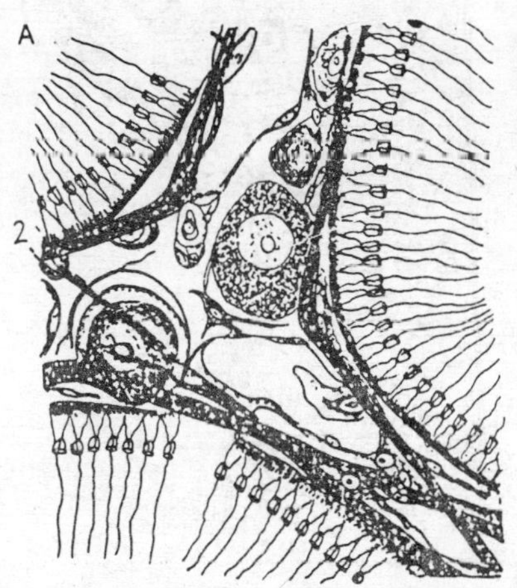

Fig. 5.1. Oocytes in the body of the sponge Sycandra.

Mature spermatozoids are expelled through the efferent canals to the exterior, and then, partly through the afferent canals and the flagellate chambers, enter the parenchyma of mature females, where they fertilise mature ova. The fertilised ova develop there, and ultimately the larvae emerge to the exterior. The genital cells of sponges are therefore widely distributed in the body, and the only special adaptations affecting them are the food-cells. The chief role in discharging genital products and in ensuring the meeting of gametes is played, firstly by the genital products' own mobility, and secondly by the irrigation apparatus, which provides almost all the communications between the sponge and the external world. Coelenterates (if we speak of individuals, not colonies) are only one step higher than sponges. We

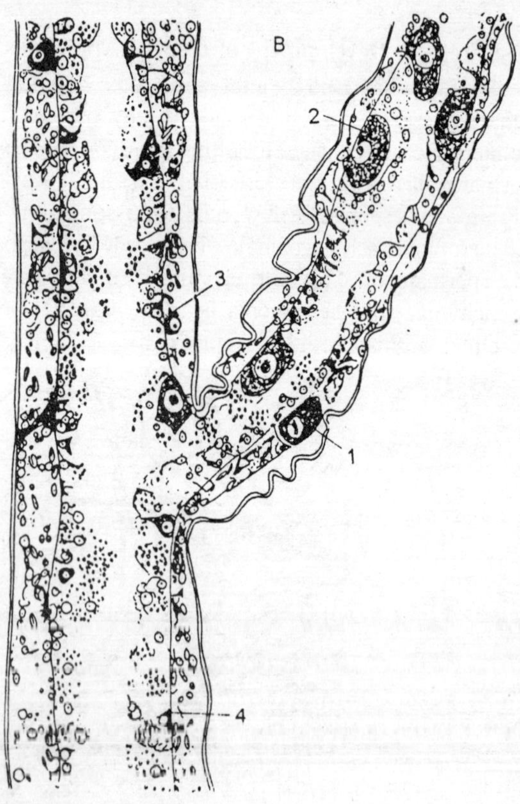

Fig. 5.2. Longitudinal section through the hydrocaulon of Eudendrium racemosum (Leptolida, Athesata): 1 and 2, wandering oogonia in ectoderm and endoderm; 3, young oogonium; 4, interstitial cell in endoderm.

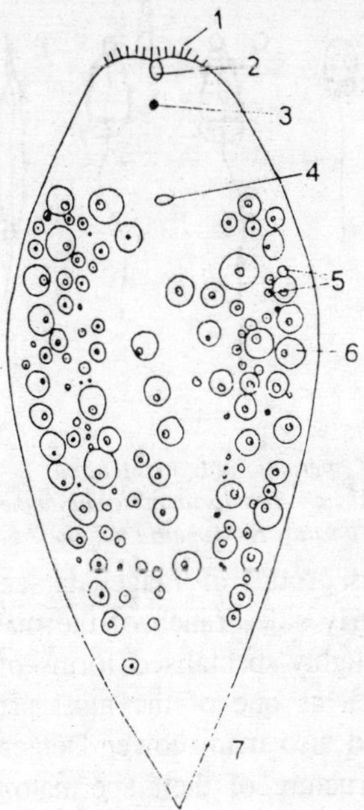

Fig. 5.3. Oxyposthia praedator (Turbellaria Acoela), location of oogonia and oocytes: 1, tactile sensillae; 2, frontal glands; 3, statocyst; 4, mouth; 5, oogonia; 6, oocytes; 7, male gonopore.

find cells resembling archaeocytes in hydroids, wandering in the interlayer between ectoderm and endoderm. These cells settle in definite parts of the body and, by multiplying, from small ovaries or testes. In hydroids these gonads lie in the ectoderm. When the genital products mature the ectoderm covering them splits and they emerge. In hydromedusae the gonads are always located near the gastrovascular apparatus, either in the walls of the proboscis or on the subumbrella, along the radial canals.

In Scyphozoa the gonads develop in the endoderm, but the ectoderm on the subumbrellar side forms a pocket beside each of them (the subgenital pit), which apparently brings sea-water, and with it oxygen, to the immediate vicinity of the gonads. Genital products are released, by splitting of the endoderm, into the gut cavity, and thence through the mouth to the exterior. In Anthozoa the gonads are also formed in the endoderm, within septa in the radial canals, and the genital products (and sometimes completely-formed larvae) reach the exterior through the mouth. The spermatozoids of Coelenterata are very similar in structure to those of sponges, which argues against the derivation of

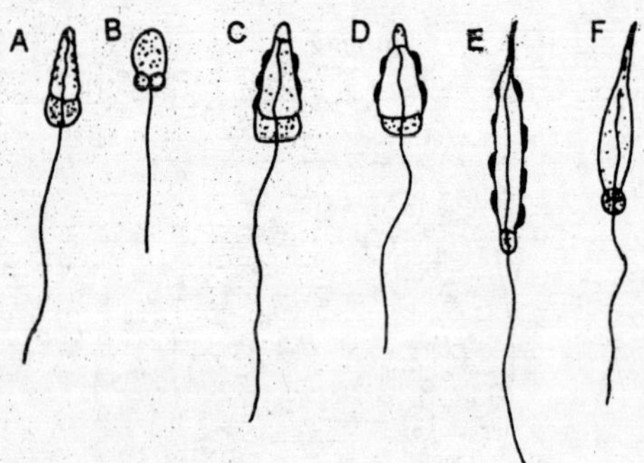

Fig. 5.4. Primitive types of spermatozoids in Metazoa. A—Grantia compressa; B—Xenoturbella bocki; C—Tubularia mesembryanthermum; D—Paracentrotus lividus; E—Styela partita. F—Anguilla vulgaris.

either from the various groups of Flagellata, and supports the view that sponges are an early side-branch of the main stem of Metazoa. Many Bilateria have highly-specialised forms of spermatozoids, but some Protostomia, such as one of the most primitive turbellarians, *Xenoturbella bocki*, and also many lower Deuterostomia are close to coelenterates in the structure of their spermatozoids. We have seen above that Cnidaria, like sponges, have almost no genital apparatus in the strict sense of the term. That refers, however, only to individuals. Colonies, at least hydroid colonies, usually have colonial genital organs in the form of gonophores, blastostyles bearing these gonophores, etc.

Occasionally the colonial genital apparatus attains considerable complexity; it is enough to mention the corbulae of Plumulariidae (Thecaphora) and similar formations. We have already mentioned the higher organological differentiation of colonies, as compared with individuals, in connection with other apparatuses. Ctenophores, unlike most other coelenterates, are hermaphrodites. The genital products develop in the endoderm of the meridional canals. It is usually believed that both ova and spermatozoa are discharged through the mouth. A. Totton (1954b) declares, however, that in most ctenophores the ova leave the meridional canals through a split in the ectoderm covering them, and not through the mouth. In Platyctenida the testes have independent seminal ducts opening directly to the exterior; T. Komai (1922) describes fine canals in *Coeloplana bocki*, opening to the exterior by eight rows of orifices and ending towards the ovaries in vesicles that sometimes contain spermatozoa. Apparently these are spermathecae containing spermatozoa of another individual. For the first time, therefore, we find copulatory organs, although still very imperfect, in crawling ctenophores.

6

BREEDING

In coelenterates a sexual reproduction takes place continuously but sexual reproduction occurs only one in a year. Most of them breed sexually between early spring and autumn, some of them reproduce during winter, whereas some tropical species may breed several times a year. In the free-swimming medusoid forms, the eggs are liberated into the surrounding water. Fertilization takes place either when they are suspended in the sea-water or when they sink to the bottom. In the sessile, polyp type of forms, the eggs are fertilized when they are still attached to the parent tissue. If the fertilization takes when the eggs are in a suspended stage, the embryo passes through a free swimming *Planula* stage before settling to the bottom and transforming to a polyp-form. This polyp which is a negative phase buds off medusae which in their turn develop reproductive organs and reproduce sexually. This type of interdependence is called *Metagenesis* alternation of generation. In the group Actinozoa, *no medusa* stage ever appears, the *polyp-stage* alone gives rise to planula larvae which attach and become polyps, producing reproductive cells which grow into planula larvae. In the Hydrozoa, the reproductive cells are *ectodermal* in nature, while in all other coelenterata, including ctenophora, they are *endodermal* in their origin.

In the group Cnidaria the *cleavage* is said to be radially symmetrical, whereas in Ctenophora it is bilaterally symmetrical. In the genus *Spirocodon*, the first cleavage furrow does not reach the vegetal pole, so there is practically no cleavage activity at that pole. As the cleavage advances the *blastula* is formed. This may be either a *coeloblastula* or a *stereoblastula*. The blastula advances to become the

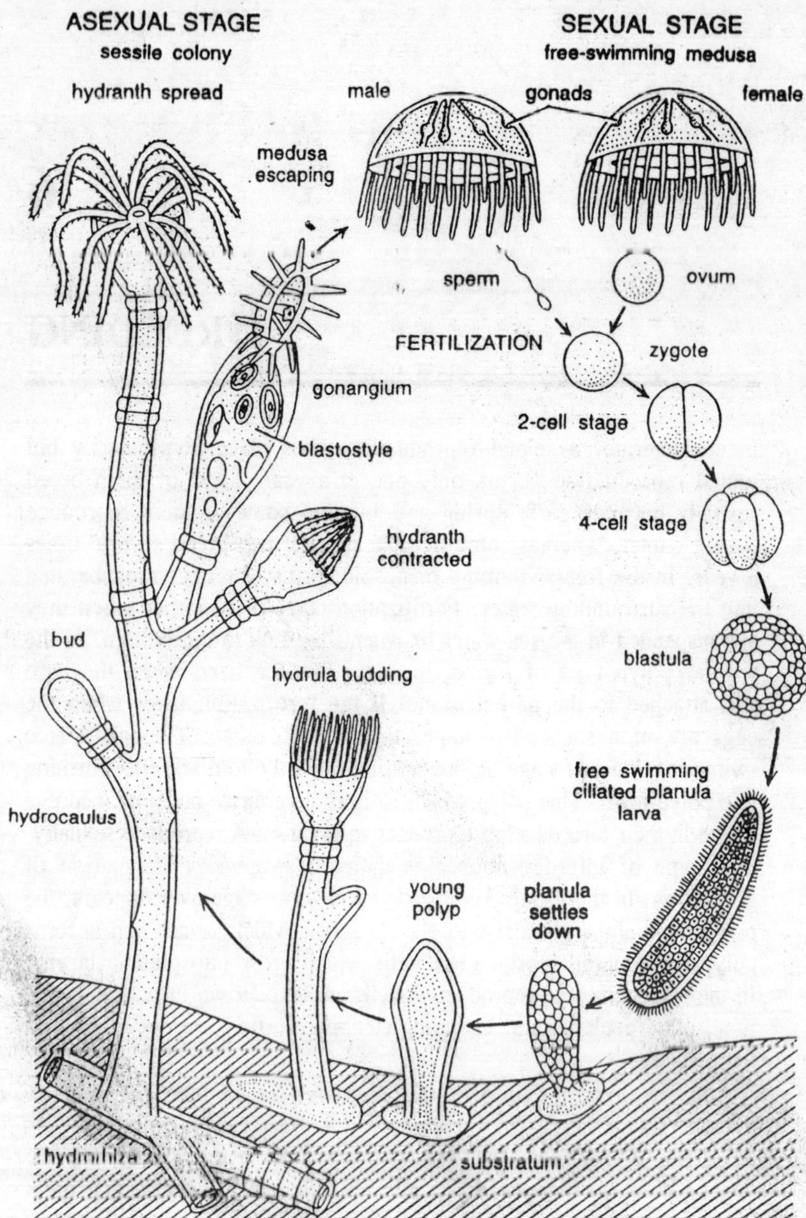

Fig. 6.1. Metagenesis.

gastrula. The endoderm of gastrula in many cases is formed by *invagination*. Invagination begins at the vegetal pole of the blastula by proliferation of cells, and these cells accumulate on the inner side of the blastular wall. In polarization, the cells of a limited area at the vegetal pole proliferate into the blastocoel. *Delamination* is a special mode of gastrulation peculiar to the coelenterates, in which all the cells of the blastula divide into the *blastocoel*. In this way the outer and inner body layers, the *ectoderm* and *endoderm*, are formed. Later, the *mesoglea* is secreted between these two layers.

In most of Actinozoa and Hydrozoa the are eggs opaque, containing large amounts of yolk. The *cleavage* appears to be *superficial*. In some cases the blastomeres separate into small ectodermal and large endodermal cells. In some forms, a *gastrula* is formed where the outer layer becomes the ectoderm and the inner layer, the endoderm. In such cases, the yolk remains in the blastocoel and gradually absorbed.

After the germ cells have been formed the ectodermal cells develop cilia, sensory and gland cells get differentiated, and a free swimming *planula larva* is produced. The planula larvae of Stauromedusae do not possess cilia and hence creep at the bottom. In Trachymedusae, Narcomedusae and Siphonophora the planula larvae are solid without a blastocoel, whereas in others there is a blastocoel.

In many cases, the planula larva becomes attached and develops into a polyp. In Zoantharia and Ceriantharia the planula larva while still swimming, transforms into a temporary swimming polyp form. The siphonophores, *Porpita* and *Valella* also develop into this special type of larva. All these larvae undergo metamorphosis to become adult forms. In Scyphomedusae after the planula has developed into a polyp, it produces a special *Strobila stage*, which asexually gives rise to the young larval medusae (*ephyra*). This ephyra passes through metamorphosis to become the adult medusa.

HYDROZOA

In Hydrozoa, each species has a polyp and medusa stage. These two stages appear alternately to give a regular alternation of generations also called *metagenesis*. In some forms the *polyp* stage may be predominant, while the degenerate *medusa* does not separate and remains attached to the polyp in the form of a *sporosac*. In some species, the whole life cycle may be *pelagic*, only the *medusa* is well developed, while the *polyp* is present in a modified or rudimentary state.

Development of Tubularia

In *Tubularia* there is *no medusa* stage and retains only the *polyp* form. It lives in shallow water on rocks and loose stones along the sea-

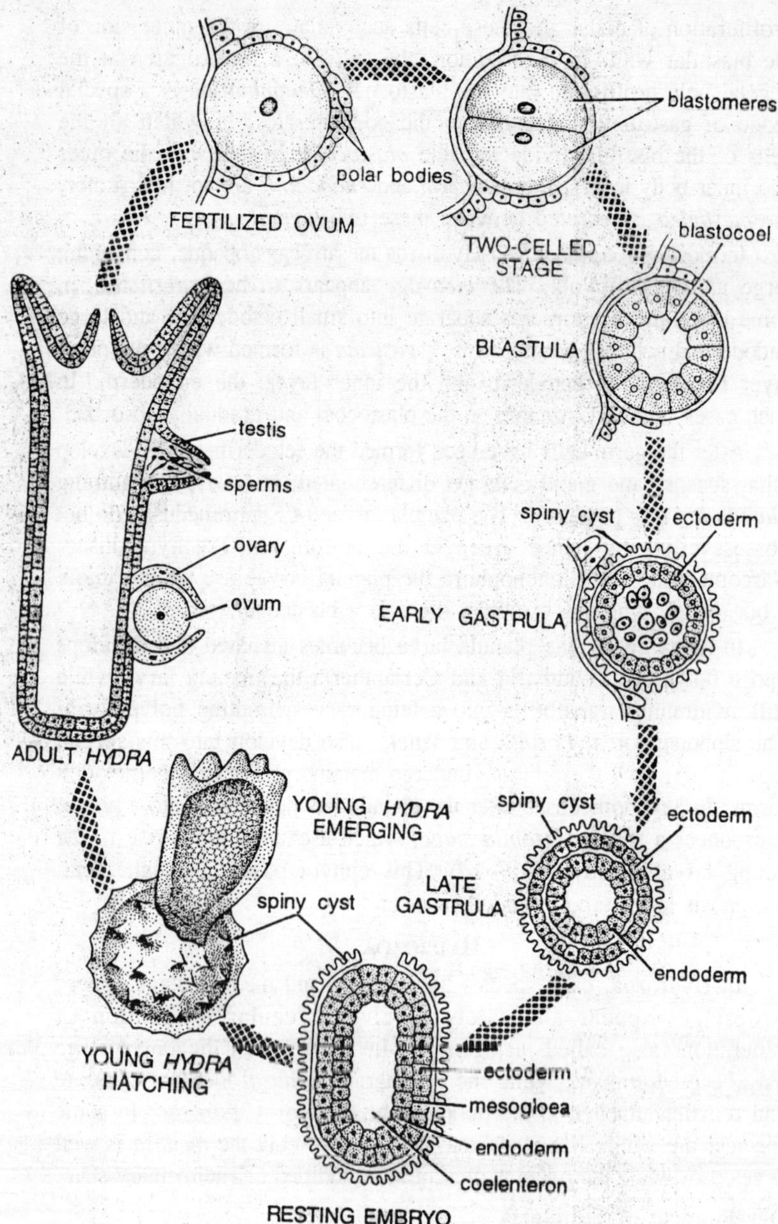

Fig. 6.2. Stages in the development and hatching.

shore or attached to eel grass. The polyp reaches to a height of 3.8 cm. It is generally found on all parts of the European and American shores.

The *gonophores* which are present on the polyp produce reproductive cells continuously by *budding*. This type of gonophore is considered to be a degenerate medusa called as the *sporosac*, which does not get detached from the polyp. The gonophores are formed in the space between the two circlets of tentacles of the hydranth. A number of processes appear at the base of the hydranth and at the end of each one of these processes is formed a *gonophore*, so that the entire structure resembles a bunch of gapes. The species of *Tubularia* are dioecious, the gonophores of a colony are either all male or all female and often exhibit sex differentiation to a certain degree. A *manubrium* is present at the centre of each *gonophore*. In the case of male gonophores, *sperm sacs* and in the case of females several *eggs* are given out from this manubrium. Fertilization is internal. The eggs are fertilized *in situ* by the spermatozoa of the male gonophores.

The eggs of *Tubularia* have good amount of yolk, hence the cleavage is rather irregular. Since the number of eggs in a single gonophore are more, this also tends to make the cleavage more irregular. The *cleavages* are somewhat *polynucleate*, the cytoplasm accumulates at the surface and the yolk towards the centre and this gives the effect of a *superficial cleavage*. The inner endodermal cells are a bit irregular in the beginning, but eventually form a definite layer. In this type of cleavage, the division of the blastomeres and the germ layer formation takes place simultaneously.

At the 16-celled stage of the regular cleavage, a small space is formed at the centre of the blastomeres. This stage is called as the *blastula* stage. As the cleavage proceeds the small *blastocoel* disappears and the embryo becomes a *solid morula*. At the time of the formation of the *morula*, the cells of a part of the blastula wall divide rapidly many times and these cells proliferate into the blastocoel. The cells of the original blastula become the *ectoderm* and the proliferated cells become the *endoderm*.

At the centre of the *solid morula*, small empty spaces make their appearance. These spaces eventually join at the centre, giving rise to a single cavity which later becomes the adult *coelenteron*. At this stage the embryo becomes flattened and plate-like. A number of small processes which are the rudiments of aboral tentacles, develop. As the development proceeds these processes bend towards the concave side of the embryo and increase in length. Meanwhile at the opposite convex side the ectoderm and the endoderm become very thin and eventually a mouth opening makes its appearance. At this stage a tendency to bend toward the aboral side becomes pronounced.

At its upper end around the mouth a number of small processes are formed which are the rudiments of oral tentacles. In the meantime, the aboral tentacles become long and slender and the aboral end of the body itself grows to about the same length as that of the tentacles. The larva at this stage is called the *actinula larva*.

This much development takes place inside the gonophore and the actinula larva escape to the outside through the mouth of the gonophore. When once they are free, they sink to the bottom, where they move about with the help of the oral and aboral tentacles and finally attach themselves to rocks or any other suitable substratum by their aboral ends. After attachment, the actinula increases in height, the part bearing the oral and aboral tentacles becomes the *hydranth*. The stalk secretes a chitinous *perisarc* and this modified actinula is called the *hydropolyp*. From the attached base a number of stolons are put out and these in turn give rise to new polyps, forming a colony. The gonophores appear for the first time during the breeding season. The development of *Tubularia* is thus completed without the inclusion of a free-swimming planula stage seen in the life histories of most other coeloenterates.

Development of Gonionema

The species of *Gonionema* is common along the pacific coast of Japan. The testes are yellowish brown while the ovaries are dark brown in colour. The eggs are brownish, opaque, with less yolk and

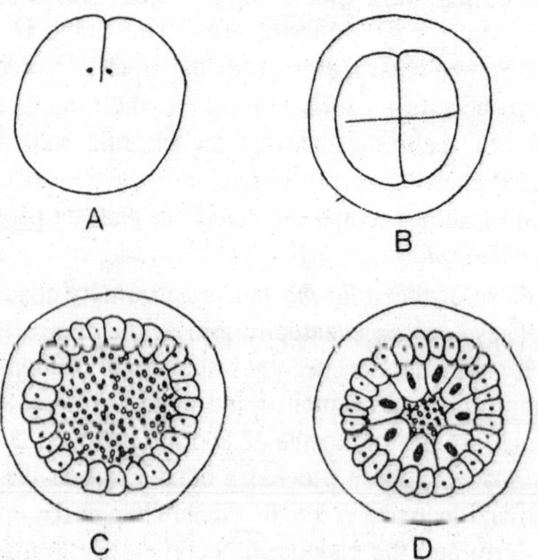

Fig. 6.3. Development of medusa, Gonionema. A—First cleavage, B—Begining of third cleavage, C—Blastula, D—Formation of endoderm.

measure about 0.07 mm in diameter. The eggs are released by the breaking of the ovarian wall, and as they fall to the bottom they are fertilized.

The *first cleavage* is completed within an hour after fertilization. In this cleavage furrow is meridional and from animal pole to the vegetal pole, dividing the egg into two *blastomeres*. The *second cleavage* is also *meridional* starts at the animal pole but at right angles to the first and takes about 50 minutes to complete. The *third cleavage* is *equatorial* giving rise to upper four and lower four blastomeres. The upper four blastomeres are turned by 45° and wedged between the four lower blastomeres. This resembles the *spiral cleavage*, but the latter divisions fail to show the characteristics of the mode of cleavage.

The *endoderm* is formed by delamination. The *ectodermal cells* develop cilia and the larva takes the form of a *planula* which starts rotating inside the chorion. After about an hour, the chorion breaks and the larva is set free. When once the larva is outside it becomes elongated, swims for several days and then settles to the bottom. The larva then loses the cilia, creeps about for a while before changing

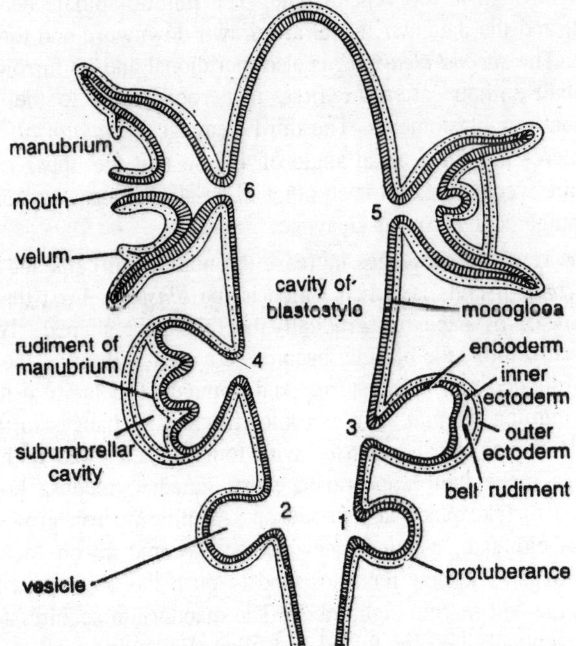

Fig. 6.4. Stages of the development of medusa from a blastostyle.

into a *polyp*. In polyp the oral operture or mouth develops first and then two tentacles appear at the opposite ends of the mouth. At a later stage two more tentacles appear at right angles to the first pair. The number of tentacles of the polyp may increase to five or six, but during the four-tentacled stage, bodies like planulae which lack cilia are budded off (fustules). This budding is repeated several times and the fustules are believed to grow into polyps. The polyps then produce one to several medusa buds, each of which develops radial canals, manubrium, tentacles, statocysts, a velum and other structures and finally set free as a free-swimming medusa.

Development of Spirocodon

The *spirocodon* breeds during the winter. The eggs of this medusa are about 0.06-0.07 mm in diameter. There is no external membrane. The cytoplasm is colourless and highly transparent. Spawning occurs during the night. The mitotic divisions are completed within the ovaries. The eggs are fertilized as soon as they are shed. The *first cleavage* is *meridional* and its furrow divides the egg into two, starting at the animal pole and reaching the vegetal pole. The cleavage furrow does not cut through at the vegetal pole. The mitotic spindle becomes V-shaped, and the daughter nuclei are drawn downward and towards the furrow. The *second cleavage* is also meridional and its furrow appears about half-an-hour after the first, is perpendicular to the first and produces four blastomeres. The third cleavage is *equatorial*, the eight blastomeres are tilted at an angle of 45° so that the upper and lower layers are wedged against each other in an alternating way, resembling the arrangement in spiral cleavage.

The repeated cleavages increase the number of cells and finally a hollow ball of cells is famed which is the *blastula*. First the blastula is elliptical in shape but gradually becomes egg-shaped. Ten hours after fertilization, the blastula becomes covered with cilia, the anterior end is rounded and the posterior end pointed. The larva at this stage is a swimming *planula*. It is not known when it changes to the polyp form. However, small *medusae* with four tentacles and four tentacle swellings make their appearance every January, medusa budding is supposed to take place at that season. As this medusa grows and its tentacles elongate, a pair of new tentacles come up on each side of the old ones. The new tentacles tend to push the older ones upward, making the bell margin eight-lobed. The manubrium acquires four lips, which eventually become folded and filled. Following such a course of development, this medusa may reach a height of 50 mm, and is one of the largest and most highly differentiated and morphologically complicated of the Anthomedusae.

General Development of Hydroidea

The members of the Hydroidea have both a polypoid and medusoid generation. These appear alternately, giving an alternation of generations. In many cases the medusoid generation is very miuch reduced and becomes attached to the polyp. Various degrees in this reduction of the medusoid generation can be recognized. These are divided into *eumedusoid cryptomedusoid*, *heteromedusoid*, *styloid* etc. In some species, the degree of reduction is different between the sexes, the female for instance, separating as a *free medusa*, while the male is represented by only a *sporosac*.

In Hydrozoa, the reproductive cells are *ectodermal* in origin and usually penetrate into the endoderm after completing maturation. The spermatozoa are flagellate. The eggs that are produced in a *sporosac* are likely to be large and yolky, while those that are formed by medusa tend to be small and have little yolk. Depending upon the amount of yolk, the latter development varies considerably. The eggs, with little yolk are shed into the sea water and fertilized there. The cleavage is regular, total and practically equal. The first two cleavage planes are *perpendicular* to each other. These are *meridional*, while the third is *horizontal*. From this point the cleavage follows the radial type with the micromeres at the animal pole and macromeres at the vegetal side. The simply organized *Hydromedusae* are also known to cleave totally but irregularly. Eggs which undergo regular radial cleavage tend to form blastulae with small blastocoels or even with none at all, and in many cases the *blastocoel* is later completely filled with *endoderm*. While the *alecithal medusa* eggs form *coeloblastulae*, the *sporosac eggs*, of such genera as *Clava*, *Gonothyrea* and *Plumularia* are likely to develop into *stereoblastulae*. In the *coeloblastula*, the endoderm is formed by invagination or polarization. While in *stereoblastula* is tends to be irregular. In the yolky eggs like those of *Tubularia* and *Enderdium*, the cleavage is irregular, giving rise to multinucleate cells, with the cytoplasm accumulated at the surface and the yolk left in the centre of the embryo. This type looks like a *superficial cleavage*, with the ectoderm cells arranged around the surface and few nuclei in the interior where endoderm cells are formed in an irregular manner.

As the *coeloblastula* proceeds to the gastrula stage it usually swims as a free *planula*. As the larva grows, it becomes oval, develops

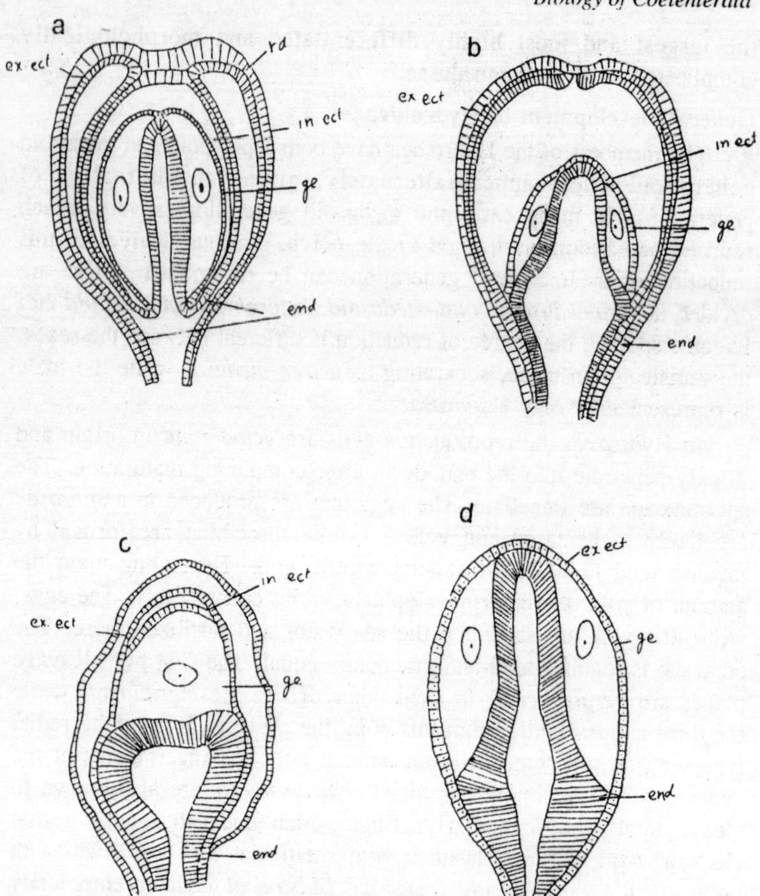

Fig. 6.5. Various types of sporosacs. A—Eumedusoid. B—Cryptomedusoid. C—Heteromedusoid. D—Styloid.

sensory and glandular cells at its anterior end. In many cases, the planula, after leading a free-swimming existence for sometime, attaches by its anterior end to a substratum. In the beginning it loses the cilia, develops pedal discs or stolons on the attached end. At the opposite end, tentacles and mouth develop as a result of which a *hydranth* is formed. The middle part becomes the *stalk*, containing the *coenosarc* which secretes the *perisarc*. The stalk gives branches whereas the stolon may form a *hydrorhiza* leading to a colony. In the colonial hydroids thee is a specialization among the individuals making up the colony, the nutritive polyp being formed first and later the zooids concerned with reproduction and other functions.

In some species of Hydroidea there is no planula stage, it being passed in the gonophore itself. In *Tubularia*, the is no planula, the embryo within the gonophore develops directly into an *actinula*. This larva has a mouth, surrounded by tentacles and constitutes a tiny polyp. In *Sertularia* and *Orthophrix*, the embryos develop in special *brood sacs* (*marsupia*) formed by the gonotheca, which contain masses of gelatinous substance.

General Development of Trachylina

There are two suborders, Trachomedusae and Narcomedusae in order Trachylina. Both these groups include species which do not form *polyp* colonies. Most of them may not have a free *planula* stage. The whole life cycle may consist of free-swimming *medusa*. However, some *Narcomedusae* species may have a larval stage which becomes a parasite on other medusae. In some cases the polyp stage is represented by the brief appearance of an *actinula larva*. They are diocious and the sexes are separate with one exception. The gonads are epidermal occurring

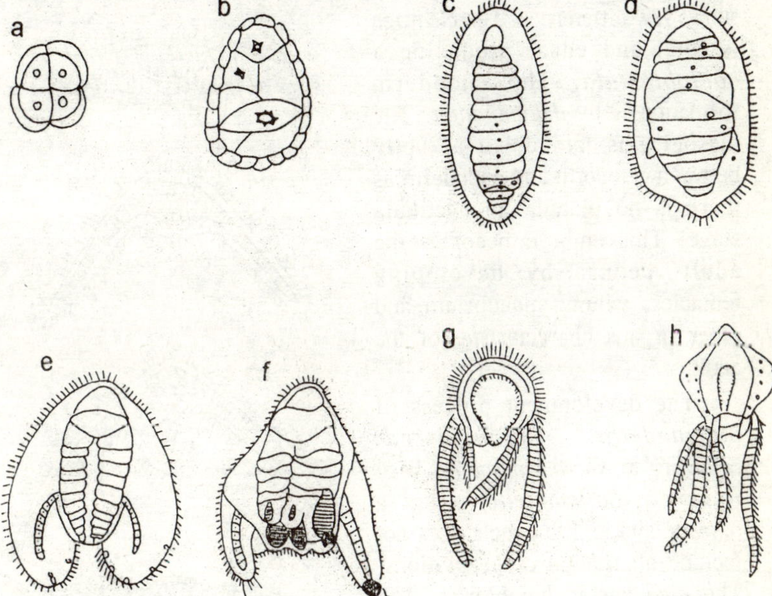

Fig. 6.6. Development of Aglaura. A—8-cell stage. B—Formation of ectoderm and endoderm. C—Planula. D—Early stage of tentacle development. E—Formation of gastric cavity. F—Increase in number of tentacles. G and H—Formation of young medusa.

as folds in *Trachomedusae* or as sacs beneath the radial canals. In the *Narcomedusae* they occur at the floor of the gastric pockets.

In Trachomedusae the eggs are small and that of *Narcomedusae* are large. The cleavage pattern varies considerably according to the species. The blastula may be a *coeloblastula* or *stereoblastula*. In the coelobastula the endoderm formation is by multipolar proliferation and in the stereoblastula it is by delamination method. A *planula* is formed at the time of gastrula and this followed by the secretion of large amount of *mesoglea*. Then tentacles and mouth appears and the planula passes into an *actinula* stage, which gradually metamorphoses into an adult.

In Trachymedusae, the cleavage is total, equal or unequal. It leads to a *morula* which becomes a ciliated *planula*. Endodermal cells at the posterior end of this larva protrude outward forming the cones of the tentacles. The tentacles increase in number and the cilia are lost as the bell of the medusa takes its shape.

In *Liriope* the development shows few differences. The *cleavage* is total and equal producing a *coeloblastula*. The endoderm formation is by *delamination*. The mesoglea is laid down abruptly between ectoderm and endoderm as there is no planula and actinula stage. This embryo becomes the adult medusa by developing tentacles, velum, manubrium and other organs characteristic of the adult.

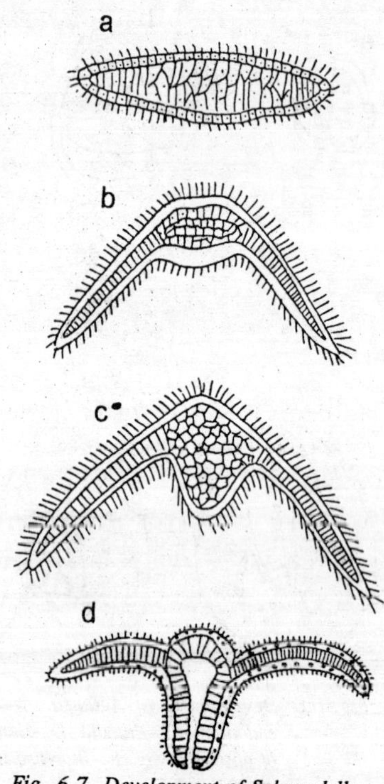

Fig. 6.7. Development of Solmundella.

The development process of *Solmundella*, *Cernoctantha*, belonging to *Nacromedusae* has been followed. *Solmundella* is not a parasite and its life cyucle does not include alternation of generations. The *cleavage* is *holoblastic*. The *stereoblastula* forms its endoderm by multipolar proliferation. A *planula* is formed, both the ends develop tentacles and the central region becomes thickened. This region later becomes the stomach with a mouth opening at its lower side. This larva slowly develops mesoglea, sense organs and becomes the adult medusa.

In the genus *Cernoctantha* a *planula* larva gives rise to an *actinula* larva with tentacles and a mouth. But there are no sense organs and no bell. This larva is parasitic in the stomach cavity or water vascular system of medusa of its own or other species. Then it produces more actinula by asexual budding. All these actinulae finally transform into medusae. In this way this animal combines a parasitic way of life with alternation of generations.

General Development of Siphonophora

Members of group Siphonophora are pelagic forming highly polymorphic colonies by means of budding. The individuals of the colony which are concerned with reproduction are called as *gonophores*. The gonads of a single colony may be either male or female but the colony as a whole may have either gonophores all of one sex or of both sexes, depending upon the species.

In general the cleavage is *holoblastic* and equal giving rise to a solid ball of cells - the *morula*. *Gastrulation* is by *delamination*. The *stereogastrula* develops cilia on its outer surface and forms a *planula*. The shape of the planula is characteristic of the species. Further development after the planula stage varies in different species.

In the suborder Calyconecta the *planula* is *ovoid*, solid with a ciliated surface. A small bud is given out from one side of this planula which later becomes the undersurface of the primary swimming bell. Tentacles are formed on the lower part of this and from the opposite end a single primary *gastrozooid* is formed when the primary *swimming bell* is completely formed. A stem grows out between bell and gastrozooid and buds extensively to produce the other individuals forming a large colony.

In the suborder Physonecta the lower or oral end of the planula also becomes the *primary gastrozooid* but before this change takes place, the aboral end forms an umbrella-like expansion. This primary or larval bract serves to *float* the larva while the *pneumatophore* is developing. The pneumatophore arises beneath the primary bract as an ectodermal invagination. Near to this are budded successively definitive bracts, dactylozooids and eventually the swimming bells.

In the suborder Cystonecta (*Rhizophysa*, *Physalia*) the development seems to be generally similar to that of the Physonecta. The *pneumatophore* develops to a still larger size and is separated from the *gastrozooid* by a septum and no swimming bells are formed. The early development of *physalia* is passed in deep water and has not been observed.

The development of such forms like *Velella* and *Porpita* show rather wide divergence. The earliest known larva termed as *Conaria* is a hollow sphere with an aboral thickening. The hollow sphere represents the primary *gastrozooid*. The aboral thickening contains a small air sac, a system of eight entodermal canals, a ring-shaped nematocyst and an entodermal plug which projects into the coeloenteron of the primary zooid. The *Conaria* larva is said to sink into deep water where it continues to develop and only appears at the surface after reaching the next stage which is called a *Rataria* larva. In this stage, the *pneumatophore* is greatly enlarged. At the junction of the float and the primary zooid a ring-like budding zone is formed from where new zooids are budded out. The *rataria larva* metamorphoses and gradually builds up the complex structure of the adult *siphonophore*.

In several species belonging to the Calyconecta, some of the units called *Cormidia* may separate from the colony and take up independent existence. These are called the *endoxides*, they develop gonads, and carry out sexual reproduction. The fertilized eggs give rise to colonies like the original one.

SCYPHOZOA

In the Seyphozoa, there are both polypoid and medusoid types, but the polypoid generation is insignificant in comparison with the large conspicuous medusoid stage. These two types appear in turn to establish a general alternation of generations. The Seyphozoa is classified into five orders:

1. *Stauromedusae,*
2. *Cubomedusae,*
3. *Coronate,*
4. *Semaeostomae,* and
5. *Rhizostomae.*

Of these, the Stauromedusae are sessile and the Cubomedusae are pelagic. These two orders have several characters in common and differ from the other three orders in many morphological and embryological characters.

In their general appearance, the Stauromedusae are medusoid in their upper part and polypoid in their lower part. Since they do not give off a free-swimming *medusa* there is no asexual reproduction and

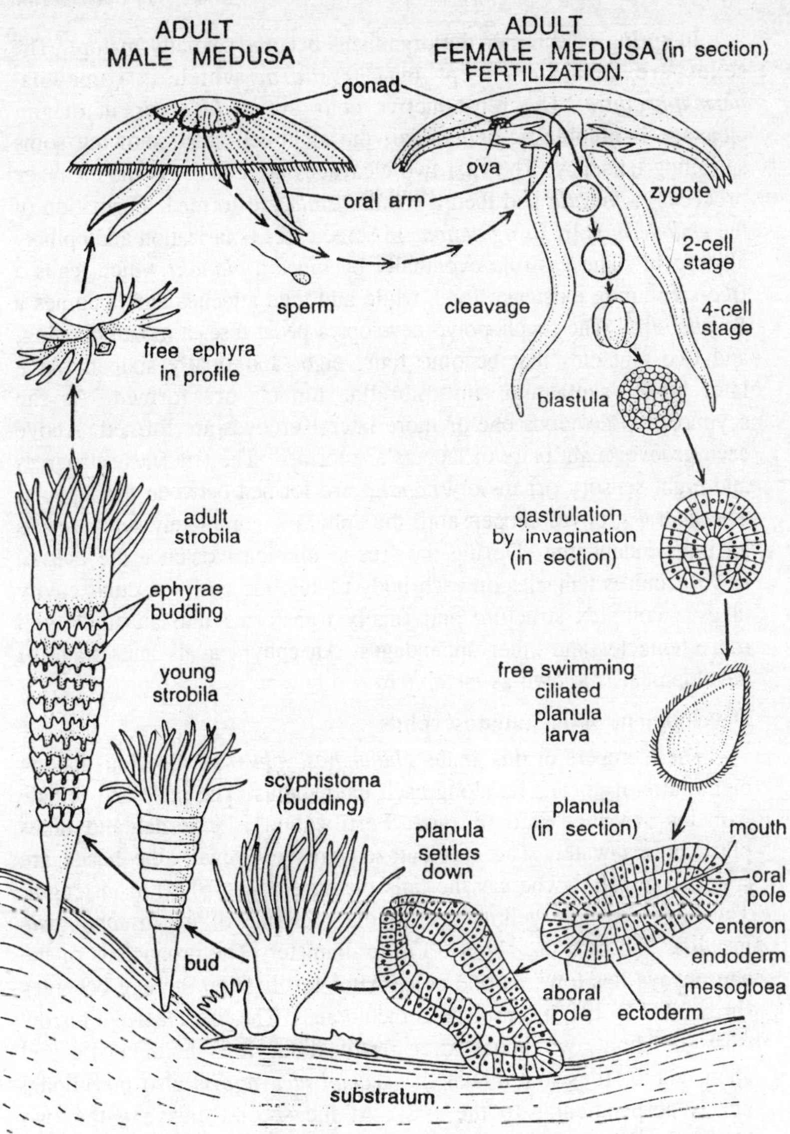

Fig. 6.8. Aurelia development.

hence no *alternation of generations*. The development of the Cubomedusae is known upto the point of the formation of the *polyp*. How the medusa is given out is not known. In the other three orders the usual alternation of generations is seen.

In spring or summer the organisms become sexually mature. The sexes are separate except in *Chrysaora*, which is somewhat *hermaphroditic*. The reproductive cells are *endodermic* in origin, cleavage is holoblastic and radial, the eight-cell stage showing some spiralling tendency. The first two cleavages are equal. As the eleavage proceeds, a *morula* and then a *coeloblastula* are formed. Formation of the *endoderm* is by *invagination*. In some cases polarization and epiboly also occur. The gastrula eventually becomes a *planula* which leads a free-swimming existence for a while and then attaches and becomes a *Scyphopolyp*. The scyphopolyp develops a pedal disc, a mouth opening, and two tentacles that become four, eight and finally sixteen. At a later stage, septa and sub-umbrellar funnels are formed. As the scyphopolyp develops one or more lateral grooves are formed. Above each groove, eight pairs of lappets are formed. The tentacles retrogress and eight sensory organs or *rhopalia* are formed between the lappets. The lateral grooves deepen until the ephyra is completely separated as an independent unit. During the free-swimming existence the *ephyra larva* acquires tentacles on each body radius, the gastrovascular cavity takes a complex structure and finally transforms into an adult with more tentacles and other appendages. An ephyra at its later stage of development is known as *metaphyra*.

Development of Thaumatoscyphus

The members of this genus *Thaumatoscyphus* are dioecious having eight pairs of gonads lie along each inter radius. The male and female gametes are shed into the sea. Fertilization is external and takes place in sea water. The spermatozoa are very small, the heads are small, triangular, whereas the tails measure about 70-80 m in length. The eggs are pale yellow due to the presence of *microscopic yolk granules* and measure about 50 m in diameter. The pronucleus of the egg always lies towards one pole. After fertilization the egg becomes surrounded by two inconspicuous membranes. The *first cleavage* furrow starts at about two hours after fertilization. It is holobla stic and about equal. It is meridional the *second cleavage* is also meridional but is perpendicular to the first. At the second cleavage the two blastomeres do not divide simultaneously but there is always a time lapse. The *third cleavage* is *horizontal*, a little above the equator. The four upper, smaller blastomeres turn horizontally to a slight extent so that they tend to slip between the four lower blastomeres. Till now the cleavage is *radial* and practically total and equal. Now it loses its irregularity and the outline of the blastomeres becomes indistinct. The

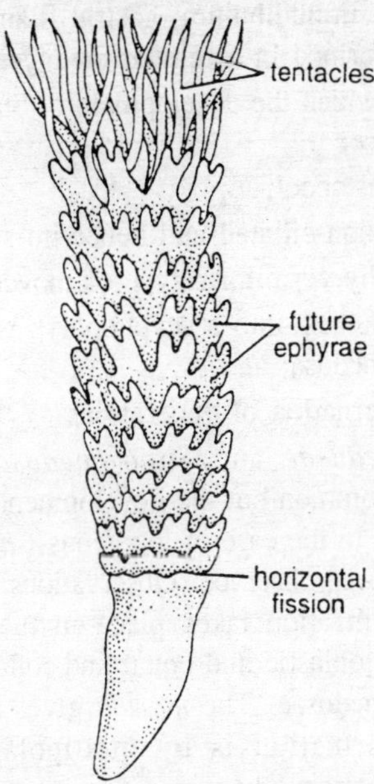

Fig. 6.9. A strobilated scyphistoma.

embryo becomes a *morula*. Instead of becoming a *blastula* the interior is filled with cells which are about the same as those of the outer layer. The endoderm formation is by unipolar proliferation. After the endoderm has been formed, the ectoderm cells flatten and become arranged in a layer around the outside of the embryo, while the endoderm cells which are now large and vasculated line up longitudinally in the interior. At this stage the embryo breaks from the fertilization membrane and becomes a long, slender, wormlike planula. This *planula larva* differs from the other planulae in not having a *blastocoel* and *cilia*. Since there is no ciliation, it cannot swim but creeps at the bottom with a vermiform type of movement. A number of such planula are found collected in masses which are surrounded by a shell made up of sand grains and debris.

The development of *Thaumatoscyphus* has been observed only to this extent, but information is available about later stagers of other Stauromedusae like etc. In *Haliclystus*, *Sesakiella* the planula finally attaches and takes on a hemispherical form and before transforming reproduces asexually by budding. The planulae produced in this way grow and form other hemispherical cell masses which finally transform into polyps. The polyp eventually forms a hypostome, pedal disc, tentacles, anchors, infundibulum, gastral filaments etc. New tentacles are subsequently formed in a regular arrangement at each per-radius.

The points in which the development of Stauromedusae differ from the other orders are:

1. there is no blastocoel;
2. the planula is non-ciliated and hence no swimming;
3. locomotion is by vermiform type of movement;
4. the planula changes into a scyphopolyp which in turn becomes a single young medusa; and
5. there is no alternation of generations.

Both *Cubomedusae* and *Stauromedusae* have a number of characteristics in common but the development of Cubomedusae is not very well=known. In the case of the genus *Charybdea* the planula and scyphopolyp have been observed. Observations on the genus *Charybdea* show that the fertilization takes place in the gastrovascular cavity. The cleavage is holoblastic and equal and follows the same pattern as that of the Stauromedusae. The *morula* gives rise to a *Coeloblastula.* The endoderm formation is by multipolar proliferation or by delamination. The embryo becomes a *gastrula*. The gastrula becomes ciliated and transforms into a small pear-shaped *planula*. After swimming for about two or three days it becomes sessile. It then loses its cilia and becomes a plate-like mass of cells from which two tentacles arise. As the tentacles elongate, the body also increases in length, a small hypostome is formed and two new tentacles make their appearance.

Development of Nausithoe

The genus *Nausithoe* is a coastal form, and hence a detailed study of its development has been done. The eggs are generally small. The *cleavage* is *radial*, equal and *holoblastic* and which eventually gives rise to a *morula* and *blastula*. *Gastrulation* may be either by invagination or by irregular, monopolar proliferation. The gast ula becomes a ciliated *planula* which detaches itself from the fertilization membranes and starts a free-swimming existence. The planula after sometime attaches by a plate-like cell mass on its underside, loses its cilia and separates into an inner and outer cell layer. The surface becomes covered by a hard transparent layer and later a mouth and tentacles are formed. The polyp is well developed and is commonly known as *Stephenoscyphus*. It grows to a height of 10 cm and starts *strobilization*. The *ephyra* larva of *Nausithoe* differ from those of the Semaeostomae and Rhizostomae in the symmetry of its lappets. The tentacles appear at a later stage. The gastral filaments occur in single numbers on the inter-radii and gradually increase in number. Finally one gonad is formed on each adradius.

Development of Aurelia

This genus is widely distributed throughout the temperate waters of the Pacific and Atlantic oceans and the Mediterranean Sea. Sexual maturity commonly occurs in spring or summer. The Sexes are separate except in *Chrysaora*. The gonads are purple in young medusae and turn brown in older ones. The gonads develop on the four inter-radii, each being much folded, forming a horse-shoe shaped structure which is pointed inward.

The spermatozoa pass through the subgenital pits into the sea water. Some of these find their way into the gastral cavities of the females where the eggs are fertilized.

The fertilized egg undergoes cleavage within the gastral cavity. The cleavages are regular, total, forming a coeloblastula. A few cells migrate from the blastular wall into the interior so that the blastocoel becomes almost full of them but later these cells disappear. The definitive endoderm is formed by a more or less typical invagination

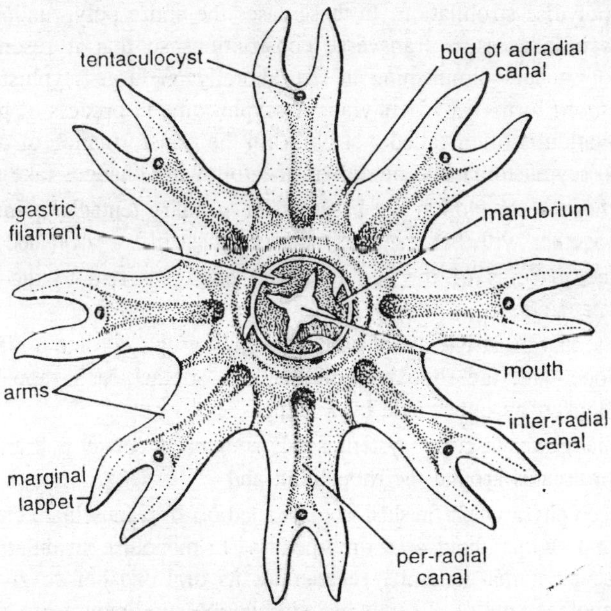

Fig. 6.10. An ephyra larva.

of the vegetal pole cells. As a result of this a hollow gastrula with a two-layered body wall made up of ectoderm and endoderm is formed. A small pore remains open where the vegetal wall is invaginated and this is the blastopore. The spherical gastrula gradually becomes oval, the anterior end is wider and the posterior more pointed. The ectoderm acquires cilia and the gastrula becomes a planula.

The planula has a small blastocoel. The ectoderm gives rise to nematocysts, sensory cells and gland cells. The endoderm contains large vacuoles. The planula swims for about four to five days, loses its cilia, attaches by its anterior end. The cells at this end secrete a pedal disc which helps in getting a firm hold to the substratum. This develops into a typical Scyphistoma larva of a trumpet form with basal stalk, pedal disc and expanded oral end with mouth, manubrium and tentacles. The mouth breaks through at the side of the closed blastopore. The first tentacles are perradial followed by interradial and then adradial ones. The Scyphistoma, in species inhabiting temperate zones remains without further change through fall and winter, feeding and producing other Scyphistomae by various asexual processes. In winter and early spring strobilation sets in. In some genera, strobilation is monodisc, with only one ephyra forming at a time, while in others it is polydisc strobilation. In these cases the entire polyp undergoes a series of successive transverse constrictions so that it resembles a pile of saucers diminishing in size aborally. A large scyphistoma of *Chrysaora* forms 13-15 ephyrae. A scyphistome in process of polydisc strobilation is often called a strobila. In the transformation of oral end of the scyphistoma into an ephyra the following changes take place:

(a) rhopalia develop at the bases of the primary tentacles which then together with the other tentacles are cast off or absorbed;

(b) the margins of the septa thicken and give rise to the gastral filaments;

(c) the margin at the site of each rhopalium grows out into a bifurcated lobe with the rhopalium in the fork so that each rhopalium is flanked on either side by a lappet;

(d) the gastrovascular system sends into each octant a branch that bifurcates around the rhopalium; and

(e) the ephyra when finished is constricted off by a muscular contraction and swims about as a tiny medusa. In monodisc stridulation, the scyphistoma generally regenerates its oral end before giving off another ephyra. In polydisc strobilation, the ephyra are released at intervals.

The scyphistomae live for several years. After a period of strobilation in winter and spring, they stop forming ephyrae and resume life as polyps, feeding and reproducing other scyphistomae asexually until the following winter when once again strobilation starts.

The newly released ephyra is a minute gelatinous creature one to a few millimetres in diameter. It has eight lobes bifurcated at their end into two lappets having a rhopalium in between them. From the centre of the subumbrella a short quadrangular manubrium hangs out. If food material is available, the ephyra expands and transforms into an adult medusa. In the process new tentacles are formed, on each adradius and their number continues to increase. More marginal lappets are also formed on the adradial regions. The manubrium develops long oral arms, the subgenital pits, gastrovascular system, gonads and other structures and increases in size. The young medusa which has completed this transformation is only about 1 cm in diameter. From this point on, it increases in size. The radial canals become more complicated, the gastral filaments increase in number and the oral arms develop elaborate frills and folds.

The development of the genus *Pelagia* differs from the other Semaeostomes. The planula develops directly into an ephyra without alternation of generations. The planula is ciliated. The mouth is on the lower side. The endoderm is formed by invagination. The blastocoel transforms into a large gastral cavity. Eight lappets are formed first on the lower side, these soon become eight pairs and rhopalia develop in between them. By this process the larva changes directly into an ephyra without passing through a scyphistoma stage.

The development of the species *Mastigias papna* belonging to the order Rhizostomae, which is found in Japan is well investigated. The early development of this species closely resembles that of *Aurelia*. It has monodisc type of strobilation. There is no stack of discs like that found in *Aurelia*. The transformation of the ephyra into adult is highly characteristic. The edges of the four lips around the mouth opening spread horizontally, developing a fringe of small tentacles. Each adradial portion becomes deeply indentated, while shallow constrictions appear in the perradii. These arms later bifurcate again, acquiring an extremely complex structure as the bifurcation is repeated many times. During this elaboration of the manubrium, the central mouth is closed off, and many small suctorial mouths appear among the fringes of the oral arms, connected by branchial canals which are subdivisions of the gastrovascular cavity formed by the repeated bifurcations of the oral arms.

Anthozoa

General Development

The Anthozoa do not have a medusoid phase in their life history, consisting only of polypoid generation. Hence the development is relatively simple than those of the other coelenterata. The gonads are *endodermal* in origin. In the eggs where the quantity of yolk is more, the cleavage is *superficial*. In the eggs with less amount of yolk the cleavage is holoblastic and radial. In most cases the endoderm is formed by invagination. The yolk remaining in the blastocoel is absorbed by the time *planula* larva is formed. The planula stage is free-swimming and lasts for several days. In the solitary forms the septa become highly developed. In free-moving forms, whether solitary or colonial, they have muscular bodies whereas in the sessile species a calcareous or keratinoid skeleton is formed. These skeletons are secreted by mesogleal cells which arise as modified ectodermal cells.

Development of Sagartia Traglodytes

Sagartia traglodytes is a European species. It breeds every year in the beginning of August, when the temperature of the sea water reaches 15°C. The perms are shed into the sea water through the pores that are at the tips of the tentacles on the pedal disc. The mature eggs measure about 0.03-0.09 mm in diameter. They are fertilized as soon as spawning takes place and form a fertilization membrane. The first cleavage furrow appears an hour after fertilization. This is equal *meridional* and holoblastic starting at the animal pole and reaching the vegetal pole. The second cleavage is perpendicular to the first forming four blastomeres. It is also *meridional*. The third cleavage is *equatorial* forming four upper and four lower blastomeres. At this eight-celled stage a *blastocoel* has already formed and it increases in size as the cleavage proceeds. The cleavage continues regularly with radial symmetry and a 128-cell stage is reached by the end of about four hours after fertilization. The blastocoel is filled with a large number of yolk cells. The *endoderm* is formed by invagination. After this is completed the ectoderm invaginates forming the future stomodaeum. Fifteen hours after fertilization and when the endoderm is still being invaginated, the embryo becomes ciliated and the *planula larva* starts swimming slowly. The body becomes thinner and longer, long cilia develop at the opposite end of the mouth. The planula swims with this end forward and later on attaches by the same end. This bundle of cilia is characteristic of sea-anemones. This stage is called the *Edwardsia* stage. The septa also become conspicuous at this stage. After six or seven days of free existence, the planula larva sinks to the bottom and becomes attached. The body increases in size, the tentacles appear in regular order, the number of septa rise and the gonads approach sexual maturity.

Development of Halcampa Duodecimcirrata

It is a sea-anemone with an elongated body which is buried in the sand. It breeds in October and November on the coast of Sweden. The eggs contain large amount of yolk and undergo *superficial cleavage*. First the males come out of the sand and shed sperms. In response to this stimulus, the females spawn. The eggs are shed from the mouth, into the sea water. Fertilization is external and takes place is sea water. After fertilization, a *fertilization membrane* is formed to which sand grains adhere. Endoderm formation takes place about 35 hours after fertilization. Endoderm formation is by invagination from one pole. The embryo is still solid even after the endoderm has been completely formed but a stomodaeum has been completely formed but

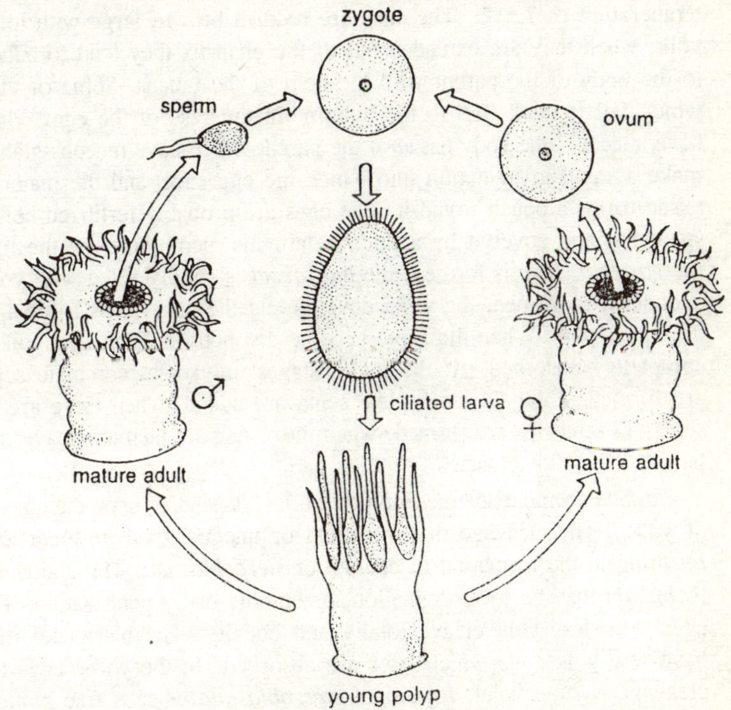

Fig. 6.11. Development of Anthozoa.

a stomodaeum is formed and the yolk cells are absorbed. The shape of the *larva* changes to oval, and becomes elongated. At one end a mouth appears and the other end becomes pointed. Up to this point the development takes place inside the jelly-like case, and now the embryo starts moving inside the case. The larva finally emerges out of the sheath with the pointed end foremost.

The larva immediately digs into the sand and continues its development there. In this species the planula has no swimming stage.

In some anemones the young develop attached to the outside of the adult, or in brood pockets or within the coelenteron. The species *Epiactis japonica* is bisexual or monoceous and have the young ones attached to the outer surface of the body. Eggs are formed on the younger septa and the sperms on the older septa. The eggs and sperms are shed through the pharynx together with mucous secretion as a result of rhythmical muscular contractions of the body wall. The spermatozoa measure about 47.5 m with the head measuring only 2 m in length. They can live for about five days in sea water at a temperature of 7.5°C. The eggs are reddish brown, large with lot of yolk; when they are extended out of the pharynx they tend to adhere to the body of the parent with the help of the mucus. Some of them which fail to stick, fall to the bottom. In the case of the eggs which have attached, the body tissue of the parent which have mucous glands, make a small invagination into which the egg sinks and the maternal tissue forms a pouch round it. The eggs are probably fertilized before spawning and covered by a thick gelatinous membrane. By the time the *gastrula* stage is formed, the *blastocoel* is solidly filled with yolk, endoderm is formed, the septa develop and the pharynx is formed on the upper part. When the septa become distinctly defined, the yolk is completely absorbed. At the initial stages only eight complete septa are formed. Later 24 incomplete septa are added, when there are 12 septa, 12 tentacles are formed, when the young seaanemone leaves the parent it has 24 tentacles.

In Sea anemone the development varies, depending upon the amount of yolk. The cleavage may be equal or unequal total or superficial resulting in the formation of coelom or *stereoblastula*. The endoderm formation may be by invagination, or epiboly or by polarization. The eggs with less yolk cleave totally and equally. The blastocoel may have few yolk cells which later are absorbed. In the yolky eggs the cleavage is superficial, forming a *stereoblastula* when a free *planula* is formed it may have a cluster of long cilia at its aboral end or they may lack cilia completely, changing into a polyp immediately. Then it passes through an *Edwardsia* stage characterized by the possession of eight complete septa. The larva then enters the *Halcampula* stage which has six pairs of septa. The number of tentacles and septa vary from one species to another.

The species *Cerianthus lloydii* belongs to the order Ceriantharia. It is a north European species, bisexual or monoceous and breeds in February. The mature ova are spawned about an hour after the sperms are shed. Fertilization takes place externally. The eggs undergo *superficial cleavage* and form *stereoblastula*. A *blastocoel* appears and gradually becomes larger after the endoderm is formed. A pharynx develops. Next, the ectoderm and the endoderm together extend out in four directions with the pharynx as centre forming a rectangular figure. These extensions are the rudiments of marginal tentacles. Two more tentacles known as oral tentacles appear at the sides of the mouth, making a total of six tentacles. Septa are formed internally and a so-called *Arachnactis* larva begins its life. The number and length of the tentacles increase, the larva elongates and sinks to the bottom and burrows into the sand where it constructs around itself a sheath of sand and slime.

Another species of Ceriantharia, *Pachycerianthus multiplicatus* has eggs containing lot of yolk. They undergo holoblastic, somewhat unequal cleavage. It is also bisexual or monoceous. The sperms are shed first and then the eggs. Since the cleavage furrows start simultaneously from both the ends it is difficult to distinguish between animal and vegetal pole. Upto the eight-blastomere stage the cleavage is somewhat spiral in nature, later radial symmetry is restored by about the 16-celled stage. The centre of the blastula is solidly filled with yolk. One pole of the embryo becomes flat and invaginates forming endoderm.

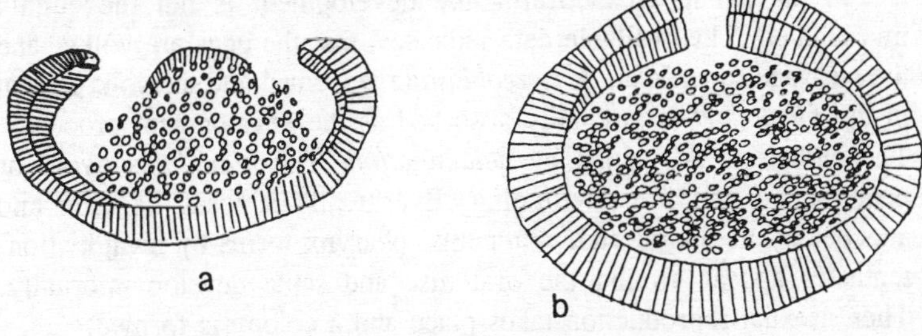

Fig. 6.12. Successive stage in gastrulation of Pachycerianthus. (a) Formaion of endoderm by gastrulation. (b) Segmentation cavity filled with yolk.

The yolk in the blastocoel, becomes gradually absorbed, and the outer ectodermal cells acquire cilia. Internally the pharynx and the septa are formed and the larva starts to swim. Then the body becomes elongated, the oral end flattens out and the aboral end narrows. The larva swims with the aboral end forward and in this position burrows into the sand with a spiral movement. The body goes on increasing in length, tentacles are formed and their number increases gradually and also the number of septa increase. In the development of this species there is no *Arachnactis* stage.

The members of the group Madreporaria live in the warm ocean currents and many of them breed throughout the year. In many cases, the eggs are heavily laden with yolk, the cleavage is superficial. Gastrulation is by invagination. The planula larva is either pear-shaped or oval, they are often brown because of the symbiotic algae (zooxanthellae) living inside them.

The planula fix by their aboral end and change to the polypoid form, developing septa and pass through Edwardsia and Halcampula stages. The characteristic feature of these species is that there is less of muscle tissue and formation of exoskeleton. Extensive colonies are built up by asexual reproduction. Formation of exoskeleton goes hand in hand with the development of septa. The young polyps have a strong tendency to aggregate which leads to the formation of colonies.

The members of the genus *Fungia* exhibit an alternation of generations. In this genus the fertilized egg cleaves and forms a planula, which attaches and develops into a branched trophozoid. This by budding gives rise to *anthoblasts*. The periphery of the anthoblasts expand and undergo repeated lateral fissions forming the larva called as *anthocyanthus*. This larva develops into the adult *Fungia*, which reproduces sexually as well as by budding.

In the group Hexacorallia the development is not thoroughly investigated. The available data indicates, that the eggs are yolky, and the cleavage superficial. A *stereoblastula* is formed and the yolk present in the *blastocoel* is gradually absorbed as the development proceeds. The *endoderm* is formed by *delamination*. The *planula* is ovoid or pear-shaped and free swimming. Everntually it loses its cilia and attaches by its aboral end. After this, pharynx forms by invagination, tentacles appear around the oral disc and septa develop internally. Then asexual reproduction takes place and a colony is formed.

The characteristic features of the development of the Anthozoa are:

1. the eggs are large and yolky,
2. the cleavage in many cases superficial,
3. a free swimming larval stage is absent and the planula develops directly into a polyp,
4. the polyp forms a pharynx and septa, and
5. in some groups musculature is predominant while in others calcareous or horny skeleton is characteristic.

CTENOPHORA

All the members of group Ctenophora are monoceous i.e., bisexual and may have two periods members of group of sexual maturity, one in the larva and a final one in the adult. The gonads degenerate between the two phases. Ovaries and testes arise along the whole length of the meridional canals. The ovaries face the principal plane and the testes on the interradial sides.

The ctenophores as well as their embryonic stages are planktonic. The eggs of *Bolina* are smaller than that of the eggs of *Beroe* which measure about 1 mm in diameter. In both the cases the eggs are transparent. The eggs of *Beroe* have been studied because of their bigger size. There is a protective this covering and a croting of jelly around egg. According to Yastu three visible concentric formations are distinguished in the egg, an extremely thin homogeneous outer layer, the ectoplasm and the endoplasm.

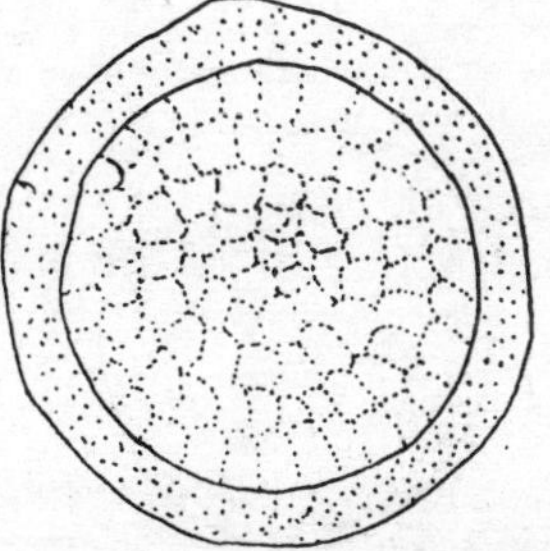

*Fig. 6.13. **Beroe ovata egg with its three concentric constituents: the thin homogenous outer layer, the fluorescent ectoplasm (fine dots) and the alveolar endoplasm.***

The outer layer is semi-fluid and orgranular and is difficult to observe. The ectoplasm is some=what thick having a fine alveolar structure. The endoplasm has a coarse alveolar structure.

In these species the great majority of the larvae are planktonic, development takes place as they float in the sea. In the genus, *Coeloplana* the eggs are ejected from the mouth and develop in the space between the underside of the flattened body and substratum. In *Lyrocties* and *Tjalfiella*, lateral chambers are formed by the distension of the meridional canals. Fertilization is internal and takes place within the brood pouch and develop to larvae within these brood pauches as in viviparous animals. These modifications have taken place in response to the various special habits of the species. Cleavage is holoblastic and two equal blastomeres are formed by the first cleavage. The first cleavage plane coincides with the future pharyngeal plane, and the second plane corresponds to the tentacular plane.

First two cleavage planes pass at right angle to one another and divide the egg into four equal sized blastomeres. The third cleavage is peculiar to the ctenophora, the cleavage plane forming diagonally between the horizontal and vertical axes and dividing the blastomeres unequally. The four more exterior cells are smaller than the interior ones. The four external cells are indicated by the letter E and the four interior ones by the letter M. The fourth cleavage is unequal, resulting a 16-celled stage. The eight micromeres derived from the interior cells are called as m_1, those deriving from the external cells as e_1. The e_1 cells divide unequally into e 1.1 and e 1.2 and almost simultaneously another set of micromeres e-2 is formed from the E blastomeres giving rise to the 24-cell stage. Shortly after-ward the m-1 cells divide into m 1.1 and m 1.2 giving the 28-cell stage and the

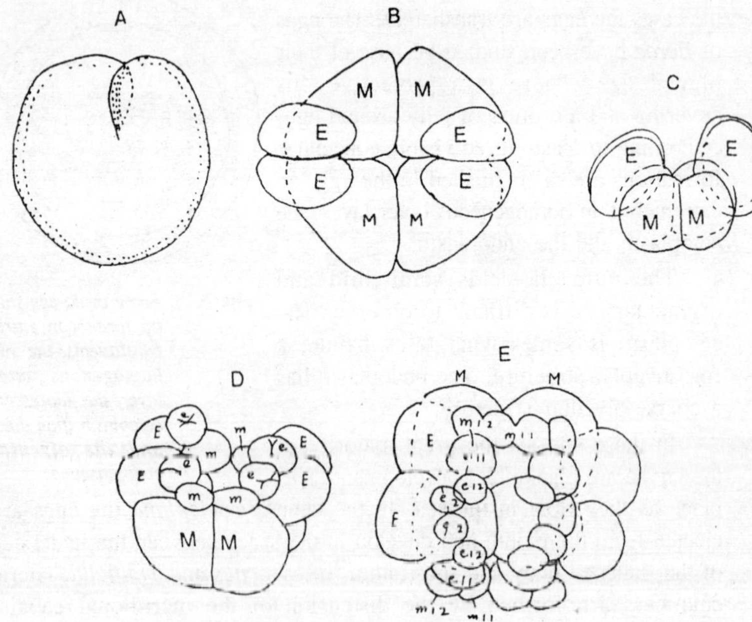

Fig. 6.14. Early development of Coeloplana. A—Mid-cleavage. B—8-cell stage from animal pole. C—8-cell stage from side. D—Transition from 16-20 cell stage. E—28-cell stage from animal pole.

m-2 micromeres are formed from the M macromeres to complete the 32-cell stage. At this stage each macromere has three micromeres. A new emission of micromeres (e3) follows but only from the E-cells. Finally the M macromeres divide into M1 and M2 and the embryo reaches the 64-cell stage. Cleavage thus proceeds with the eight macromeres giving rise to successive generations of cells in an orderly fashion, although there is a tendency for the micromeres derived from the smaller E macromeres to divide earlier than those from the larger M macromeres.

Prospective Destiny

The destiny of the various blastomeres can be traced in different ways. The most simple method is to mark the blastomeres with coloured granules and to determine the position of these granules in the developed larva.

Suppose a granule is placed at the animal pole in the immediate vicinity of the polocytes. At the end of development this granule can be seen at the bottom of the gastral cavity in the larva. This indicates, that the blastopore and later the larval mouth forms at the animal pole. A granule on the opposite pole does not move much. It remains on the surface and is finally found on the apical organ which forms at the vegetal pole of the egg. A granule which has been put on an E cell is found at the larval stage in a gastro-vascular canal, while a granule on an M-cell is found in the gastric sac. According to these results the E gives rise to the longitudinal canals, the M to the gastric pockets.

At the 16-cell stage, a granule placed upon an e_1 micromere is found on the ciliated combs of the larva. A granule placed on an m_1 micromere is found on the apical organ. These observations suggest that m_1, contrary to the opinion held in the past, have nothing to do with the formation of the ciliated plates. One granule placed on an e_{11} and another on an e_{12} are found in the larva on ciliary plates but on different rows. This suggests that every e_1 is involved in the development of the rows of ciliated plates.

The micromeres given from macromeres at gastrulation at the animal pole have a completely different destiny. The derivatives from M give rise to mesenchyme and those of the E to the larval musculature.

The fragmentation has been performed by Yastu (1911). Sectioned of the unfertilized egg (*Beroe*) into two parts and then fertilized the fragments. Each fragment gave rise to a small but normal larva. La

Spina (1963) also obtained corresponding results after breaking the egg by centrifugation. The centrifugal fragment containing the ectoplasm segmented typically, gastrulated and gave rise to a larva having abundant ciliature, whereas the centripetal, larger fragment containfing the endoplasm segmented irregulary and gave rise to an undifferentiated cellular mass lacking in cilia. These results emphasize the importance of the ectoplasm for the development of the ciliary plates.

Chun (1880) isolated the blastomeres. He collected fresh eggs from the sea after a storm and sometimes noticed two half larvae in the same capsule each having only four rows of ciliary plates instead of the usual eight. Chun stated that these half larvae had derived from the two mechanically isolated blastomeres.

Yastu (1912) also isolated the blastomeres from eggs at the eight cell stage. From a simple E-cell he obtained a larva with only one row of comb-plates. From two isolated cells he obtained a larva with two rows of comb-plates. From two M-cell he obtained larvae of various types. In one of them two comb-rows were formed, in another embryo a very small group of comb-plates was developed. In three no comb-plates were formed at all. According to Yastu the developmental potencies of the E and M cells at least with regard to the formation of the comb-plates are the same. Each cell is responsible for one row of comb-plates. The combination, 1E + 2E, according to the hypothesis should yield larvae with three comb-rows.

At eight celled stage, *Faraglic* isolated all E-cells of the same egg and followed their development. The development was good, the cells continued to segment as though *in situ* and finally gave rise to four partial larvae each having two comb-rows.

The isolated M-cells do not develop easily. This is probably due to the fact that the M-cells, produce fewer micromeres than the E and so remain uncovered. In any case, the larvae which developed did not have any combs.

Isolation of 1E + 4M : the larvae has only two comb rows.

Isolation of 1E + 2M : the larvae obtained from this combination has only two comb-rows.

Isolation of 2E + 1M : resulted in larvae with four comb-rows.

These results suggest that only the E-cells and not the M-cells, are involved in the formation of the comb-rows, each of them giving rise to two comb-rows. This conclusion is confirmed by the results obtained after killing cells of the egg at the 16-cell stage. Since the m_1 are not involved in the genesis of the comb-rows, what is their function? Ortolavi (1964) removed all the m1 cells and obtained larvae with an apical organ, defective in respect of some structures, the long cilia forming the protective dome were still present, but the four pillars supporting the otholite and the short cillia of the pit were absent. Probably the m_1 are responsible for formation of the ciliary pillars, each m would be responsible for one pillar.

Gastrulation

During gastrulation as the number of micromeres increases and form a layer which spreads over the macromeres and finally covers them. This process is also called the *epiboly*. By this time the macromeres are arranged in the shape of a bowl, with the macromere layer as a lid over its upper surface while lower surface is exposed. On the lower surface in this stage, a small cell is budded off from each macromere. These small cells are believed to correspond to the mesodermal cells found in the platyhelminthes and other groups. The micromeres continue to multiply, finally spread over the aboral pole and grow down as a one-layered sheet over the macromeres. The macromeres also invaginate into the interior. This part of the gastrulation is called emboly. Thus the gastrula arises by the combined processes of epiboly and emboly. When the micromeres cover the embryo to become the epidermis, four interradial bands of particularly small, rapidly dividing cells become noticeable and these differentiate into the comb-rows. The aboral ectoderm differentiates into the *statocyst* and related parts. The ectoderm at the oval pole invaginates extensively to form the *stomodaeum*, which pushes against the entoderm in such a way as to constrict it into four pockets, two on each side of the tentacular plane. From these the gastrovascular canals arise by active entodermal outgrowth. The tentacle sheaths that posses them originate as ectodermal invaginations from whose base the tentacle sprouts. At this stage, the embryo escapes as a free-swimming *cydippid* larva. This cydippid larval stage occurs in the development of all the ctenophores, being common to all the orders except that in the Beroidea. The tentacular and pharyngeal axes are of approximately the same length and the pharynx is circular in its sectional view.

The larva swims in the sea by moving its comb-plates. During metamorphosis the changes that occur are, the number of comb=plates increase, the cilia grow longer, the tentacles lengthen, branch and become freely movable in and out of their sheaths. The water vascular system develops completely and the general shape of the body gradually changes.

In the other group Lobatea, the whole body of the cydippid larva shortens in the tentacular axis and lengthens in the pharyngeal direction. The lower parts of the body in the pharyngeal plane are prolonged into conspicuous sleeve like structure and the water canals grow into this region and make complicated ziz-zag revolutions there. At the bases of the sleeves ear-shaped projections are formed. The main cones of the tentacles retrogress and their branches elongate to form the so-called secondary tentacles.

In the group Cestida the body takes the shape of a ribbon the difference between the pharyngeal and tentacular axis is still more pronounced. The four rows of comb-plates adjacent to the pharynx grow very long, forming a border along the upper edge of the ribbon while the plate rows next to the tentacles become vestigeal consisting of a few comb plates situated near the centre of the upper margin. The secondary tentacles hang from the lower margin of the ribbon.

In the *Coeloplana*, the mouth of the cydippid larva widens, and the metamorphosing larva moves over the substratum by means of the cilia on the stomodaeal surface. Then after the comb plates are shed, pigment cells develop and the whole body becomes flattened so that it resembles a turbellarian. The body elongates in the tentacular axis and the tentacle sheath regions extend upward like chimneys, and thus the general shape of the body becomes somewhat like a harp. In most of the species the water canals form a highly ramified network.

7

MUSCULATURE

First Stages of its Development Coelenterata

All the components of the muscle system of any animal, located within its body in a definite order, form its contractile-motor apparatus. As well as contractile elements, that apparatus always contains non-contractile formations that fulfil passive mechanical functions, especially the functions of uniting the contractile elements and of resilient opposition, playing the role of antagonists to the musculature contractions. In addition, any energetic contraction of the musculature requires increased solidity at the points where the increased force is applied; therefore the contractile-motor apparatus includes skeletal elements in the wide sense of the term. Subsequently the latter often assume the role of levers, increasing the speed, the power, and the precision of movements, and the role of implements acting on objects in the external environment.

The structure of the contractile-motor apparatus depends to a great extent on the level and type of the animal's organisation, and in turn profoundly affects its entire organisation.

The function of uniting the separate contractile elements is well shown by the boundary membranes of Actiniaria; the contractile parts of the epithelial-muscle cells are attached to these membranes, and thus are combined into functional units of a higher order—muscular areas; these membranes are unusually elastic and have a certain amount of resilience, which enables then to return to their original state after stretching. The forces involved, however, are not great enough for these formations to play the role of antagonists to the musculature. A clear example of a resilient formation acting as antagonist to the musculature is the mesogloea of the umbrella of Scyphomedusae. The entire musculature of the umbrella is located on its subumbrellar side, and by its raid contraction reduces the subumbrellar cavity, expelling water from it and moving the animal forward jerkily. The slow return of the umbrella to its former shape is caused by the resilience of the mesogloea. The economy of energy obtained by substitution of resilient formations for antagonistic muscles has led to wide distribution of such formations in the animal world.

In medusae resilient resistance is produced by the mesogloea, which is so firm that it gives the animal's body a stable form. For that reason the muscles are attached to it only on one side; they merely bend the mesogloea layer, which later resumes its shape through its own resilience.

Unlike medusae, such animals as Hydrida, Actiniaria, and most flatworms do not have a constant body form; on the contrary, their body becomes shorter and thicker or longer and thinner, as a whole or partially, depending on the amount of contraction of different groups of muscles. All of these animals possess musculature in the form of a sac formed by their longitudinal and annular muscles and filled with fluid. In polyps that is the fluid of the gastrovascular cavity (practically water), and in flatworms it is the semi-liquid tissues of their body. In both cases, when the mouth is closed the fluid filling the body acts as a support; muscle tension in the sac gives the body rigidity, and different amounts of contraction of the longitudinal and annular muscles produce changes in body shape (Actiniaria). Further improvement of that type of motor apparatus is shown by polychaetes and some other coelomate animals (see below).

A third structural principle of the contractile-motor apparatus arises after the appearance of hard skeletal parts. Skeletal elements include formations of varied origin, structure, and composition, their only common characteristic being increased mechanical stability: boundary membranes, cuticles, solid epidermal secretions, different kinds of internal skeletons such as the skeletal plates of echinoderms, the cartilages of cephalopods, the fibrous and cartilaginous-fibrous endosternites of Chelicerata, etc. Usually all these formations arise in connection with a supporting or protective function, and only secondarily and partially do they begin to be used for muscle attachment, being thus brought into the contractile-motor apparatus. The role of skeletal components in the structure of that apparatus becomes much greater as their number and solidity increase and as the musculature develops into a constantly-growing number of separate, individualised muscles. These two processes are intimately linked together. Among invertebrates they reach the highest level in arthropods, with their complexly-segmented transversely-striated musculature and external skeleton. Parallel improvement takes place in the sensory and motor innervation required to set in motion a complex motor apparatus, and also in the development of adaptations for the nutrition and respiration of energetically-acting muscles.

What was the original structural plan of the contractile-motor apparatus, and what was its original function? It has already been observed that the question of the first appearance of neural and muscular apparatuses coincides with the question of the origin of the type of organisation characteristic of Enterozoa, and the origin of that type of organisation is clearly linked with the transition of primary Metazoa to nutrition by the swallowing of large prey. Therefore the primary function of the contractile-motor apparatus was not locomotion by the animal itself, but the seizure and swallowing of food: the prehensile and swallowing functions, not the locomotor function, were primary. The original locomotor apparatus of Metazoa was a ciliary, not a muscular, apparatus. Ctenophora, as representatives of primary, free-swimming coelenterates, provide an example of the level of organisation at which the animal travel almost entirely by means of a ciliary locomotor apparatus, but uses its musculature mainly to seize and swallow prey. If we assume, with A. A. Zakhvatkin (1949), that in the adult state primary Metazoa were sessile colonial animals, the connection between the origin of musculature and the prehensile and swallowing function becomes still more evident. Sessile coelenterates, in fact, especially hydroid polyps, represent in their whole organisation the fullest embodiment of the prehensile and swallowing function to be found in the animal kingdom.

The muscle system of hydropolyps corresponds fully with their thoroughly-polarised structure, and is formed exclusively of epithelial-muscle cells. The contractile-motor apparatus of hydroids is characterised by the fact that the muscle processes of these cells are longitudinal in the ectoderm and annular in the endoderm; in the terminal region of the hydranth the musculature is more developed on the hypostome, and around the mouth the annular fibres form something like a sphincter. On the other hand, there are practically none in the coenosarc. The body of Hydrida, which has no perisarc, is contractile along its entire length. Freshwater Hydida represent the simplest variant of a motor apparatus supported by fluid enclosed in a muscular sac. When the mouth is closed the hydra may stretch into a long thin thread or contract into a short thick sac, depending on whether the annular or the longitudinal muscles have more tension. By expelling water it can contract into a small pellet. Metagenetic hydroids have a perisarc closely apposed to the branches of the coenosarc and to the stalks of the hydranths, and consequently these parts are no contractile and lack musculature. The head of a hydranth is usually contractile in all its parts, but sometimes a firm supporting tissue of endodermal origin develops in its basal part, giving the head a certain amount of constancy of shape (e.g. *Tubularia*). The musculature is best developed in the terminal part of the head, in the hypostome, which makes the movements of swallowing.

The hollow tentacles of Hydrida have ectodermal longitudinal and endodermal annular musculature, and are capable of great extension in length; the tentacles of marine hydroids (Leptolida) are solid, and their endoderm has no annular musculature. These tentacles are able to bend, but are incapable of any great extension. The same types of tentacle muscle occur outside the order Hydrida. Actiniaria have hollow and very extensible tentacles, furnished with annular and longitudinal muscles; Bryozoa and Serpulimorpha have tentacles without annular musculature and of more or less constant length. The elongation of hydroid and sea-anemone tentacles is achieved by increase in the pressure of the cavity fluid, produced by increased tension in the body musculature; contraction of the annular musculature of the tentacles prevents them from expanding in diameter, and relaxation of the longitudinal musculature enablesd them to increase in length.

Besides the functions of seizing, swallowing, and retaining food, the motor apparatus of hydroids has another function, that of shortening, drawing back, and pulling in the tentacles and the head in response to unfavourable stimuli, or, more briefly, that of concealment. All sessile animals are able to contract in one way or another in response to unfavourable stimuli, and probably, together with prehension, that was a primary function of the contractile-motor apparatus. Primitive organs and apparatuses are usually multifunctional.

Muscle System

When Hydrida are not sessile the motor apparatus also has a locomotor function. *Hydra*, as is well known, has several methods of locomotion. *Protohydra* burrows in the sand by pumping fluid from one end of the body to the other, like many worms.

The motor apparatus of Anthozoa is distinguished by development and partly by differentiation of adaptations designed to fulfil the same basic functions, food-capture and concealment. In addition, solitary forms (Actiniaria), like Hydrida among Hydrozoa, are capable also of progressive movement. The ectodermal musculature there also is longitudinal. In Cerianthatharia it is strongly developed and enables the body to contract; accordingly the longitudinal muscles of their dissepiments are relatively weak. In Actiniaria, on the other hand, the ectodermal longitudinal musculature of the stalk is poorly developed, and in the majority it is absent. Functionally it is replaced by the endodermal longitudinal muscles of the newest septa. That substitution became possible in Actiniaria because of the regular location of new septa along the whole circumference of the cross-section of the body. In Ceriantharia all new septa are concentrated on one side of the body, and therefore they cannot replace the strictly-longitudinal muscultature of the stalk. On the oral disc the longitudinal muscle fibres are arranged radially and act as dilators of the mouth. The endodermal musculature of the trunk and the pharynx is annular; in many sea-anemones it forms a sphincter beneath the edge of the disc, as a result of which the edge of the body moves over the disc and covers it in a concealment reaction. In the septa (mesenteries) the endodermal musculature forms powerful longitudinal muscles (retractors), the arrangement of which is very important in the morphology of Anthozoa. At both ends of the body the fibres of each bundle diverge fanwise and are attached to the basal and oral discs along their radii. On the reverse sides of the septa the muscle fibres run radially: in 'complete' septa (i.e., those attached to the pharynx) the radial fibres act as dilators of the pharynx. In the lower parts of the septa they become oblique and are attached by their ends to the basal disc: they are parietobasal muscles. The basal disc of sea-anemones possesses radial and annular musculature of endodermal origin. The musculature of the basal disc itself and also the septal muscles attached to it are used for suctional attachment to the substrate and for crawling.

The septal musculature is very important in digestion, by closing the mesenterial filaments with their digestive epithelium around swallowed prey. The longitudinal septal muscles also play a major role in protective contraction of the animal. Rapid contraction of all the longitudinal muscles of the body and of the radial muscles of the oral disc produces retraction of the disc and of the whole front end,

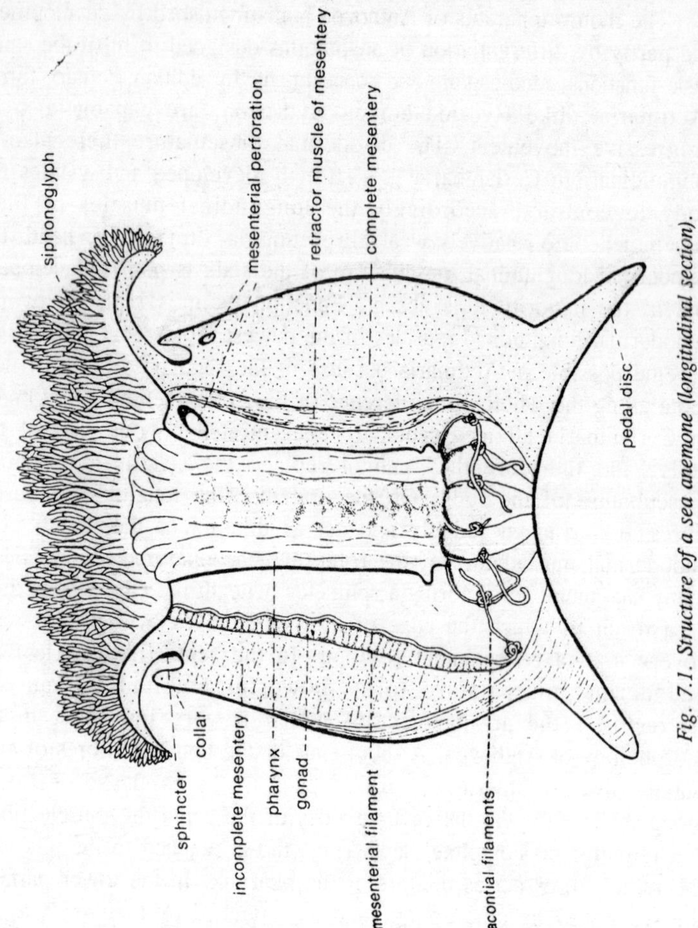

Fig. 7.1. Structure of a sea anemone (longitudinal section).

followed by contraction of the disc sphincter, which covers the retracted head with a fold of the body wall. As a result the whole animal may be transformed into a dense flat cake, presenting a minimum of surface to enemies and the elements.

The physiology of sea-anemone movements has been described in a number of works. On the protective contraction of the animal all the cavity fluid is expelled through the open mouth. The opposite process of filling the gastral cavity with water, without which the sea-anemone could not expand, is performed by the activity of the cilia of the siphonoglyph, which force water into the body. It is also possible to swallow water by peristaltic movements of the pharynx. In any case

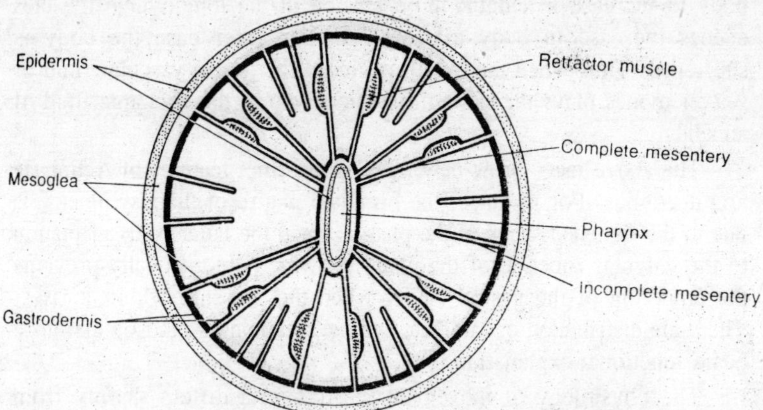

Fig. 7.2. Sea-anemone—cross section at the level of the phrynx.

the expansion of the sea-anemone takes place slowly; its body does not have enough elasticity for automatic expansion. When a sea-anemone (*Metridium senile*) is in a state of rest, the pressure of the cavity fluid is equal to that of a 2-3 mm column of water and is due entirely to the activity of the siphonoglyph cilia. When the parietal musculature contracts the pressure in the gastral cavity increases and compresses the pharynx, which because of its flat shape makes an excellent valve. When the mouth is closed, the tension of the longitudinal and annular muscles of the stalk causes increased pressure within the cavity, which in *Metridium senile* reaches that of 5-6 cm of water. When full of water a sea-anemone can make great changes in its body shape and tentacle length by muscular action. When there is more tension in the annular muscles the pressure of the gastrovascular fluid forces the

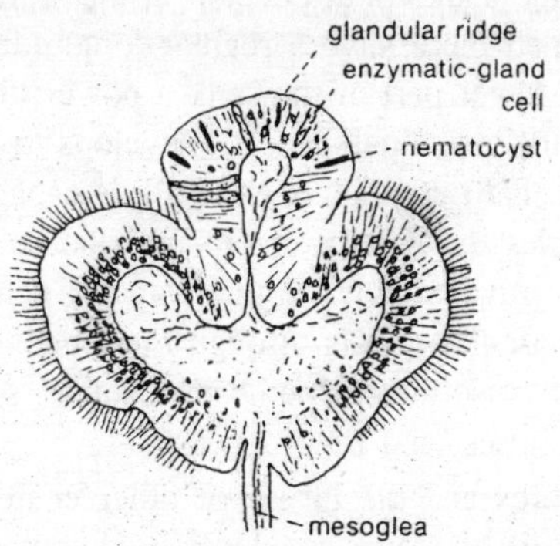

Fig. 7.3. Cross section of mesenterial filament of a sea anemone.

body to increase in length; more tension in the longitudinal muscles causes increase in body diameter. In the latter case the body is shortened. Thus when the mouth is closed the gastro-vascular fluid of sea-anemones plays the same mechanical role as the coelomic fluid of annelids.

The above facts throw new light on the chief features of Actiniaria architectonics. For instance, the biradiate nature of their symmetry is due to the flattened shape of the pharynx, and the latter is an adaptation to the valvular function of the pharynx, which automatically prevents the emptying of the gastral cavity when the pressure within it rises. The wide distribution of biradiate symmetry among Anthozoa therefore has a functional explanation.

The physiology of movement in Actiniaria differs sharply from that in higher animals such as arthopods and vertebrates. The musculature of Actiniaria consists mainly of undifferentiated muscle layers that form part of the body wall and its extensions (tentacles and septa). Their contractions determine the shape of the animal. Every movement of the body calls for co-ordinated action by many muscle groups. That is partly true also of Hydrida. The most primitive structure of the neuromuscular apparatus therefore not only does not exclude, but demands, the most complex co-ordination in the work of its parts; and the most differentiated parts of the motor apparatus of Actiniaria, such as the front-end retractors and the disc sphincter, have the simplest movements. Progressive evolution necessarily includes simplification of organs and functions as well as complexity.

Sea-pens (Octocorallia Pennatularia) have a well-developed colonial muscular apparatus. It consists of the external longitudinal and internal annular muscles of the stalk. Both develop in the wall of the perisarc canals (solenia), which there have a regular longitudinal and annular arrangement. In the lower part of the stalk a powerful sphincter, able to compress even the principal canals, develops from the annular muscles. Because of that musculature the stalk is capable of peristaltic movement that enables the colony to dig into the soil with its basal element and even to travel slowly; in this way sea-pens, by collecting water in the gastrovascular canals through the siphonozoids and then forcing it back, partly use the hydraulic mechanism so widespread in fossorial worms, molluscs, and Enteropneusta.

Unlike ctenophores and the larvae of other coelenterates, which swim by means of cilia or ctenes, medusae (and some siphonophores) have advanced to swimming by means of muscular contractions of the bell, which is the expanded margin of the basal part of the head of a polyp. Rapid contraction of the bell forces water out of its cavity and thrusts the medusa backward. Medusa movement is therefore reactive. This method of movement is as secondary as is the organisation of the medusa itself. In Scyphomedusae the bell contracts by means of annular and radial muscle bands formed by the ectoderm on the subumbrellar side; the elasticity of the bell mesogloea restores its shape; on the exumbrella there is little or no musculature. Possession of elastic mesogloea gives the medusa stability of shape. The locomotor apparatus of medusae therefore differs radically in its organisational principles from that of Actiniaria. The total amount of locomotor musculature in the body of Scyphomedusae is very small: in *Cyanea arctica*, according to I.A. Vetokhin, it constitutes only 1.2% of the body mass. The prehensile function is fulfilled by the musculature of the marginal tentacles and the arms, which is fairly complex in its nature.

As we have seen, however, many Scyphomedusae (e.g. *Aurelia*) in the adult state feed on small plankton brought by cilia to the orifices that remain after the mouth is overgrown, and the arms and tentacles lose their prehensile function. We shall not dwell on the peculiarities of the musculature of Hydromedusae and Siphonophora.

The musculature of ctenophores is all embedded in the mesogloea. It consists of longitudinal and annular fibres beneath the skin, the same kind of fibres around the gut, radial fibres running from the gut to the skin, and the powerful musculature of the tentacles. Sometimes there are sphincters around the mouth and the aboral pole. The aboral sphincter, by its contraction, may cover up the aboral sense organ and thus have a protective function, resembling the role of the sphincter of the oral disc in sea-anemones. In the general principle of its organisation (a musculo-cutaneous sac enveloping the connective tissue of the body together with its enclosed internal organs) the muscular apparatus of ctenophores is therefore closer to that of flatworms than to that of Cnidaria. Most ctenophores are predators that swallow their prey; some of them seize it with their tentacles, and others directly with the mouth; the chief role of their musculature is prehension and swallowing, and in addition it has a protective function; in most ctenophores it does not have a locomotor function, but it does acquire that in some specialised groups. *Cestus veneris*, for instance, which has an elongated ribbon-like body of considerable size (over 1 metre) swims by undulating its body, and its musculaure (which consists of longitudinal, transverse, and diagonal muscles) is comparatively well developed. In Lobifera the musculature of the lobes is composed of two systems of bundles arranged fanwise and interesting each other rapid contraction of these muscles produces flapping of the lobes and retrograde motion of the animal, e.g., when it meets an obstacle, a curious resemblance to Cephalopoda Decapoda. In both cases there are two types of progressive movement: smooth movement, mouth-foremost, by means of fins (Cephalopoda) or ctenes (Ctenophora), and jerky backward movement.

8

MESENTERIES IN ANTHOZOA

Anthozoans are the coelenterates belonging to the class— Anthozoa. In Anthozoa only the polyps are seen, medusa stage is absent. The polyps have mouth, phary, gullet or stomodaeum. The stomonxdaeum hangs into the gastrovascular cavity. The gastrovascular cavity is divided into chambers by mesenteries. The mesenteries are defined as the radiating partitions that extend from the body wall and the stomodaeum. Usually these are present in pairs.

Structure

Mesenteries are more or less evenly spaced around the circumference. The upper margin of each mesentery is attached to the oral disc, the lower margin to the pedal disc and the outer margin to the wall of column. The free-edges of the mesenteries are thickened to from twisted cords, the *mesenteric filament*. The filaments are trifid in the region of gullet but below gullet they are simple in sub-class zoantharia. The middle lobe is called cnido-glandular band, it contains cnidoblasts and gland cells. The lateral lobes bear cilia and are called ciliated bands. The lower ends of mesenteric filament are produced into free threads, the *acontia*. These are beset with thread cells. These are organs of defence. They can be protruded through the mouth or through the *cinclides* (these are special cells of body wall).

Intermesenteric chambers between the mesenteries of same pair are called *endocoels*, where as chambers present between the mesenteries of the adjacent pairs are termed the *exocoels*. The chambers are separate in the pharyngeal region but they open into the central part of gastrovascular cavity. The mesenteries near bear their upper or oral side apertures called *ostia*. Two such ostia are found on each mesentery, the *oral ostium* lies near the body wall. The stomata put the adjacent intermesenteric chambers in communication with one another to permit flow of water between them.

Gonads are located along the inner free edges of the mesenteries just behind the filaments.

Histology

A mesentery consists of a middle supporting lamina of mesogloea covered on both the sides with a layer of gastrodermal cells. The gastrodermis is composed of long columnar supporting cells with sensory cells, enzyme secreting gland cells and nematocysts. Nematocysts are present only in cnidoglandular band. The mesenteries are highly contractile due to presence of three types of muscles: the *longitudinal* or *retractor muscles* which runs as a narrow band from the base of oral disc; the *parietal muscles* that runs obliquely across the lower outer angle; and *transverse muscles* which run parallel to the oral disc. The longitudinal and parietal muscles are so thick so as to form

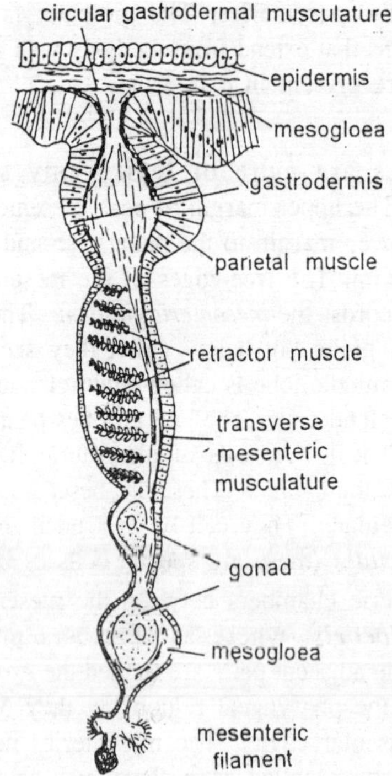

Fig. 8.1. T.S. mesentery of Metridium.

a bulging on one side of mesentery. This bulging is called *exocoelic*. Here the bulging face away from each other. If the bulging face one another it is called *endocoelic*.

The muscles form the chief means of contraction and extension of the body.

Typical Type of Mesenteries

The typical type of mesenteris are present in *Metridium*. Six pairs of more broad mesenteries extending from the body wall to the gullet. These are called *primary mesenteries*. A little short and incomplete *secondary mesenteries* are found in between the Paris of primary mesenteris. More small pairs of *tertiary mesenteries* are also found between them. The secondary and tertiary mesenteries are attached to the body wall but they do not extend upto gullet.

Type of Mesenteries

Based on the position and its development, the mesenteries are groups into following:

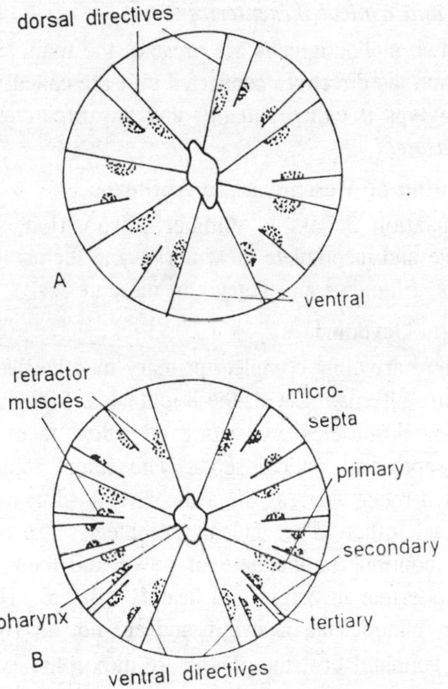

Fig. 8.2. T.S. through pharynx. A—Gonadinia; B—Protanthea.

Complete and incomplete mesenteries

The complete mesenteries are those connecting the body wall and stomodaeum *e.g.*, primary mesenteries. The incomplete mesenteries are those extending mid-way between the body wall and the stomodaeum *e.g.* secondary and tertiary mesenteries.

Couples and pairs

The couples refer to all the mesenteries which occur in one half and similar to the other half. Whereas on either side of the animal; the mesenteries facing each other in two's are called *pairs*.

Macrosepta and microsepta

The large sized septa are called *macrosepta* e.g. primary mesenterises. The small-sized septa are called *microsepta* e.g. tertiary mesenteries.

Directives

The pharynx commonly bears a ciliated groove, the *siphonoglyph*. The septa that are connected to the siphonoglyph are called directives.

Sulcal and asulcal mesenteries

If two siphonoglyphs are present, the main siphonoglyph is called *sulcus* and the directives connected to it are called *sulcal*. The secondary siphonoglyps is called *sulculus* and the directives connected to it are called *asulcal*.

Disposition of Mesenteries in Anthozoa

Variation occurs in number, disposition, relative numbers of complete and incomplete mesenteries and the arrangements of retracter muscles, filaments and gonads in mesenteries.

Subclass Alcyonaria

There are eight complete primary mesenteries. They are unpaired. These are alternating with the 8 tentacles. Siphonoglyps is single. The longitudinal muscles are strong on sulcal faces. There is a pair of sulcal septa and asulcal septa. The sulcal septa (ventral directive) face each other whereas the asulcal septa (dorsal directive) look away from each other. The asulcal mesenteries are often longer than the others, continue to the base of polyp and their mesenteric filaments are ectodermal in origin and heavily ciliated. The remaining six are smaller, endodermal in origin and are not heavily ciliated.

In Pennatulacea, the polyps are dimorphic. Mesenteries are found only in siphonozooids.

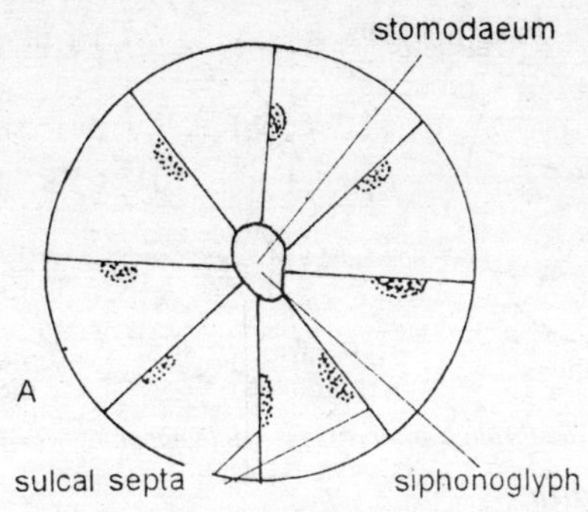

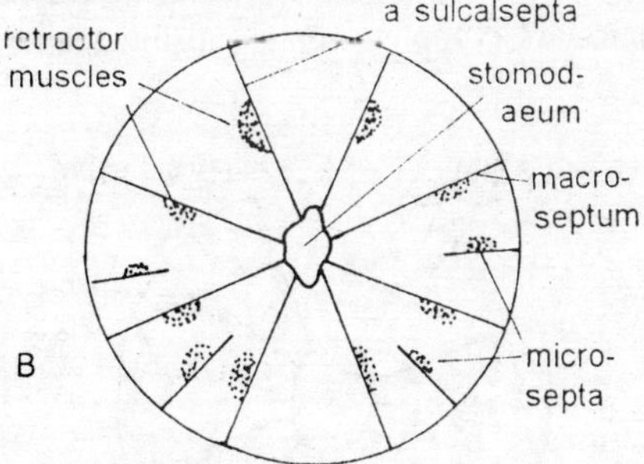

Fig. 8.3. T.S. through pharyngeal region. A—Alcynarium; B—Edwardsia.

Subclass Zoantharia

The number, kind and disposition of mesenteries are so variable among the different orders or within the same order that no generalised condition can be described. The differences in various orders are summarized:

Order—Actinaria

In *Edwardsia* there are eight macrosepta. But there are only four microsepta facing the sulcus. The sulcal retractors face away. Besides four microsepta are also found near the oral disc, pairing with single microsepta in such a manner that their retractors face each other. In *Gonactinia*, in the exocoels, between these six pairs of mesenteries, are usually present additional two pairs of microsepta. There are six pairs of microsepta in *Protanthea*. *Halcampoides* shows a similar condition in mesenteries as in sea-anemone. In *Halcampa* there are six pairs of microsepta along with six pairs microsepta. *Haloclava*, is usually decamerous i.e. with 10 pairs of complete septa.

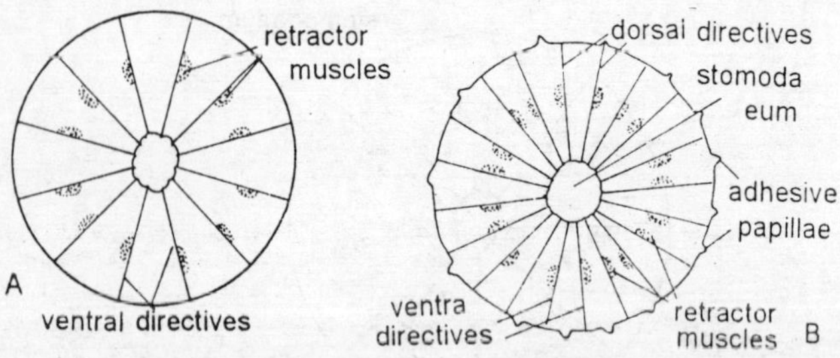

Fig. 8.4. T.S. through pharynx. A—Halcampoides; B—Haloclva.

Order—Madreporaria

The mesenteries are arranged in the hexamerous plan. In *Acropora* and *Porites* there are twelve mesenteries out of which two pairs are directive, four pairs are complete septa and four pairs are incomplete septa.

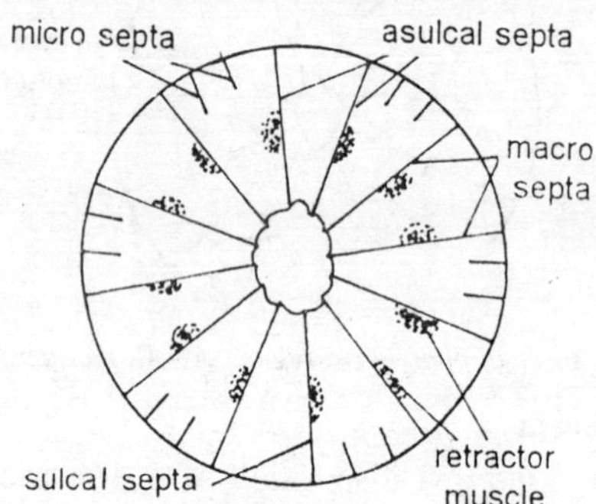

Fig. 8.5. T.S. through pharynx. Halcampa.

Order—Zoanthidea

The number of mesenteries is six or multiple of six. The arrangement is different from any living Anthozoa but somewhat resembles that of extinct Terracoralla. The mesenteries are paired and the arrangement of septa is called *brachyncemous* in *Zoanthus*. This includes one pair of macrosepta, one pair of microsepta and one pair of ventral directives. Some zonathids like *Epizoanthus*, however, shows the *macrocnemous* arrangement. This includes three pairs of macrosepta, the ventral directives and the fourth and fifth septa on each side counting from the dorsal directives.

Order—Antipatheria

There are ten complete mesenteries, arranged in couples, but there may be six or twelve. Of these six are primary mesenteries including

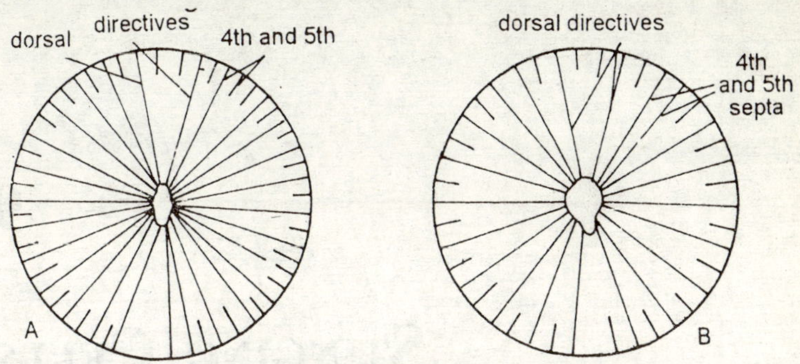

Fig. 8.6. T.S. through A—Zoanthus; B—Epizoanthus.

four directives and two transverse. The transverse are largest and bear gonads and filaments. The septal muscle is poorly developed.

Order—Ceriantheria

The septal arrangement is quite different from that of other anthozoans. *Cerianthus* shows numerous, complete but coupled mesenteries. The number is not definite. During growth new mesenteris added continuously. The ventral directives are very small. The retractor muscles are absent. Three most dorsal couples, including the directives and two couples to either side of them develop first in the embryo hence called *protosepta* and the rest-too short to extend ventrally are called *metasepta*.

9

STINGING CELLS

These are diagnostic of coelenterates for defense and offense. They develop in special cells called *nematoblasts*, or *cnidocytes*. They are also known as *stinging cells* or *nettle cells*. Actually these are not cells but cell-organoids and are made up of a substance similar of chitin.

Distribution

Nematocysts are located all over the body except on the basal disc or pedal disc. These are lodged between or invaginated within epitheliomuscular cells. They are especially abundant on the tentacles. Here they are arranged in groups. These groups are called *batteries*. A battery may consists of one or two large nematocysts surrounded by small nematocysts, each in its own cnidocyte.

Size

Nematocysts are minute structures. The capsule measures 5.50μ in length. The tube in several mm in length.

Structure

A cnidocyte is a rounded or oval cell with a basal nucleus. One end of the cell contains a short, stiff, bristle-like process called *cnidocil* in hydrozoans and scyphozoans. The cnidocil has an ultrastructure similar to that of a flagellum which is exposed to the surface. A cnidocil is usually absent in anthozoans. The interior of the cell is filled with nematocyst. The nematocyst consists of a rounded, oval pyriform or elongated soluble-walled *capsule* containing a coiled, usually *pleated tube*. The base of the tube or thread may be swollen into a *butt*. The tube on its outer surface has three spiral ridges, each bearing a row of spines. The spines may be present throughout the tube or only on a part of it. The end of the capsule that is directed toward the outside is covered by a *lid* or *operculum*. The capsule is full of a poisonous fluid, the *hypnotoxin* which contains proteins and about 75% as virulent as cobra venom. In hydrozoans and scyphozoans, *support rods* run the length of cnidocyte, and the base is anchored to the lateral extensions of an epitheliomuscular cell. The base of the cnidocyte may also be associated with a neuron terminal.

Discharge of Nematocyst

In an undischarged nematocyst, the butt lie invested in the capsule with the spines on its inner surface and the tube coiled round it. The discharge mechanism apparently involves a change in the permeability of the capsule wall. Under the combined influence of mechanical and chemical stimuli, which are initially received and conducted by the cnidocil, the operculum (hydrozoans and scyphozoans) or apical flap (anthozoans) of cnida open. Hydrostatic pressure within the capsule everts the tube and the entire nematocyst explodes to the outside. Discharge can apparently be affected by nerve impulses from the associated neuron terminal, and perhaps neuronal connections serve to bring about coordinated firing by a large number of nematocysts.

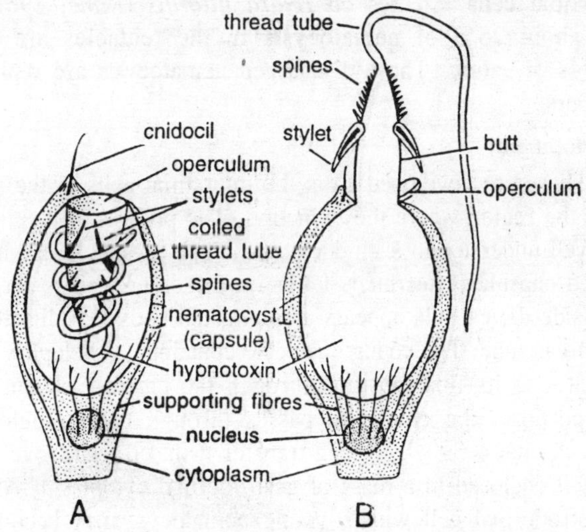

Fig. 9.1. A cnidoblast. A—Undischarged; B—Discharged.

Nematocyst may be used but once and are released from the cnidocyte after discharge. New cnidocytes are formed from nearby

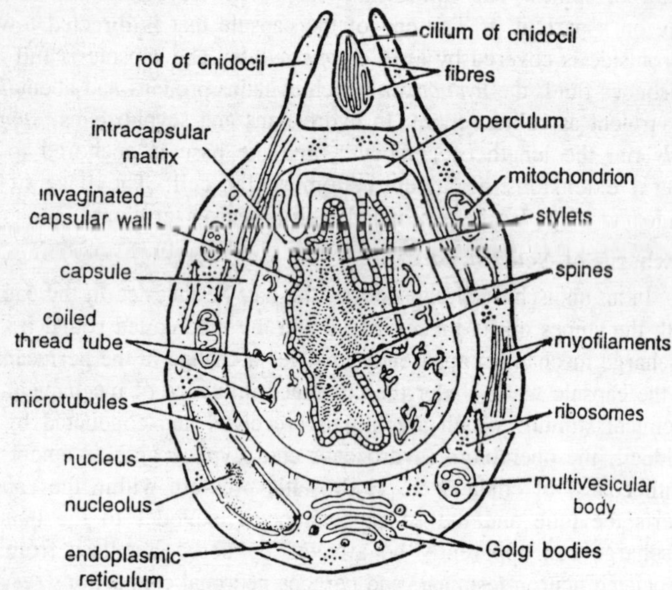

Fig. 9.2. A cnidoblast as seen under electron microscope.

interstitial cells. Studies on *Hydra littorals* (*Kline*, 1961) indicated that about 25% of nematocysts in the tentacles are lost in the process of eating. These discharged nematocysts are replaced within 48 hours.

Development

These are developed from the interstitial cells of the region away from the region where they function. The process of their formation is not well understood. A cnidocyte develops a vacuole in which a single walled capsule is secreted. Later on the outer wall is formed around it. Inside the capsule appears an elongated body that differentiates into the thread tube. The young cnidocyte containing developing nematocysts migrates to its final position through the gastro-vascular cavity. For this purpose, the cnidocyte passes through the mesogloea into the gastrodermal cells which then transfer it into the gastrovascular cavity as such enclosed in a mass of gastrodermal cytoplasm. Alternatively, a gastrodermal cell with a young nematocyst may become free and move in an amoeboid manner in the gastrovascular cavity. From gastrovascular cavity, the cnidocyte is ingested by the gastrodermal cells, which transfer it once again into the epidermis. Here, it complete its differentiation.

Types of Nematocysts

Weill (1964) described seventeen different types of nematocysts based on the diagnostic characters of the tube or thread. He studied about 119 species of different coelenterates. Recent classification has been suggested by *Weill* (1964), *Werner* (1965), *Deroux* (1966) and *Mariscal* (1972); *I. Astomocnidae*. Tube closed at the distal end.

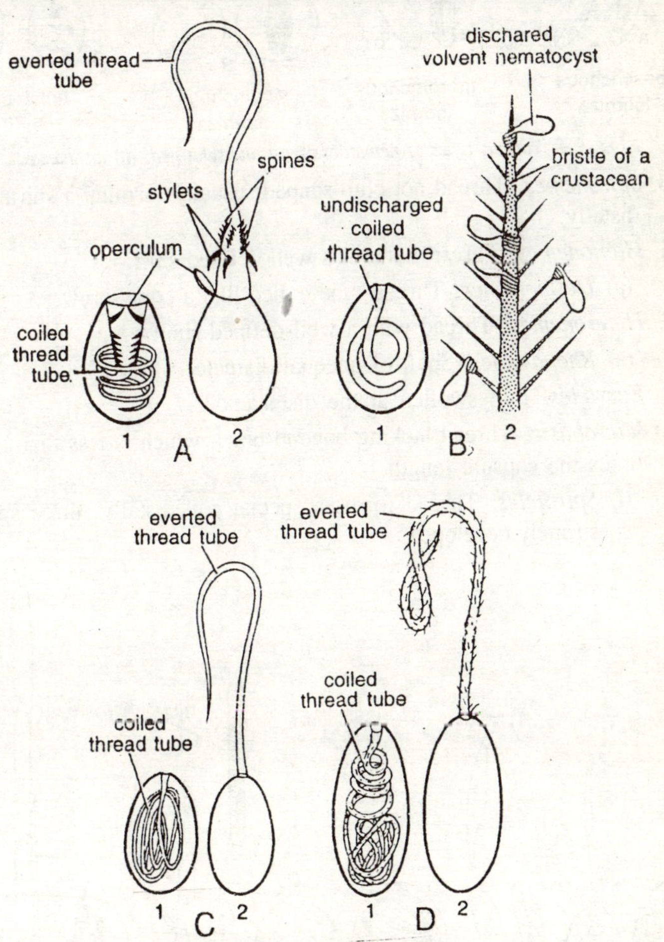

Fig. 9.3. Types of nematocysts. 1-Undischarged. 2-Discharged. A—Penetrant. B—Volvent. C—Stereoline glutinant. D—Streptoline glutinant.

A. *Rhopalonemes*. Tube club-shaped and much greater in volume then the capsule.

1. *Anacrophores*. Tube without an apical projection.
2. *Acrophores*. Tube with an apical projection.

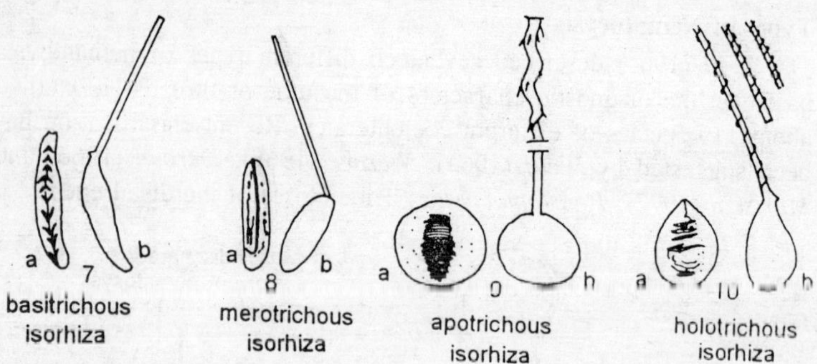

Fig. 9.4. Various types of nematocysts (a) undischarged; (b) discharged.

B. *Spironemes*. Thread not club-shaped, usually forming a spiral coil distally.

1. *Haplonemes*. Thread without a well-defined shift.
   (a) *Desmonemes*. Thread-tube coiled like a cork-screw.
2. *Heteronemes*. Thread with a well-defined shaft.
   (a) *Rhopaloides*. Shaft of unequal diameter.

1. *Euryteles*. Butt swollen at the distal end.

(a) *Microbasic*. Thread lacking beyond butt., which is less than three times the capsule length.

   (I) *Spiroteles*. Thread forms a special coil distally, three spines strongly developed.

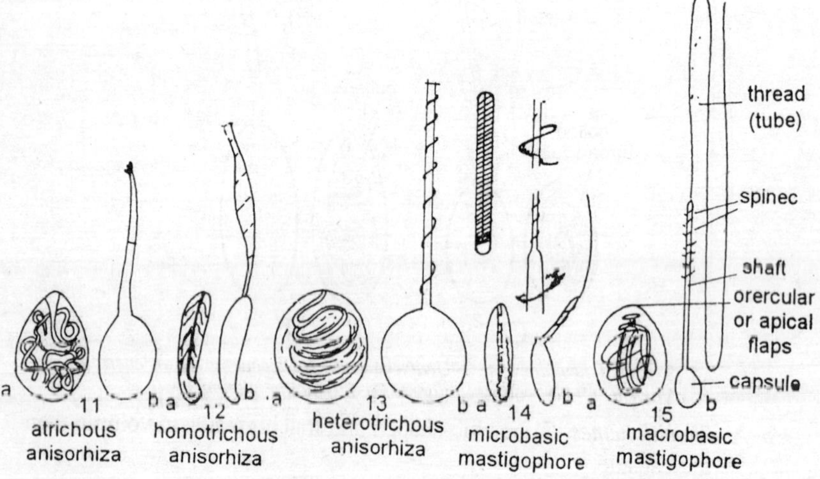

Fig. 9.5. Various types of nematocysts (a) undischarged; (b) discharged.

(II) *Aspiroteles*. No thread beyond the shaft, three spines strongly developed.

(III) *Stomocnidae*. Thread open at distal end.

A. *Haplonemes*. Thread without a well-defined shaft.

1. *Isorhizas*. Thread of the same diameter throughout (glutinants).
   (a) *Atrichous*. Thread without well-developed spines.
   (b) *Basitrichous*. Thread with well-developed spines only at the base.
   (c) *Merotrichous*. Thread with well-developed spines on the intermediate part only.
   (d) *Apotrichous*. Thread with developed spines on distal part only.
   (e) *Holotrichous*. Thread with well-developed spines along whole length.

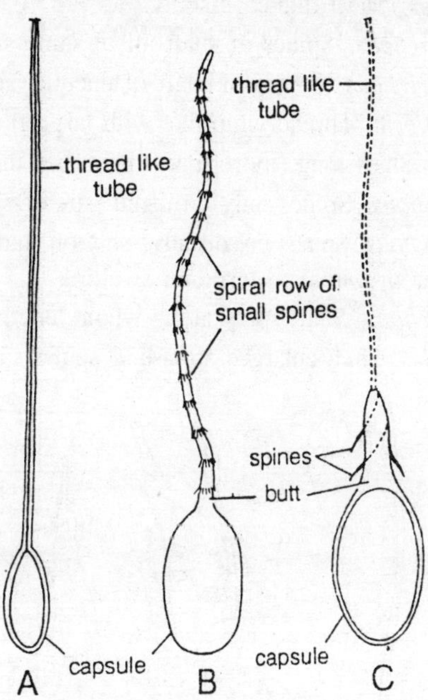

Fig. 9.6 Types of nematocysts. A—Atrichous isorhiza; B—Holotrichous isorhiza. C—Microbasic eurytele.

2. *Anisorhizas*. Thread slightly dilated towards the base.
   (a) *Atrchous*. Thread with well-developed spines.
   (b) *Homoterichous*. Thread spiny throughout, spines all of equal size.
   (c) *Hetesotrichous*. Thread spiny throughout, spines larger at the base of thread.

3. *Heteronemes*. Thread with well defined shaft.

1. *Rhabdoides*. Shaft aglindrical; same diameter throughout.

(a) *Mastigophores*. Thread continuous beyond the shaft.

(1) *Microbasic*. Shaft short, less than three times of capsule length.

(a) *Microbasic b-mastigophore*. Shaft tapers slowly into thread.

(b) *Microbasic p-mastigophore*. Shaft tapers abruptly into thread.

(2) *Macrobasic*. Shaft long, more than four times of capsule length

(b) *Amastigophores*. Thread absent beyond the shaft.

1. *Macrobasic*. Shaft short, less than three times of capsule length.

2. *Microbasic*. Shaft long, more than four times of capsule length.

3. *Rhopaloides*. Shaft of unequal diameter.

(a) *Euryteles*. Shaft dilated distally.

(b) *Homotrichous*. Spines of shaft all of same size.

(c) *Heterotrichous*. Spines of shaft of unequal size.

(d) *Semiophoric*. Thread whip-like with large flat spine medially.

2. *Macrobasic*. Shaft long, more than four times the capsule length.

(a) *Telotrichous*. Spines only on distal part of shaft.

(b) *Merotrichous*. Spines not distally, only on shaft area of uniform diameter proximal to terminal swelling.

(c) *Holotrichous*. Shaft spiny along whole length.

(d) *Stenotales*. Shaft enlarged at base, 3 spines strongly developed.

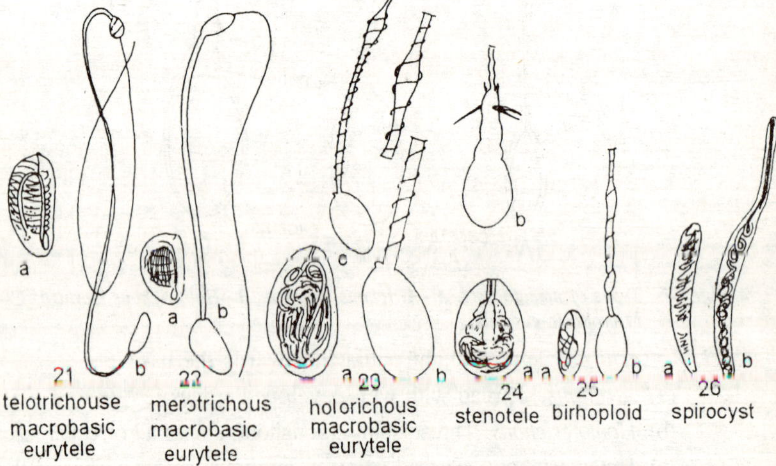

Fig. 9.7. Various types of nematocysts (a) undischarged; (b) discharged.

Birhoploides. Shaft to unequal diameter at distal and proximal end.

Spirocysis. They have a thin capsule containing a thin, long, spirally coiled thread without spines.

Function of nematocysts

Different functions are performed by different nematocysts. Rhopalonemes and desmonemes serve for holding the prey. Atrichous isorhizas are adhesive. Function of holotrichous isorhizas is not yet clear. A few nematocyst like stenoteles, mastigophores etc. penetrate and anchor in the tissue of prey.

Nature of the discharged fluid is not well-known. The discharge causes paralysing action on the prey and cause burning sensation in the human skin in same cases.

10

SENSATION

Coelenterates and the First Appearance of Nerve Cells and of a Neural Apparatus

As we have seen, sponges have no neural elements; their incipient contractile and glandular cells are independent effectors, reacting directly to external Stimuli. At the same time, sponge tissues possess an incipient capacity for conducting stimuli: if one pricks a sponge at some distance from one of its oscula, after some time the latter contracts by the action of its myocytes. That conduction of stimuli, however, which takes place without the help of specialised neural elements, is exceedingly slow, at the rate of 1 cm per minute. The function of conduction of stimuli therefore appears in Metazoa before the development of a special apparatus to perform it. A neural apparatus first appears in coelenterates.

The absence of neural and muscular apparatuses in sponges and possession of these by coelenterates is no accident: the difference has the closest connection with the types of feeding and the methods of procuring food in the two groups. Sponges feed on small particles suspended in sea-water—ultra and nanoseston; food is brought by the activity of the flagella in the choanocytes, and the food particles are absorbed by separate body cells, choanocytes and amoebocytes. Co-ordination in the activity of the separate cells is not required in either the bringing or the absorbing of food. Sponges require a definite degree of integration of all cells in the body into a single organism only in the creation and maintenance of a specific structure, with the proper relationship of parts, to ensure the regular functioning of the irrigation apparatus. Although in many respects sponges are very far from other

Metazoa, their method of feeding is clearly primary: in fact, the colonies of Flagellata from which Metazoa are most probably derived must originally have lived on food that could be entrapped and digested by separate members of a colony, by separate cells.

In contrast to sponges most coelenterates, including all the primitive ones, feed on large prey, which they take by means of the mouth. Here the seizure and swallowing of prey requires harmonised action by all the epithelial-muscle and muscle cells of the body; that is necessary if the animal is to act as a whole, as a single organism, and to act quickly. Such integration and rapidity of action can be attained only by possession of a neural apparatus and musculature. I think it almost beyond question that in the evolution of Metazoa both the neural and the muscular apparatuses arose because of transition from primary feeding on fine dispersed food to the taking of large prey, i.e. because of the appearance of predation. In other words, the transition to predation served as a stimulus also for development of the organisation of Enterozoa, and among the latter predation was the primary, original method of feeding.

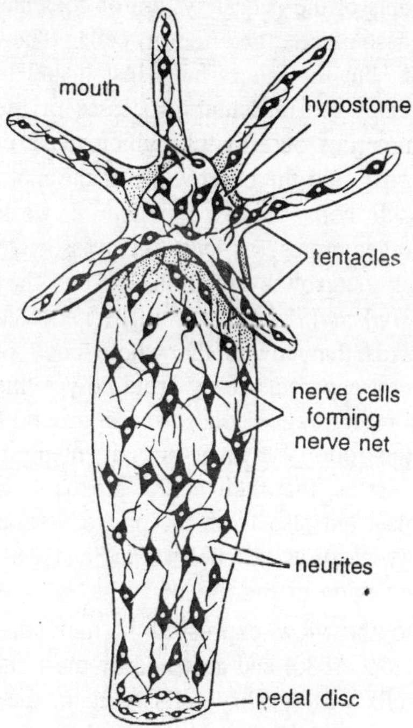

Fig. 10.1. Nerve net in Hydra.

In coelenterates both kinetoblast and phagocytoblast contain neurosensory as well as other types of cells.

N. Kleinenberg (1872) has suggested that the epithelial-muscle and neuro-sensory cells of coelenterates were formed by splitting of original epithelial-muscle elements that were independent effectors: a muscular outgrowth of the original cell formed a muscle cell, and its epithelial part formed a neurosensory cell. A.A. Zavarzin (1941) also holds similar views.

From the structural-morphological point of view, of course, the epithelial-muscle and neurosensory cells are the result, not of splitting, but of differentiation of original multifunctional ectodermal cells: some of them lost sensitivity and became dependent effectors, epithelial-muscle cells, while others lost their contractility, retaining and increasing their sensitivity, and became neurosensory cells. Their phylogeny also must have been of that kind. N.G. Khlopin (1946) also believes divergent differentiation to have been the most probable method of original differentiation of neural and muscular elements in Metazoa, and substantiates that view in detail.

The components of the neural system of coelenterates are of three types: sensory, associative, and motor cells. The first retain their epithelial location, but the others have lost it and lie subepithelially, on the same level as the muscular processes of the myoepithelium. They all form numerous outgrowths, which make contact with other nerve cells (neurons), and the outgrowths of the motor cells also make contact with muscle cells. Unlike the neurons of higher forms, the nerve cells of coelenterates (at least their associative cells) are not polarised; all their outgrowth are uniform, and they are not divided into a neurite (axon) and dendrites. All the nerve cells are combined into a single plexus that covers the whole body of the animal and constitutes its neural apparatus. In hydroid polyps this plexus is diffuse and forms only a rather vague concentration around the mouth, and in *Hydra* also on the foot. We have already mentioned a remarkable feature in coelenterates, that their neural plexus is developed not only from the kinetoblast but also from the phagocytoblast, the epithelium of the gut. The two plexuses, ectodermal and endodermal, interpenetrate each other in the region of the mouth.

According to the view expressed in their time by O. and R. Hertwig (1878, 1879, 1880) and accepted by most authors, the primary form of nerve cells was sensory cells lying in the epithelium, from which both associative and motor cells are derived. It is most natural to regard as the prototype the neural system of coelenterates in the form of a network of cells in the epithelium, each having a receptive device at its apex, and at its base outgrowths whereby it made contact both with cells similar to itself and with muscle cells. The direct link between sensory and muscle cells would be the simplest conceivable reflex arc—monomeric. This type of link has actually been described in the tentacles of Actiniaria. Mutual linkage of all the cells in the neural plexus results in reaction to purely local stimuli by wider areas of the body. Let us now imagine that some of these universal cells lose their direct connection with the muscle cells, but retain their epithelial location and their connection with other nerve cells; that other lose their direct connection with both the surface of the body and the musculature, but retain their connection with other nerve cells; and, finally, that a third group retain their connection with the musculature and other nerve cells but lose direct connection with the external world; in such a case the first cells would become sensory, the second associative, and the third motor cells, and their assemblage takes the form of the neural apparatus of modern coelenterates. This picture of the course of differentiation of types of nerve cells in coelenterates (which differs somewhat from the usual formulation of O. and R. Hertwig's theory) also explains the absence of polarisation

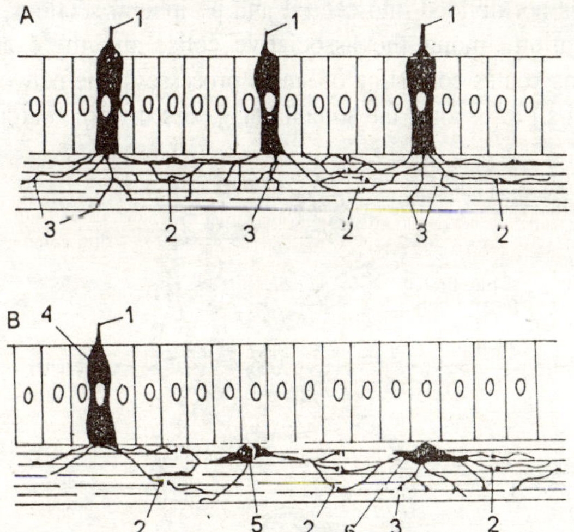

Fig. 10.2. Diagram of origin of sensory, associative, and motor nerve cells in coelenterates (B) from universal nerve cells of the prototype (A). 1, sensory ending; 2, synapses at places of contact of different neurons; 3, motor endings on muscle cells; 4, sensory; 5, associative; 6, motor.

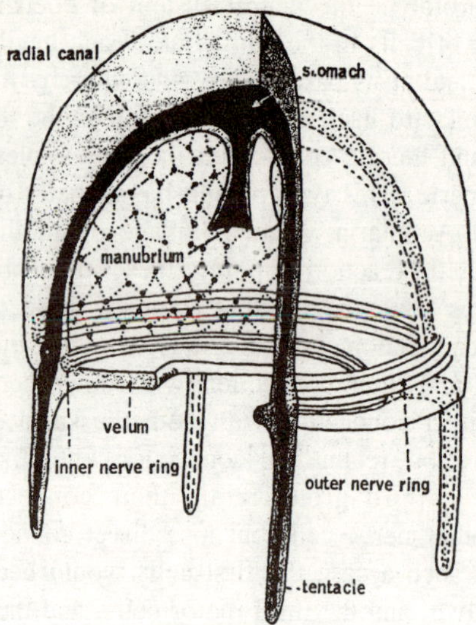

Fig. 10.3. Nervous system of a hydromedusa.

in their associative cells, since from the above point of view all the outgrowths of these cells are equivalent in nature.

The diffuse neural apparatus of coelenterates is a very primitive type. It is not divided into central and peripheral sections, there is no specialisation among the associative cells, and there are no long conducting routes consisting of single processes. The network conducts stimuli in all directions, the stimuli being passed from neuron to neuron.

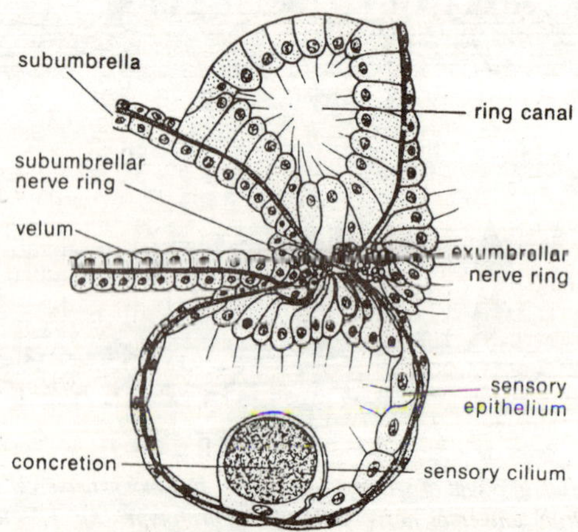

Fig. 10.4. Statocysts of hydromedusae.

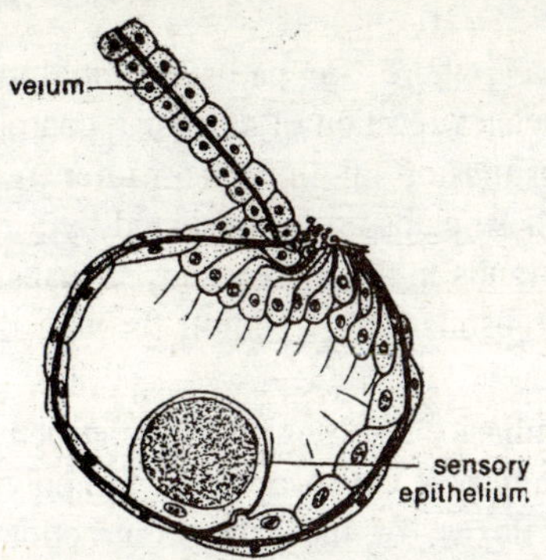

Fig. 10.5. Tilted structure of statocysts.

At the same time each of them is also connected with motor cells. Therefore a wave of cxcitation spreading through the neural plexus from any point is accompanied throughout its journey by a wave of muscle contraction. A polyp is incapable, as a rule, of reacting to stimulus arising at any point of its body by contraction of muscles at a distance from that point without previous reaction by all the muscles between the two points. Stimuli are transmitted through the neural plexus far more rapidly than in sponges, where neural elements are absent, but far more slowly than in higher Metazoa, where they do not have to surmount synapses (i.e. points of contact between the processes of different neurons) located along their path. Even in

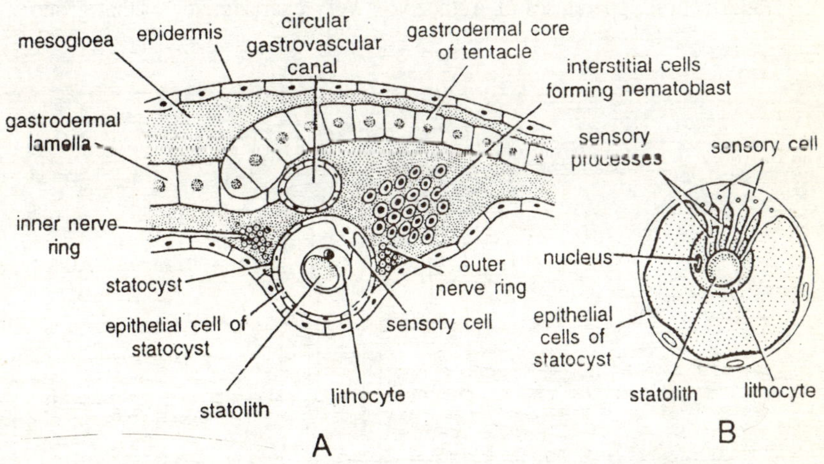

Fig. 10.6. Obelia. A—Base of an adradial tentacle in L.S. showing statocyst. B—Statocyst magnified.

Actiniaria, however, there are paths of rapid transmission, which produce, for example, a reaction of defensive contraction by the animal. The rate of transmission in these is rather more than 1 m/sec. Morphologically these paths are represented by a ring on the oral disc and longitudinal strands in the mesenteries. Synapses probably exist on these paths, but transmission of stimuli through them is permanently faciliated.

A second peculiarity resulting from the structure of diffuse neural apparatus is that any part cut away from the polyp's body is capable of independent reflexes, of muscular contractions in response to a stimulus, unlike, for instance, the amputated leg of an insect. This is because, with the diffuse, absolutely-uncentralised neural apparatus of polyps, each small part of the body always retains the whole neural mechanism necessary to complete a reflex—sensory, associative, and motor cells and an effector muscle, all linked together.

Anthozoa, especially Actiniaria, differ from hydroid polyps in their greater concentration of the neural plexus in the circumoral-disc region; basically, however, their neural apparatus remains diffuse. A second important difference between Anthozoa and hydroids consists in the great development of the parts of the neural apparatus located in the mesenteries of Anthozoa, which are generally lacking in hydroids. The outstanding role of peripheral phagocytoblast in creating the muscular and neural apparatuses of Anthozoa is very characteristic of that group.

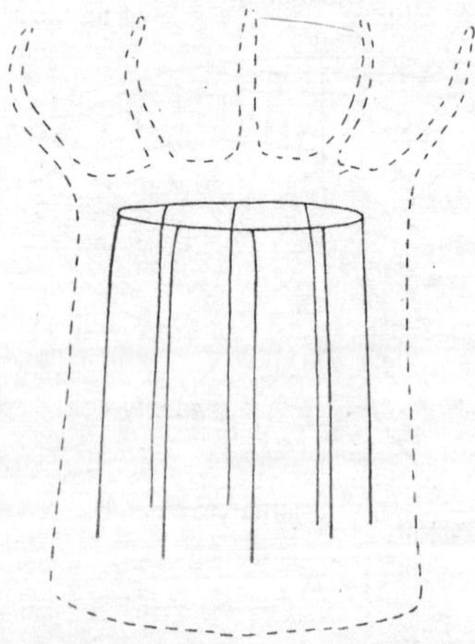

Fig. 10.7. Diagram of arrangement of paths of accelerated transmission of stimuli in body of Actinia (annular thickening in oral disc and radial nerve cords in mesenteries).

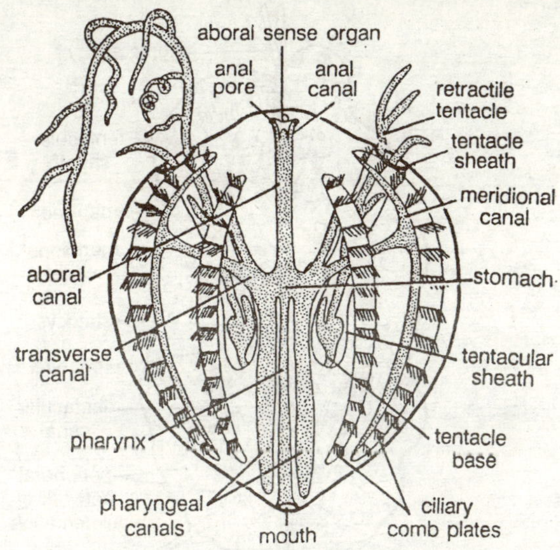

Fig. 10.8. Pleurobranchia.

In this respect they show a very interesting resemblance to the lower Deuterostomia (see below). The endodermal neural plexus of Anthozoa combines with the ectodermal not only in the region of the mouth but also throughout the mesogloea.

All medusae differ from polyps in possessing specialised sense-organs and in making considerably more complex movements. As a result the neural plexus of hydromedusae forms two annular thickenings at the edge of the umbrella, one at the subumbrellar and the other at the exumbrellar side of the annular canal. Possession of a subumbrellar ring has also been demonstrated in several Scyphomendusae (*Aurelia*, *Rhizostoma*), in which it is joined by radial strands to all the marginal sense-organs. At the base of each of the latter there is a considerable accumulation of nerve cells—a marginal ganglion. According to physiological research data, rhythmic impulses arise in these marginal ganglia, are transmitted by the entire neural plexus, and produce rhythmic contractions of the bell. The marginal ganglia of Scyphomedusae represent a first step towards creation of a central section of the neural apparatus. It is instructive that they originated because of the formation of the first complex sense-organ, whose origin was in turn due to the mobile mode of life of medusae.

The neural apparatus of ctenophores is of exceptional interest to comparative anatomists, but because of the extreme delicacy of these animals it is difficult to study. According to K. Heider (1927), the

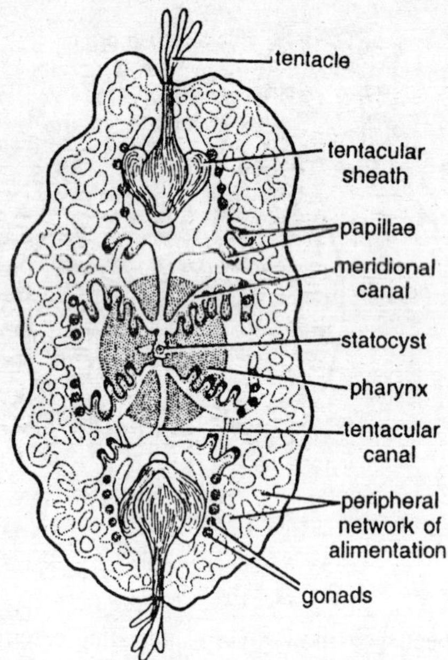

Fig. 10.9. Coeloplana

general sub-epithelial plexus of *Beroë* forms a thickening (circumoral ring) around the external mouth. Along all the rows of ctenes the plexus also thickens, thus forming eight meridional condensations. These meridional condensations extend still farther aborally, becoming fused in pairs and converging along the course of the four ciliated bands towards the base of the 'springs' of the aboral sense-organ. *Coeloplana* has there four ganglion-like concentrations of neural tissue. As physiological experiments indicate, the aboral sense-organ and adjacent parts of the neural plexus certainly exert a tonic and regulating influence on the activity of the ctenes, which are, however, capable of reflex responses to stimuli affecting any other parts of the neural plexus.

Thus a co-ordinating centre linked with the aboral sense-organ, eight meridional strands, and a circumoral ring are developed in the neural plexus of ctenophores, a structural plan very similar to that which we find in polychaete trochophores and partly in Scolecida. B. Hanström (1928) suggests that in ctenophores, as in Scyphomedusae, the appearance of balancing organs (which obviously resulted from a free-swimming mode of life) was the stimulus to the first creation of neural centres.

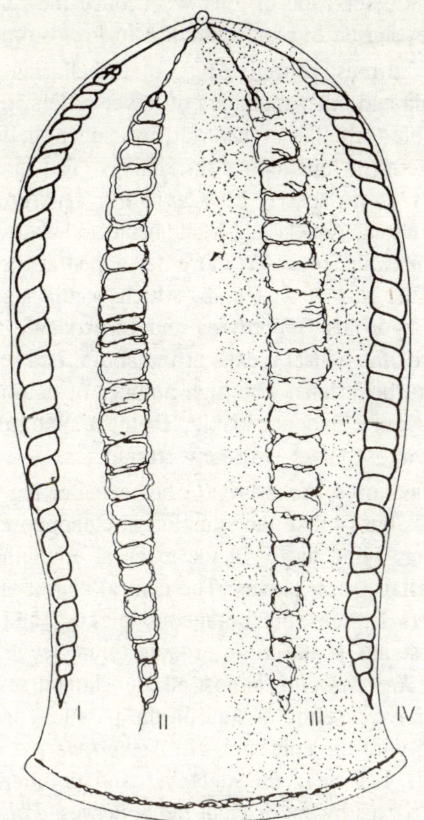

Fig. 10.10. Neural apparatus of the ctenophore (diagrammatic). The nerve cords along the ctenes, the circumoral plexus, and part of the diffuse cutaneous plexus are shown. In radius II the course of the nerve fibres, and in radius III and interradius III-IV the nuclei of nerve cells, are shown.

H. Korn (1959) describes a neural plexus beneath the epithelium of the stomach of *Pleurobrachia pileus*; judging by his data Ctenophora, as well as Cnidaria, possess not only an ectodermal but also an endodermal section of the neural apparatus, a fact of great comparative-anatomical significance.

Lower Worms, and General Principles of the Evolution of the Neural Apparatus

The neural apparatus of lower turbellarians is at the same level as that of coelenterates, but it reaches higher levels in higher turbellarians. Unlike coelenterates, however, turbellarians no longer show any traces of an endodermal neural system; their neural system is entirely of kinetoblastic origin. We deducc the reasons for this by studying the evolution of phagocytoblast in Protostomia.

We find the most primitive type of turbellarian neural apparatus in some Acoela and other members of lower orders, e.g. *Xenoturbella*. This remarkable turbellarian has a diffuse subepithelial neural plexus, without any stems or longitudinal thickenings. The same type of diffuse neural plexus is possessed by *Convoluta stylifera* and *Otocoelis gulmariensis* among Acoela and a number of other lower turbellarians. In all these forms the plexus is more developed at the front end of the body than in the rest of its length, which occurs also in coelenterate planulae (the hydroids *Gonothyrea* and *Clava*). We ascribe the same condensation of the neural plexus at the aboral end to the coelenterate prototype of turbellarians. The high number of neural elements at the aboral pole is quite understandable, since in swimming it is the first part to come into contact with new stimuli.

Like ctenophores, *Xenoturbella* has a balancing organ located at its aboral pole, but (unlike the situation in ctenophores) its statocyst is a closed vesicle lying partly in the external epithelium and partly in the sub-epithelial neural plexus. The central neural apparatus contains no other parts besides the cutaneous plexus, and in this respect *Xenoturbella* stands at the same organisational level as coelenterates. In contrast to *Xenoturbella*, almost all Acoela and several other lower turbellarians have a statocyst sunk in the parenchyma, which has lost its connection with the epidermis. In *Nemertoderma* (Acoela) a group of nerve cells lies near the statocyst and communicates with the subcutaneous plexus by six or eight radial nerves. *Polychoerus* (Acoela) has four such nerves, lying in the same planes as the four ciliated bands connecting the aboral organ of ctenophores with the ctenes. In the embryo of *Otocoelis chiridotae* (Acoela) four brain rudiments form in the same planes. We may regard the group of nerve cells serving the statocyst in *Nemertoderma* as a brain, even if an incipient one.

It was in consequence of the statocyst—a sense-organ at the aboral pole, sunk beneath the skin—that the brain, or cerebral ganglion, arose in turbellarians; the cerebral ganglion is a part of the cutaneous neural plexus that has separated from the rest and sunk into the body behind the statocyst. We have already seen that in ctenophores and Scyphomedusae the first rudiments of neural centres arise in connection with the service of the balancing organs. In view of the location of the cerebral ganglion of higher turbellarians within the cutaneous plexus, on the body axis, E. Reisinger (1925) calls it the *endon*.

Developing from the above-described prototype, the neural apparatus of flatworms passes through a series of successive stages of improvement. That improvement takes five main directions: (*i*) centralisation of the neural apparatus; (*ii*) migration of its elements into the depth of the body; (*iii*) increase of the role played by the cerebral ganglion (endon); (*iv*) external architectonical simplification of the neural apparatus; (*v*) greater complexity of its internal architectonics, causing greater complexity and plasticity in the animal's behaviour. The latter question is too complex for discussion within the limits of a general course in comparative anatomy. A detailed study of it may be found in the works of B. Hanström (1928) and A.A. Zavarzin (1941); we can touch only on a few points that are relevant here.

11

Polymorphism

Amongst the coelenterates, hydrozoans provide very good examples of polymorphism. The phenomenon is essentially for division of labour. Division of labour is first seen in the cells of *Hydra* where the cells are specialized to perform different functions of individual as a whole. Physiological differentiation of this type had its effect upon the morphology of cells which led to cell specialization and give rise to cells of different structures. In *Obella* this specialization is carried still further. In it not only cells are specialised but individuals get specialised to perform different functions. The polyp perform different functions. The polyp performs vegetative function such as feeding, respiration, etc. and the free swimming medusae are reproductive in nature.

There are different types of polypoids and medusoids specialized for different functions. There are three types polyploid and four types of medusoids individuals as given below:

Polypoids Zooids

Gastrozooids

The gastrozooids or the siphons are the nutritive or food-ingesting individuals of the colony. Each gastrozooid is a tubular or seccular structure with a large mouth. A single, long, contractile and hollow tentacle arises from the base of the gastrozooid. It bears numerous 'ateral contractile branches called the *tentilla*, each ending into a knob or coil of nematocysts.

Dactylozooids

These are the protective polyps of the colony and are variously known as *palpons*, *tasters* or *feelers*. Typically they resemble the gastrozooids except that they lack a mouth and their basal tentacle is unbranched. In *Vallela* and *Porpita* the dactylozooids arise from the margin of the colony in the form of long, hollow and tentacle-like fringing bodies called bodies called the *tentaculozooids*. When associated with gonophores, the tentacle-like dactylozooids are known as *gonopalpons*. In *Physalia* the dactylozooids become excessively long.

Gonozooids

They are reproductive zooids which are also known as *blastostyles*. They are without mouth and tentacle. They reproduce asexually by budding and form medusae. In *Vallela* and *Porpita*, they resemble a gastrozooid and possess a mouth. Usually, the gonozooids take the form of branched stalks, called the *gonodendra*. These bear grape-like clusters of gonophores and are often provided with gonopalpons as in *Physalia*.

Medusoid Zooids

Swimming Bell

The swimming bells which are also known as *nectocalyes*, *nectophores* or *nectozooids* are medusoid form with a bell, velum, four radial canals and a ring canal. But these are devoid of mouth, manubrium, tentacles and sense organs. It shape is variable and may be bilaterally symmetrical, prismatic, elongated or flattened. Due to well developed musculature, swimming bells act as excellent swimming organs and help in the locomotion of the colony.

Pneumatophores

The pneumatophores or the *floats* are bladder or vesicle-like structures filled with gas, and keep the colony floating. Each pneumatophore represents an inverted medusa bell, devoid of mesogloea and consisting of an external exumbrellar wall, *pneumatocodon*, and an internal subumbrellar wall, the *pneumatosaccus* or *air-sac*. The walls of both these are double-layered and are highly muscular. The space between the two walls is known as *gastrovascular cavity*.

A great degree of variation in shape and size is observed in different siphonophores. In *Agalima*, the float is simple and its air sac is lined by a layer of chitin secreted by the epidermis.

The shape of pnematophone is variable and way or may not be divided into a number of concentric chitinous chambers arranged in one plane. These communicate with each other and with the central chambers by pores in their walls. The air sac may open or closed. The air sac may be perforated by a single or several pores. In some

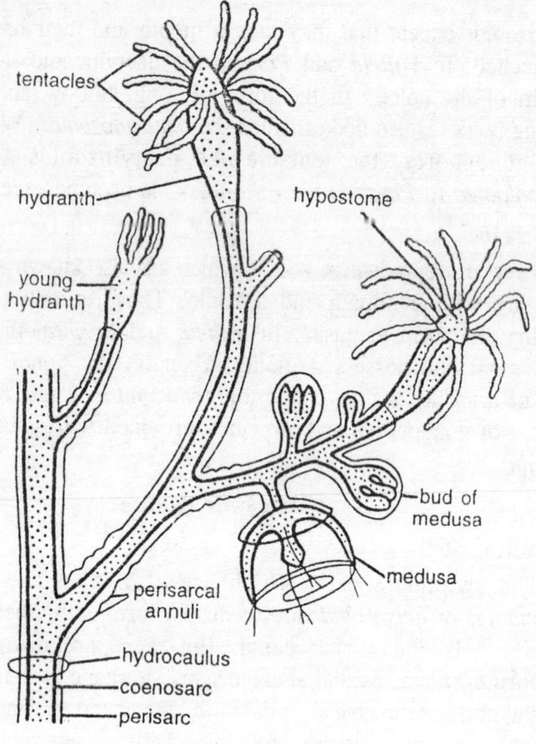

Fig. 11.1. Bougainvillea. A portion of colony.

cases a portion of the float is partly constricted off and assumes the form of an ovoid medusa-like, called *aurophore*.

Bracts

The bracts which are also known as the *phyllozooids* or *hydrophyllia* are thick, gelatinous and curved plates of mesogloea. These may be prism-like, leaf-like, shield-like or helmet-like in appearance. They are unlike medusae and contain a simple or branched gastrovascular canal.

Gonophores

The gonophores or the reproductive medusoids occur singly on separate stalks or in clusters on polypoid gonozooids as in *Vallela* or on simple or branched gonodendra. The gonophores may be medusa-like with bell, velum, radial canals and a manubrium bearing gonads. But the mouth, tentacles and sense organs are always absent. In number of hydrozoans e.g. *Physalia*, the female gonophores are medusa-like

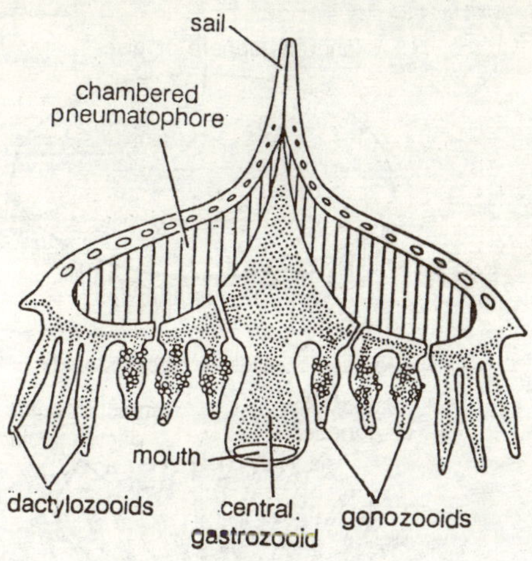

Fig. 11.2. Velella in V.S.

while the male ones are sac-like. In animals like *Physalia* (male), the gonophores may remain attached to the colony or are set free as in female *Physalia*, *Porpita* and *Vallela*. Since they cannot feed, they perish after the discharge of sex-cells. The gonophores are decisions but the colonies are hermaphrodite bearing both types of gonophores in the same or separate clusters. Gonophores may be budded off from the pedicel of the gastrozooid as in *Diphys*, or from a blastostyle as in *vgallela* or from coenosarc as in *Agalmopsit*.

Grades

Some coelenterates posses only two types of zooids, others have there types and still others several types. These forms are respectively described as dimorphic, trimorphic and polymorphic. There are, thus, three grades of polymorphisms:

Dimorphic forms

The dimorphic forms have two types of individuals, namely, *polyps* or *hydranths* and *medusae*. These types are regarded as the fundamental types from which the additional types found in the trimorphic and polymorphic colonies are derived by modification.

(a) *Polyps*. The Polyps have a cylindrical form, are fixed by aboral end, enclose a wide gastrovascular cavity, and bear mouth and tentacles at the free oral end. They serve to feed the colony and are, therefore, also known as the *gastrozooids* or *trophozooids*.

(b) *Medusae*. The medusae have a bell-, bowl- or saucer-shaped body, lead free-swimming life when mature, enclose gastrovascular cavity

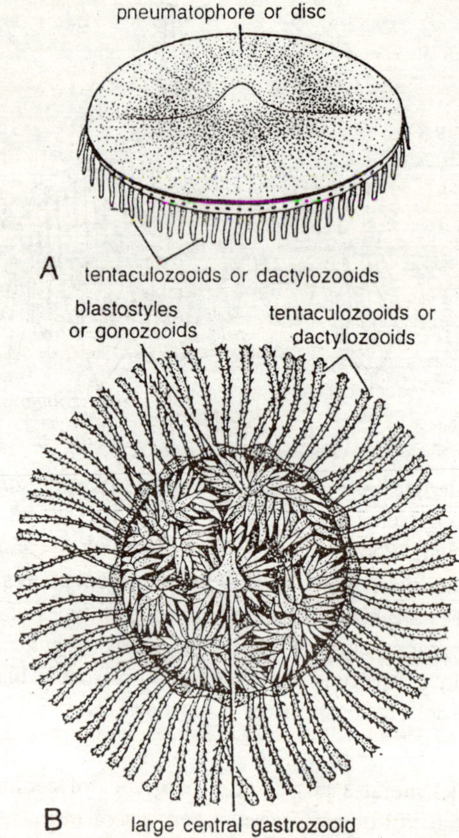

Fig. 11.3. Porpita. A—Colony in dorsal view; B—Colony in oral or ventral view.

in the form of narrow radial and circular canals, and bear mouth and marginal tentacles. They bear gonads and bring about sexual reproduction. They are, therefore, also known as the *gonophores* or *sexual zooids*.

Though diverse in form and function, the polyps and medusae have a similar basic plan so much so that they can be derived from each other.

Bougainvillea

(Class Hydrozoa) and *Corallium* (class Actinozoa) are examples of dimorphic colonies.

(a) *Bougainvillea*. The polyps are stalked and uncovered, have an elongated manubrium and bear a single circled of solid filform tentacles. Medusae have four per-radial tufts of simple marginal tentacles and four groups of branched oral tentacles. They arise by budding from the hydrocaulus, polyp stalk and other medusae.

(b) *Corallium*. The zooids of *Corallium* are called *autozooids* and *siphonozooids*. Both are polypoid. The autozooids have tentacles and four groups of branched oral tentacles well developed mesenteries and a normal Siphonoglyphe. They serve to feed colony. The siphonozooids are small, lack tentacles, have reduced mesenteries and very large siphonoglyphe. They serve to drive a current of water through the cavities of the colony and bear gonad.

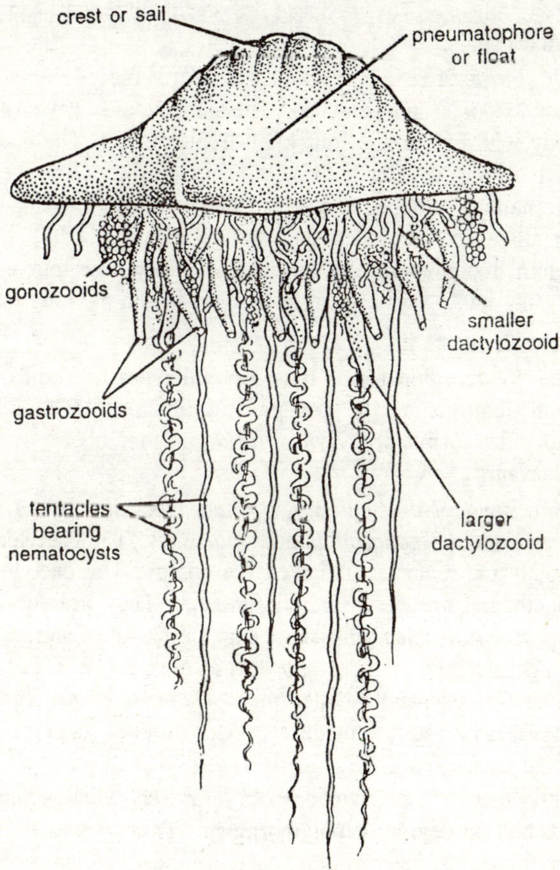

Fig. 11.4. Physalia. A colony.

Trimorphic forms

The trimorphic forms have three types of zooids, i.e., *polyps*, *medusae* and *gonozooids* or *dactylozooids*. The polyps and medusae are similar to those of dimorphic forms. The gonozooids are modified polyps. They lack mouth and tentacles. They serve to bud off medusae or their morphological equivalents. The dactylozooids are also modified polyps. They are mouthless and serve to protect the colony.

Examples of trimorphic colonies are many. *Obelea* and *Millepora* are well-known:

(a) *Obelea*. The zooids of *Obella* include polyps, gonozooids and medusae. The polyps are surrounded by hydrothecae and gonozooids by gonothecae. The medusae are saucer-like and bear gonads on radial canals.

(b) *Millepora*. The zooids of *Millepora* include gastrozooids, dactylozooids and medusae. The gastrozooids have short plump body with mouth and four knob-like tentacles. The dactylozooids have a long, slender body without mouth but with several, short, alternating, knobbed tentacles. The medusae develop in pits or ampullae of the colony and are greatly reduced, being without velum, mouth, tentacles and canals. They bear four nematocyst-bearing knobs on the margin and gonads on the long manubrium.

Polymorphic forms

The polymorphic forms have several types of zooids. All these are modifications of the polyps and medusae. The best known polymorphic forms are *Hydractinia* and members of the order *Sinphonophora*.

(a) *Hydractinia*. *Hydractinia* develops four types of zooids: *gastrozooids*, *dactylozooids*, *gonozooids* and *sporosacs*. The dactylozooid have mouth and tentacles and feed the colony. The dactylozooid lack mouth and are defensive in function. They are further of two types: *spiralzooids* with short capitate tentacles and *tentaculozoid* which are long, slender and devoid of tentacles. The gonozooids retain short tentacles called the *nematocyst heads*. The sporosacs are reduced sac-like medusae. They produce gametes, either ova or sperms.

(b) *Siphonophora*. The siphonophores form free-swimming colonies with the highest degree of polymorphism. Their polyps occur in three modifications: gastrozooids, dactylozooids, and gonozooids—

(i) *Gastrozooids*. The gastrozooids are also called the siphonozooids, hence the name of the order. They feed the

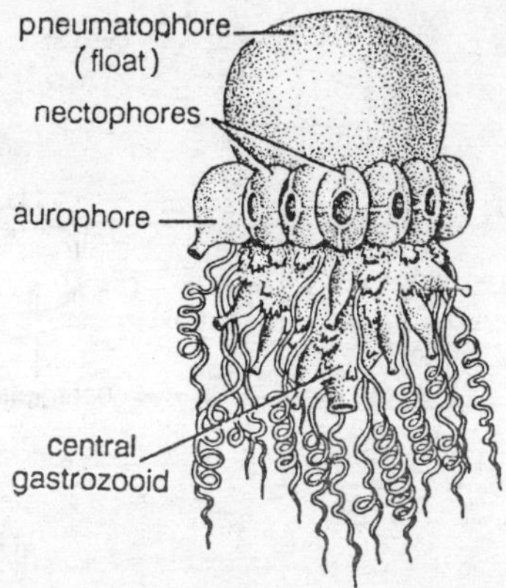

Fig. 11.5. Stephalia showing swimming bells and aurophore.

colony. They have the usual polyp form but lack the usually located tentacular ring. Instead, they bear a single, hollow, long and contractile tentacle at or near the base. The tentacle gives off lateral branches, the *tentilla*, each ending in a knob or coil of nematocysts.

(ii) *Dactylozooids*. The dactylozooids are also called the *palpons*, *feelers* or *tasters*. They lack mouth and their basal tentacle is unbranched. They may be long, hollow and tentacle-like, when they are termed the *tentaculozooids*.

(iii) *Gonozooids*. The gonozooids may look like the gastrozooids and even have a mouth, but lack a tentacle. Usually, however, they are long, slender and branched and are called *gonodendra*. The may bear tentacle-like dactylozooids called the *gonopalpons*. The gonozooids produce clusters of gonophores on them.

The medusoid individuals exist in four modifications: swimming bells, bracts, gonophores and pneumatophores.

(i) *Swimming Bells*. These are also termed the *nectophores* or *nectocalices*. They are medusae with velum and radial and circular canals, but without mouth, manubrium gonads and sense organs. They are very muscular and bring about locomotion of the colony.

(ii) *Bracts*. These are also called the *hydrophyllia* or *phyllozooids*. They are thick, gelatinous individuals with simple or branched gastrovascular cavity. They are protective in function.

(iii) *Gonophores*. The gonophores may be medusa-like but lack mouth, tentacles and sense organs. They may be reduced to rounded sacs.

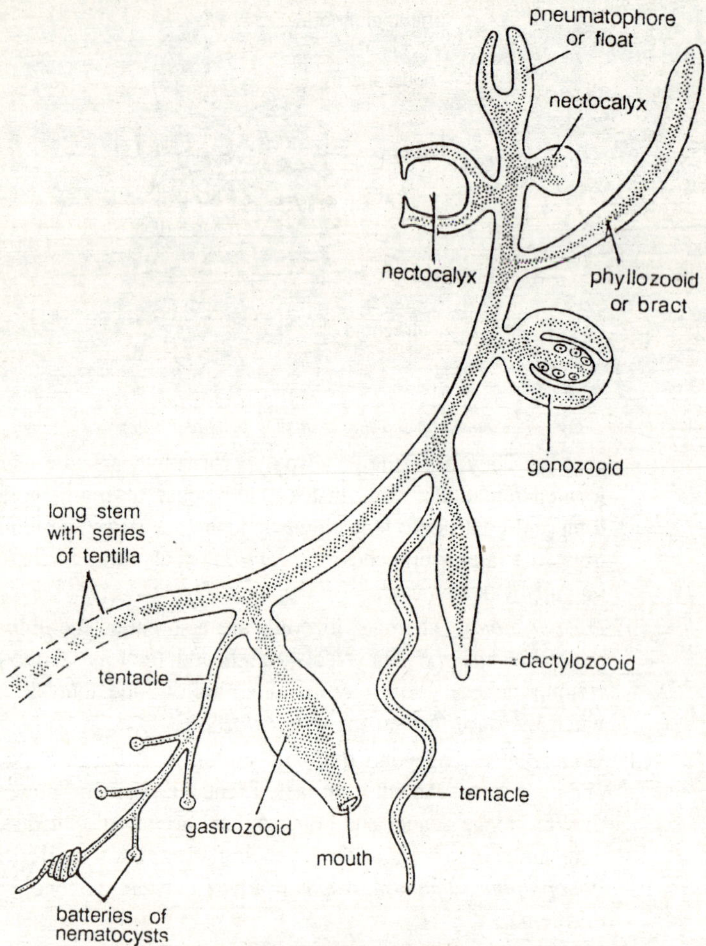

Fig. 11.6. Generalized calycophoran Siphonophora showing a single cormidium.

In some cases, the female gonophores are medusa-like and male gonophores are sac like. The gonophores produce gametes.

(iv) *Pneumatophore*. It is also known as the float. It is an inverted medusa without mesogloea. Its outer (exumbrellar) and inner (subumbrellar) walls are respectively called the *pneumatocodon* and *pneumatosaccus* or *air-sac*. The opening of the air-sac directed upwards, is reduced to a small pore, the *pneumatopore*, and is guarded by a sphincter muscle. At the bottom (original roof) of the air-sac, the epidermis is modified into a *gas gland* that secretes gas having composition similar to that of air. The float keeps the colony affloat.

All types of zooids arise by budding from a common *stem* or *coenosarc*. The latter may have the form of a tube or a disc and bears zooids in groups called *cormidia*.

The common polymorphic siphonophores are *Halistemma*, *Physalia*, *Velella* and *Porpita*.

(a) *Halistemma*. *Halistemma* possess a long slender stem with a small float at the top, several closet set swimming bells below the float and numerous cormidia lower down. The cormidia arise from nodes and each comprises a gastrozooid or dactylozooid and several sporosacs. Bracts arise from the internodes and partly cover the sporosacs.

(b) *Physalia*. *Physalia* has a disc-like coenosarc with a large sail-bearing, baloon-like float over it and several cormidia hanging from it. A cormidium consists or gastrozooids, small and large dactylozooids, gonodendra bearing gonophores, gonopalpons and peculiar gelatinous zooids of unknown function. The colony is poisonous.

(c) *Velella*. *Velella* has a rhomboidal float divided internally into a number of gas filled chambers. It bears an oblique sail above and a disk-like coenosarc beneath. There is a single, large, central gastrozooid surrounded by gonozooids which are, in turn, surrounded by dactylozooids. The gastrozooids lack a tentacle. The gonozooids have mouth and bear gonophores at the base. Dactylozooids are tentacles-like.

Origin of Polymorphism

After studying the polymorphism in coelenterates the question arises whether the metagenesis is a direct consequence of polymorphism or the life-cycle of primitive coelenterate has led to polymorphism. According to one view, the original coelenterate was a polyp and through specialization the sexual function was relegated to the secondarily developed medusoid form and this led to metagenesis. According to another view, the ancestral coelenterate was a medusoid form and the polypoid generation is persistent larval form, thus leading to polymorphism.

Polyp Origin Theory

This theory was proposed by *Huxley*, *Eschscholtz* and *Metshnikoff*. According to this theory component zooids, are really organs of a single medusoid individual, whose manubrium, tentacles and umbrella have become multiplied and migrated from their primitive positions. According to this theory, the various zooids are organs that have not yet attained the grade of polymorphic individuals, so that siphonophores are the most primitive existing coelenterates. In short, a siphonophore, in possessing a marked degree of power of vegetative increase of its parts, resembles a plant more than an animal. The theory errs too much in denying the colonial nature of siphonophores.

Polyp Person Theory

This theory was proposed by *Leucart*, *Vogr* and *Gagenbaur*. According to this theory siphonophores are free-swimming polymorphic colonies of highly specialized polyps with the power to produce medusae. This theory maintains that the parts of a siphonophore are either polyps or medusae or both, but the primitive zooid of the colony is of the polyp type.

Sedgwick, *Haeckel* and *Balfour* hold the view that the colonial theory (polyp-person theory) is more appropriate but that the primitive zooid of the colony was a medusa which has produced other medusa by budding and that the parts of these medusae possess the power of becoming discrete. Then they are removed from the bud to which they belong and in some cases, multiply secondarily. The many organisms of the colony, which according to the old colonial theory are modified polyps are, according to this view, nothing more than the parts of the medusiform individuals which have shifted their attachment and are therefore real organs. For instance the structures called palpons are to be looked upon as mouthless manubria of medusoid; the tentacles are to be looked upon as only surviving amongst the marginal tentacle of the medusoid. The theory agrees in asserting the colonial nature of Siphonophora but admits that there has been a vegetative repetition and specialization of certain organs which is demanded by the old colonial theory.

Recently, *Moser* has revived the polyp organs theory of *Huxley* and *Metschnikoff*. According to *Moser*, the various individuals of siphonophore colony are organs that have not yet attained the grade of polymorphic individuals; and, therefore, the siphonophores are the most primitive existing coelenterates. Although the theory has not gained general recognition, yet it cannot be doubted that the siphonophores early diverged from the coelenterate stem.

We may, therefore, conclude that the identification of the component structures (especially in the siphonophores) as organs or individuals is a problem which rather defies an easy solution and may best be left as such.

Polymorphism and Alternation of Generations

Polymorphism is intimately associated with the life-history of organisms. In monomorphic forms as in ***Hydra***, the life-cycle is simple and without any larval stage. It may be represented by the formula *polyp-egg-polyp*. With the advent of polymorphism, the reproductive powers of the organisms are divided among the different individuals of the colony. In these organisms the polyps reproduce asexually to form medusoid form, the *gonophore*, and the gonophore reproduces sexually to form the *polyp*. The life-cycle of such organisms may be represented by the formula: *medusa-egg-planula-polyp*. Thus the *alternation of generations* or *metagenesis* comes into existence in the life-cycle the asexual polypoid generation alternates with the sexual medusoid generation.

12

Coral, Reefs and Attols

In this chapter, the surface views of coral reefs will be described along a cross-section from the outer slope to the inner side. This description will include the most common features; the peculiarities producing the different reef types. Some general problems, such as those related to the formation of beach rock and the influence of hurricanes on reef morphology, will also be discussed here.

Generalized Cross-Section

Outer Slope

The outer slope of the reefs is, as a rule, very step, although the gradient differs according to sites. The differences concern principally the length of the slope, which varies in relation to the depth of the sea-floor upon which the reefs have been built. Some frameworks rise from a sandy sea floor where the depth does not exceed a few metres, but even there the reef slope is steep: the corals rise abruptly as living colonies forming a kind of staircase, as, for example, off Noumea in the New Caledonia lagoon, where reefs are built on a sandy floor 9 to 22 metres deep.

More significant, however, are the slopes surrounding oceanic reefs, which rise several thousands of metres from the deep sea-floor. At Bikini Atoll, the outer slope descends to 3600 or 4500 metres, the maximum and average slopes being 46° and 36° respectively. On Rongelap, Rongerik, and Ailinginae Atolls, also in the Marshall Islands, the same authors reported average slopes of 31°, 31°, and 35°. The leeward slopes at Bikini are steeper (37°) that those to windward (33°). At Mururoa Atoll in the Tuamotu Islands (Chevalier *et al.*, 1968), the slope averages 45° from 20 metres to 300-500 metres depth,

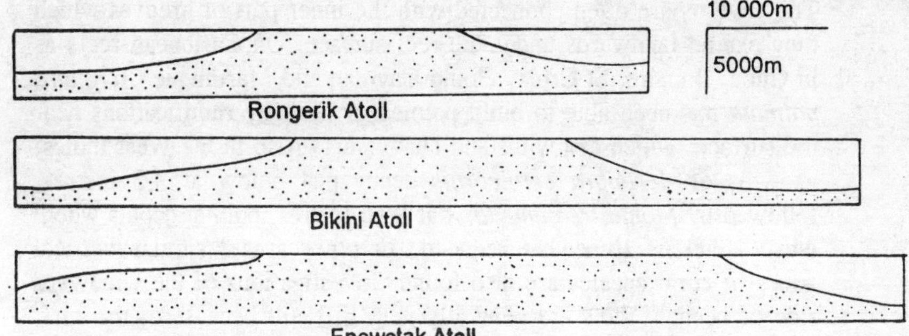

Fig. 12.1. Cross-profiles of three atolls in the Marshall Islands. Heights and lengths at the same scale.

and then decreases to 30°-35° in the north and 15-25° in the south, the relation with wind direction being less evident than at Bikini. On the barrier reef of the Gambier Islands, Eastern Polynesia, the slopes, exceptionally gentle as far down as 30 metres, are between 12° and 40° from 30 to 100 metres and between 16° and 32° from 100 to 1000 metres.

In many cases, terraces interrupt the slope in its uppermost section. A terrace is of general occurrence in the Marshall Islands where it occurs between 15 and 21 metres, and has been referred to as the 10 fathom terrace. Its greatest width is 3.6 kilometres, and it contains shallow irregular depressions. Terraces also exist at Mururoa in the Tuamotu Islands, at 8 and 20 metres depth, with some irregularities and they have also been recognized at the same levels at Raroia in the same archipelago. Thus the precipitous slopes may not start from the reef edge. At Johnston Island in the Central Pacific, terraces or platforms are equally reported at 9 or 18 metres depth, with numerous ridges of knolls, and many sink holes in the upper one. Such terraces, however, are not found on all reef slopes. In the Red Sea they do not occur on the reefs of the northern part of Farsan Bank, where they have not been found at Shab Suleim, Malathu and Marmar. At Shab Suleim, at least in the upper part, the eastern, inner side, is steeper than the western, windward side, as at Bikini. But on all these Red Sea reefs the slopes average 45-50° down to 200-500 metres.

A feature common to the upper few metres of outer slopes is the occurrence of many overhangs and caverns, inhabited by brilliant and numerous schools of fish, the overhangs testifying to the active growth of coral outwards. The most beautiful pictures taken by divers to portray coral reefs are from this part of the hermatypic frameworks.

The caverns are often connected with the inner part of grooves which they extend landwards under the reef surface. On Caribbean reefs as in Cuba, Bonaire, St Croix, Grand Cayman and Martinique, *Acropora palmata* has been able to build palmed or branchy ramifications near the surface, which can withstand heavy surf. Also in the West Indies, growths of *Acropora cervicornis* occur just below at ±5 metres, followed by *Montastrea annularis* at 6-10 metres, both at depths where wave attack is somewhat reduced. In other areas, similar vertical zones of coral species are also found. Investigations of the windward side of oceanic reefs are generally very difficult because of the surf; nevertheless, where divers have explored these environments, as on Aldabra Atoll in the Indian Ocean, they have confirmed a very marked decrease in branching corals and Alcyonariae and a sharp increase of foliaceous and encrusting types below a depth of 30 metres.

Outer Edge: Spur and Groove Systems and Algal (*Porolithon*) Ridge

The outer edge of the reefs on the windward side, from the surface down to a few metres, is generally cut by systems of more or less spaced grooves, separated by spurs. The classic description is from Bikini Atoll, where 'the seaward slopes are cut by well-developed grooves normal to the reef front, 50 to 100 feet long, 3 to 10 feet wide, and 6 to 25 feet deep. The grooves are relatively straight but in some places are forked, usually downslope. The spurs separating the grooves are flattened algal ridges 25 to 50 feet or more in width. The tops of the ridges are covered by living algae, and the sides also appear to be algal...Reefs to this type are best developed on the windward side of the atoll. The algae which coat and partly form the spurs and grooves are pink in colour and belong to the genus *Porolithon*, which includes *P. onkodes*, smooth or slightly undulating, and living especially is heavy surf, and *P. craspedium*, with an aspect of cauliflower, thriving mostly in more sheltered corners or caverns. The upper part of the system forms a ridge or broad arch, commonly from 5 to 15 metres wide, and rising some 0.3 to 0.6 metre above the lowest part of the backing reef flat. This *porolithon* ridge is constantly beaten and washed by the surf, which is necessary for *Porolithon* to live.

This *Porolithon* ridge, shaped in spurs and grooves, is found in the other Marshall Islands; it is not found on Tarawa and Abemama at Kiribati perhaps because these atolls lie in the doldrums so that the surf is moderate (but spurs and grooves occur there too). This feature appears on many other Pacific archipelagos, especially at Funafuti.

Tuvalu; at Tutuila, Samoa; in the Society Islands at Mopelia and Scilly Atolls and at Bora Bora barrier; on many Tuamotu atolls as Mataiva, Rangiroa, and Mururoa. In these islands the *Porolithon* ridge rests generally upon a more or less eroded older reef. The windward reef at Kure Atoll, Hawaii, is also made of red-brown living algae with a spur and groove system.

As the spurs are prograded seawards, the intervening grooves are progressively coated by *Porolithon* in their inner parts, so that they tend to become surge channels, in which the water pushed by the uprush of the swash circulates in caverns and is compressed through blow-holes in the inner part of the ridge; this is the room-and-pillar structure described by American authors, which is well represented at Rangiroa, Mopelia, Bora Bora, and others. There are variations from place to place in the width of the grooves and the height of the ridges, but they are only details.

A curious feature in several places in Indonesia and at Tongatapu, Tonga Islands is that, around blow holes fed by surge channels, calcareous algae have built 'shallow, rimed polls resembling the terrace formations around geysers and hot springs; the pools are circular and arranged in successively lower terraces as sections of a circle around the over flow from the higher levels'. The low rims are built by algae, but resemble the erosional rims found as residuals on calcareous outcrops around similar pools in Morocco and Madagascar, also in the intertidal zone. It is an interesting convergence phenomenon. Incipient rimmed pools, also with edges built by *Porolithon*, exist on the southern part of the reef edge at Mataiva Atoll, Tuamotu Islands, and at Mopelia Atoll, Society Islands.

On ST Croix, Guadeloupe and elsewhere in the Caribbean, algal ridges are even more important outer parts of reefs than in the Pacific. They often form only a veneer over or beside a fossil Holocene or Pleistocene reef edge, a structure especially evident on Mopelia Atoll in the Society Islands, and at Mataiva, Rangiroa, Kaukura, Hikueru and other atolls in the Tuamotu Islands.

In the Pacific, living hermatypic corals are comparatively scarce on the upper reef edges, but they generally thrive in front of and just below the spurs, as, for example, on Mopelia Atoll, where the outer slope is covered by living colonies down to at least 25 metres, and at Mataiva Atoll where *Acropora*, *Porites*, *Montastrea*, *Pocillopora* and other corals form many small colonies accompanied by corallinaceae and brown algae on the reef front.

In some cases, as at Taiaro Atoll, Tuamotus the *Porolithon* ridge is not well developed on sheltered sectors, where scleractinian corals are abundant in the spur and groove system, but it reappears in areas exposed to strong surf.

Outside the Central and South Pacific, spurs and groves have been reported from many reefs. They are well represented on the Mayotte barrier reef in the Western Indian Ocean, where the length of the spurs varies from 40 to 80 metres, and the intervening distance from 12 to 24 metres. In places where the barrier reef is now drowned under several metres of water, the length increases to 200-900 metres and the spacing to 23-50 metres. Well-developed spurs and grooves are also present on Farquhar Atoll, on Grand Récif off Tulear on the southwestern coast of Madagascar. At Shab Suleim on the Farsan Bank in the Red Sea they occur on the western, windward side of the reef, and are absent from the leeward side, much as at Bikini. Off Tulear, Weydert showed how the grooves, which extend down to 18 metres, display an intersecting pattern in their middle sectors. But, if algal ridges to exist on some reefs in the Indian Ocean, as at Diego Garcia Atoll and on the Maldives, they are generally missing from reefs lying outside of the northeastern part of the marine area, at Mayotte. Madagascar, Mauritius, and are poorly developed in the Seychelles. A consequence is that spurs and grooves on reefs in the Western Indian Ocean and on Farsan Bank are neither build of, nor coated with, *Porolithon*, but are of large hermatypic corals. The bottom of the groves is often, at least partly, covered with coralline sand.

Spurs and grooves have also been described from reefs in the West Indies. Stoddart (1962) reported them from Belize in Central America, again on the windward side, where the spurs were made of vigorous branching corals, chiefly *Acropora palmata*, with no indication of algal blanketing. Roberts (1974) described them around Grand Cayman Island, south of Cuba, as systems associated with two fore-reef terraces found respectively at 5 metres and 15 metres depth above a steeper slope. As elsewhere, they characterize the reefs on coasts exposed to high wave energy, and are poorly developed on the more sheltered west coasts. There is an angular disparity between the spurs and grooves on these two terraces as a result of adjustment to variations in wave refraction. At Tulear, too, where the southwesterly swell is oblique to the reef edge, the orientation of the spurs is a resultant between the alignment of the reef edge and the direction of the swell.

So far, we have mentioned only spurs and grooves on the edges of reefs rising from great or moderate depths, but similar features have also been described from the edges of reefs rising from shallow sandy areas. Such is the case along the northwestern coast of Madagascar, for example at Nosy Foty, where the depth in front of the reef does not exceed 14-16 metres, and at Nosy Valiha, Radama Islands, where the near shore depth is similar. A close examination of the reef at Nosy Valiha has shown the same pattern as at Tulear, on the Farsan bank, and elsewhere, with vigorous *Acropora* colonies covering the spurs, and sand and coral debris flooring the grooves, which are 5 to 6 metres deeper.

At Bikini Atoll, Munk and Sargent (1948) explained the spur and groove systems as an adaptation of the reef front to the surf: the spurs disperse the energy of the waves and allow the organisms to grow, utilizing oxygen and nutrients from the constantly stirred water. Newell (1954), describing the system at Raroia, Tuamotus, wrote that it 'clearly combines both erosional and depositional processes, and it is probable that it is also true of reef in general': as growth occurs on spurs, 'scour by sand is the chief agency of erosion along the floors of the grooves... that are gradually extended headward, in many places reaching well into the reef flat as very shallow, more or less bare, furrows, embryonic surge channels'. This statement, made for an oceanic reef, is equally true for a coral reef in shallow water, as at Nosy Valiha in Madagascar, where the same relation between growth of spurs or buttresses, and erosion of grooves or furrows, appears valid. At Grand Cayman Island, the grooves 'provide avenues through which sediments produced in shallow water environments can be transported to deep off-shelf sites of deposition' and this has also been observed at Tulear.

Detrital Outer Ridge

Many fore-reefs bear on their outer edge detrital accumulations instead of, or beside, growing coralline or algal structures. Such accumulations stand slightly (one metre or less) above the inner reef flat which will be described below. Both types of ridges are found off northwest Madagascar, a detrital ridge existing, for example, on reefs rising from shallow depths (around 10 metres) at the entrance to Baie du Courrier. Huge blocks, which Matthew Flinders and others called negro-heads or nigger-heads, are common, eroded from the living reef and thrown on to the reef flat during tropical hurricanes. Such blocks can even be pushed as far as the inner side of coral reefs by exceptionally strong cyclones. At Grand Récif near Tulear, which receives a heavy swell originating in the Southern Ocean, blocks form either ramparts or series of elongated domes perpendicular to the general direction of the reef front, along which the grain size decreases toward the inner ends. Some authors, quoted in Battistini *et al.* (1975), have called them 'crags and tails', by analogy with the shape of glaciated structures in Scotland, but the 'crags' here are not made of *in situ* rocks, but of big blocks. Detrital outer ridges are also common the smaller reefs south of Tulear, but at Itampolo their place is exceptionally taken by a small algal ridge. Many subtypes exist, and sometimes the ridge is flat, consisting of sparse blocks of various sizes scattered over the outer reef, as a mayotte.

Reef Flat

The reef flat generally forms the widest part of a coral reef. Its width can vary from less than 100 metres to several kilometres. An example of wide reef flat is found on the fringing reef of Rodriguez Island in the Indian Ocean, which reaches 2000 metres in width in several places. As a rule, the reef flat consists mainly of a pavement cemented by calcareous algae and incorporating various detrital elements derived from the coral community, such as coral fragments, articles of Halimeda, and foraminifera. The reef flat is generally exposed to the atmosphere at low spring tides, but there are many pools of various sizes, often with overhanging sides, some decimetres deep, which provide habitats for soft algae, gastropods, lamellibranches, echinoderms, starfishes, and small colonies of living coral. Loose elements such as sand or gravel formed from the branches of broken *Acropora* partly floor these pools.

The most interesting feature of reef flats is their *striation*. They currently bear series of long stripes, consisting alternately of consolidated reef flat and of sand and gravel. These stripes have been described and illustrated from Tulear Grand Récif, smaller reefs in northwest and southwest Madagascar and Mayotte barrier reef. They run at right angles to the general trend of the reef and appear quite distinctly on air photographs as white and dark bands, the while being sand and gravel. At Mayotte they clearly continue across the reef in the alignments of the grooves at the outer edge, and the difference in height with the solid stripes varies from 10 to 50 centimetres. As with the grooves, the sand and gravel stripes are controlled by wave dynamics: where the outer edge of the Mayotte reef shows re-entrants, and refracted waves arrive in a modified pattern, the direction of the stripes shifts accordingly. At Tulear 'there is a balance, on the reef flat, between the total area covered by reef builders as a whole, and

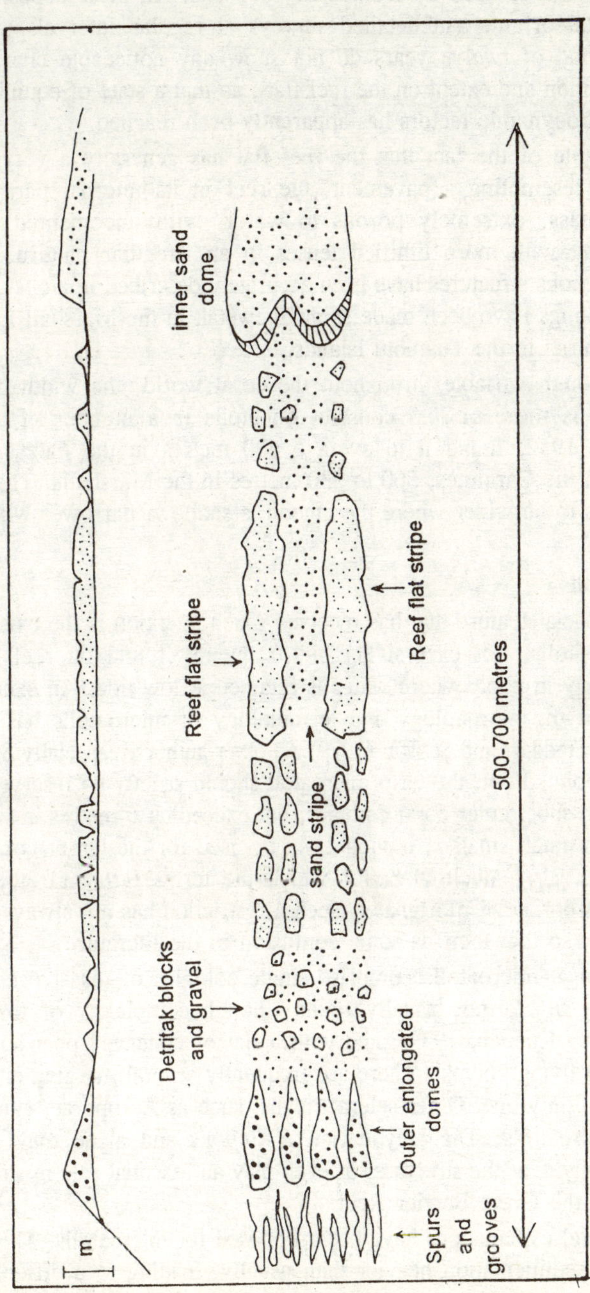

Fig. 12.2. Reef pattern across the central part of Toliara barrier reef, southwest Madagascar.

the total area covered by sedimentary deposits. All these deposits are of skeletal origin, and detailed surveys at regular intervals over a total period of twelve years do not show any noticeable change in their position and extent on the reef flat', so that a state of equilibrium with hydrodynamic factors has apparently been reached.

In spite of the fact that the reef flat has generally a very hard surface, resembling a pavement, the reef in its internal parts is a looser mass, extremely porous to water, with uncemented areas alternating with more lithified lenses in an irregular pattern. Such heterogeneous structures have been accurately described in atolls where deep drillings have been made as at Enewetak in the Marshall Islands and Mururoa in the Tuamotu Islands.

Although variable throughout the coral world, the width of the reef flat is more or less constant on atolls in a number of areas. Tayama (1952) found it to average 600 metres in the Palaus, 500 metres in the Carolines, 500 to 530 metres in the Marshalls. The reef flat tends to be wider where the lagoon is shallow, narrower where it is deep.

Microatolls

A special feature that has received much attention is the existence of microatolls, not excessively but frequently found on reef flats, particularly in pools where water is retained at low tide. An excellent review of the morphology and terminology of microatolls has been made by Stoddart and Scoffin (1979) As other authors, especially Scheer (1971), pointed out, the term microatoll should strictly be reserved for individual subcircular coral colonies, not exceeding 6 metres in width, and often much smaller; it should not be used for the larger corallian structures, also subcircular, for which the terms faro and atoll are more appropriate. Unfortunately such a restriction has not always been observed, so that there is some confusion in the literature.

'Typical microatolls comprise single colonies of massive corals, especially of *Porites*, usually round (though complexity of form is stressed by Kuenen, 1933), and with a flat or concave upper surface devoid of living polyps'. There are frequently several merging circles, instead of only one. Other scleractinians, such as Acroporae, can also form microatolls. The alcyonarian *Heliopora* and algae may grow sporadically over the structure, or even play an essential role in shaping it, as on the Great Barrier Reef.

Several explanations have been proposed for microatolls. The first one is a sedimentation heavier than usually, leading to a differential growth: fine particles and suspension in sea water settle on the top of the colonies and kill the central polyps, which continue to grow on the outer vertical surfaces where the ciliae prevent deposition. This explanation was put forward, for example, by Wood-Jones (1912) for Cocos-Keeling Atoll in the Indian Ocean, and by Krempf (1927) for frameworks off the Vietnamese coast. It is also valid at Mayotte as a result of fine sediment derived from erosion of lateritic soils on the adjacent mainland.

A second explanation is that colonies are limited in upward growth by the level of low tides, so that lateral growth prevails, and the central area dies. Kuenen (1933) and Umbgrove (1947) give both explanations for Indonesian reefs. A third explanation, suggested for alcyonarian atolls in American Samoa, is that nutrient supplies are greater round the periphery than in the centre. Dead microatolls standing at abnormally high levels may have grown in former polls which have been drained, or they may have formed during a phase of higher relative sea-level.

Microatolls are especially frequent on the Great Barrier Reef, where their origin(s) were discussed by Hopley (1982). In the coral world as a whole, the present writer thinks that the first explanation is generally valid.

Back Reef; Boat Channel

The area behind the reef flat differs on fringing reefs, barriers reefs, and atolls. In the second and third type, is replaced by a lagoon, usually several tens of metres deep, which will be discussed. Behind fringing reefs, there is frequently, but not always, a depression, up to a few metres deeps at low tide, between the reef flat and the mainland shore, which has only poor coral growth, probably because of deposition of terrigenous sediment from the continent, or the island on which the reef is anchored. This depression, which was mentioned and explained in this way by Darwin (1842), is called the boat channel because it allows a coastal traffic with small boats: an example is found near Mombasa in Kenya. A boat channel existed just north of Jeddah in Saudi Arabia before it was artificially infilled during the seaward extension of that city. In the southern and southwestern parts of Rodriguez, Indian Ocean, the bat channel can exceed 5000 metres in width in places. Boat channels also exist in Madagascar.

Parts of these shallow depressions are colonized by meadows of marine phanerogams, which are also sometimes found on the inner parts of reef flats and lagoons connected with barrier reefs, as at

Tulear where *Cymodocea* and *Thalassia* spp. from prairies and at Mayotte. In northwest Madagascar and near Mombasa in Kenya, these meadows are partly associated with sets of crescentic pools the phanerogams growing on the sides of the ridges between the pools. Such patterns result from wave action at high tide, when the rides and polls are submerged. They seem to require to tidal range of at least 2 or 3 metres at spring tides. They have not been reported from areas outside the Western Indian Ocean.

In some places, the back reefs bear extensive meadows of *Halimeda*, whose small discoid plates, released after death, provide an important part of the carbonate sedimentation. This is especially true in the northern region of the Great Barrier Reef of Australia, where these meadows cover at least 2000 km^2. *Halimeda* platelets are frequently extremely important in the sedimentation of other parts of reefs, for example in Brazil where they accumulate in great numbers. *Halimeda* is extensive on the inner side of the barrier reef at Guadeloupe Island in the Caribbean, at depths of up to 3-4 metres, and on the deep fore-reef of Enewetak Atoll in the Marshall Islands, where it is a major carbonate producer, as it is also on the fringing reefs of Jamaica. In other places, algae beckoning to the genus *Caulerpa* have a wide distribution, for example at Takapoto Atoll in the Taumotu Islands, where they cover the outer slope all round the atoll from 40 metres to more than 70 metres depth.

On the coastline behind a fringing reef, in the innermost parts of the boat channels where the surf is sufficiently weakened, mangroves are able to grow, usually belonging to the genera *Avicennia* or *Rhizophora* (e.g. in the Gulf of Elat or Aqaba), but including sometimes *Lagguncularia* and *Conocarpus*, as in Paraiba on the northeastern coast of Brazil.

SANDY ISLANDS: VEGETATION

Several types of sandy islands exist on reefs, or in association with them. They will be described in details but it is necessary to characterise their vegetation cover before the effects of hurricanes are considered.

Summarizing the facts, the natural vegetation is usually quite dense except just behind the shoreline. In the society and Tuamotu atolls for example, where the islands are directly exposed to surf and spry, the outermost fringe of plants consists of a salt-and surfaces, becoming taller and much thicker just behind, where they are associated with *Scaevola* and *Suriana*. Farther inland was originally and inextricable broad-leaved, mesophytic forest, several metres high, including *Tournefortia*, *Pandanus*, *Pisonia*, *Guettaria*, and other species. This forest still exists in places, but generally it has been replaced by coconut plantations, in which the undergrowth deriving from early stages in the revival of the former vegetation is periodically cleared. On similar islands in the Indian Ocean and in the West Indies, the natural vegetation was the same broad-leaved forest, but there too coconuts have been widely planted. Although the climate is very dry, the islets scattered on the Farsan Bank in the Red Sea are covered by a thick spiny growth, more than 2 metres high. In more sheltered positions, such as on islands situated in lagoons, the natural vegetation is extremely dense just behind the beach.

Beach Rock

Beach rock is commonly found on sandy cays and in beaches behind coral reefs. According to a definition adopted in a recent symposium, beach rock which occurs on beaches as a series of flagstones sloping seawards, is a sedimentary formation indurated at the mediolittoral level by a calcareous cement, consisting initially of aragonite or calcite. There are fossil beach rocks, lying now at higher or lower level. Beach rock does not exist only in association with coral islands: it occurs in many calcareous beaches around the Mediterranean, and perhaps at even higher latitudes. Numerous references on beach rock are included in Dalongeville, 1984. The origin of beach rock remains largely a matter of discussion. Russell (1962) said that cementation consisted of the deposition of calcite, but in many exposures, especially in coral areas, aragonite has been found to be the primary cement, and in other places both occur. Dalongeville and Sanlaville stated that, around the Mediterranean, beach rock was exposed on retreating shores, and is still forming within unconsolidated Holocene deposits which have appeared on the surface as a result of coastal retreat; such a view has seemed partly true but too exclusive, especially for tropical areas where coral reefs thrive. The action of ground water has been invoked, but beach rock is present in sites where there is not ground water table because of aridity, or for topographic reasons. A debated problem is the relative age of inner and outer slabs: at Aquaba in the Red Sea, the most recent slabs occur on the outer side, and at Tulear, Madagascar, the upper ones are the most recent; but the Great Bahama Bank the inner ones appear to have been cemented subsequently; at Hao in the Tuamotus, the inner slabs are less consolidated, but may not necessarily be younger. It is not yet clear whether the flagstone pattern existed as beach layering prior to the consolidation, or whether it develops during the hardening.

Where it is exposed on the shore, beach rock is currently being undercut by wave erosion, and as this washes to the underlying loose sediment, the layers are undermined, and fractured into slabs which can either slump to subtidal levels, where beach rock does not form, or be thrown by storm waves or tsunamis to supratidal levels and piled up as a mass of broken rock. Evidence for former sea-levels can only be derived from *in situ* beach rock and not from isolated fragments, even if these are very large. On the other hand, disintegrated fragments of beach rock are frequently incorporated into new beach rock layers. In a seminar held in Tahiti, some participants have warned against the use of beach rock or determine former sea-levels, while others through it may be a good indicator, especially in areas of low tidal range.

Undoubtedly there are places where beach rock is still being formed, but beach rocks have been recorded from very remote times, perhaps as early as the Proterozoic. Beach rock patterns are useful in the study of variations of sandy beach outlines, and can be used to determine the trend to changes in progress, as shown by Battistini and Guilcher in Dalongeville; (1984); Stoddart (1975), from Tongatapu; and Stoddart (1969e), from Guadalcanal in the Solomons. In northeast Brazil, parallel outcrops of Holocene beach rock, now slightly below sea-level, have formed basements of coral reefs which consequently occur in parallel rows. These unusual Brazilian structures, already described by Darwin at the end of the second edition of his book on coral reefs, are locally known as *arrecifes*, a Portuguese world which thus designates sand-stones rather than true coral reefs, although the latter are connected with the former.

Beach rock is associated with sandy deposits on coral reefs, but can also incorporate pebbles, sometimes large, and not exclusively coralline: pebbles of volcanic origin have been supplied to coral beaches in Madagascar and cemented into beach conglomerates.

Beach rock should not be confused with what has been called *cay sandstone* by Kuenen (1933) in the East Indies. Both can occur on coral reef islands, as on Belize in the Caribbean and at Diego Garcia Atoll in the Indian Ocean, but cay sandstone consists of horizontal strata, while beach rock is characterized by seaward dipping layers. Like beach rock, cay sandstone occurs outside the regions of coral reefs, for example in the sandy Tiger Peninsula in South Angola where coral growth in impossible because of the Benguela cold current. It seems to result from vertical percolation of freshwater through the beach sand during heavy occasional rains, causing solution and then deposition of carbonates.

Calcium Carbonate Budgets

Such budgets have been established for particular reefs. Discussion and results of studies of annual energy budgets and calcification mechanisms may be found in Gladfelter (1985). An example was based on a study of a fringing reef on the west coast of Barbados in the Caribbean by Scoffin *et al.* (1980). On this reef, measuring 7 hectares, the overall balance is favourable, with a calcium carbonate production of 206 tons/year, against a destruction of 123 tons/year. In this case algae, including *Halimeda*, play only a minor part in carbonate production, while elsewhere they have a more important role. Corals provide more than one half of the carbonate sands (43 to 93 per cent), followed by Corallinaceae, cryptocrystalline particles, molluscs, foraminifera, and echinoderms. Terrigenous elements on this reef never exceeded 3.3 per cent, but they are more abundant on many fringing reefs, as in Kenya. Mechanical erosion is very small, since hurricanes are rare in Barbados, but again this factor is quite different on reefs in other areas. Bioerosion, on the contrary, is important in that it controls the grain size, the shape and the texture of the particles. Sponges are the most efficient boring organizms, but *Cliona* spp., molluscs, sipuncules, polychaetes, and crustaceans also contribute. As a whole, erosion by boring amounts to 25 tons/year. Grazing by fishes is here quite negligible: 0.4 ton/year, but elsewhere many observers, including Darwin, have stressed this kind of erosion. Conversely, urchins are extremely active especially *Diadema antillarum* which destroys 97.5 tons/year of epilithic algae.

Another example of active boring by echinoids has been noted at Enewetak in the Marshall Islands, where they accomplish bioerosion of 80 g/m^2/year on corals in the lagoon and 325 g/m^2/year on the outer reef flats. This author considers, like Scoffin *et al.* (1980), that the sponges are probably the main bioeroding agents on tropical coral reefs, but it is necessary to stress the variability of these budgets from one reef to another, and regionally.

At Lizard Island on the Great Barrier Reef, Kiene (1985) found that the rate of bioerosion was markedly different on the reef flat, where it averaged 0.41 kg/m^2/year, on the lagoon floor where it averaged 1.95 kg/m^2/year, and on the reef slope where it averaged

2.71 kg/m^2/year. There were variations from year to year, but the differences between the three sites remained: in other words, the slope is always being more rapidly eroded. Kiene found also that fishes were important in grazing, in addition to boring by molluscs, sponges and worms.

Randa (1974) has reviewed the effects of fishes on coral reefs in many different areas. Plectognath fishes (triggerfish, filefish, puffer) have powerful jaws and specialized dentition, which make them quite able to bite off corals. The polyps appear to regenerate, so that the food resource for these fish is not destroyed. Though parrotfish has been cited as feeding extensively on live coral, most studies have shown that it lives mainly on algae which it scrapes from dead coral rock, cutting into the surface as it does so. The amount of sediment produced is enormous. The surge on fish also contributes to sedimentation by feeding in part on coralline Algae, and ingesting inorganic sediment with its algal food. Fishes that eat coral are among the most highly evolved of modern teleost fishes, a features which suggests that this mode of feeding has developed relatively recently in geologic time. In Hawaii, the grazing of coralline algae by parrotfish and surgeon fish is so intense that it changes the algal succession by allowing only certain forms to persist, while the succession in other areas follows a relatively undisturbed sequence.

Calcium carbonate production in reef environments does not decrease steadily with increasing latitude, as might be supposed. At Houtman's Abrolhos in Western Australia (29°S), average carbonate production amounts to 32 mg/m^2/day, with seasonal fluctuations, a figure not much lower than in equatorial areas.

Bioerosion in Emerged Reefs

When reefs have emerged to form raised fringing reefs, barriers or atolls, the living reef backs against a cliff cut into the dead reef. Such situations exist in a large number of places including Kenya, Papua new Guinea, Vanuatu, the Solomon Islands, and the Farsan Bank in Saudi Arabia. In these cases, it is common to find the old coral reef cut into by cliffs with deep notches, sometimes with more than one notch in the same cliff, with spectacular over changes or visors. The surface of the rock is always extremely jagged, as are also the residual, sharp pinnacles (the so-called feos) of old corals, as seen on Rangiroa and nearby atolls in the Tuamotu Islands. The nature of this corrosion has been investigated by Hodgkin (1970) in the emerged coral reefs of the Langkava Islands and Samporia Bay in Malaysia, where deep notches and residual mushroom stacks are common. Hodgkin has shown that these erosional features are due to the action of endolithic algae, which are forced to penetrate deeper into the rock as they are rasped by chitons and other browsing molluscs, the boring sponge *Cliona*, the bivalve mollusc *Lithophaga*, and other boring organisms. In other worlds, the intertidal and supratidal erosion leading to such notches in mainly *bioerosion*, which is by no means restricted to coral areas since it has been also studied in detail on the calcareous shores of the Mediterranean Sea. However, the studies carried out by Trudgill (1976) at Aldabra Atoll in the Indian Ocean have indicated, in addition to bioerosion, solution by sea-water, a process that has not been generally acknowledged in recent times.

Infestations of Reefs: Acanthaster Planci

This 'plague' of coral reefs is a starfish, also called the Crown of thorns, which kills corals as it feeds on them. It is now widespread throughout the Indo-Pacific region, including the Red Sea, but has not been recorded from the Caribbean. It has been known since the eighteenth century, when it was named by Linnaeus, and its habits have been long documented. Although it is reported as common in a number of scattered areas, such as Christmas Island, the Solomon Islands, Java, and known to have various impacts on corals in many parts of the South West pacific, its predations have drawn special attention to the Great Barrier Reef since the nineteen-sixties, when reefs near Cairns (lat. 17°S) became infested. Subsequently *Acanthaster planci* was reported from other areas between Townsville (19°30'S) and Cooktown (15°30'), and, in spite of apparent recoveries, renewed outbreaks have occurred in the central region of the Great Barrier Reef, and caused until 1984 heavy destruction of massive as well as branching corals (especially Endean and Cameron in Bradbury *et al.*, 1985).

The recoveries have been explained by the fact that, even in areas which appear devastated on superficial examination, a significant and diverse stock of surviving corals remains. There is, however, little viable *Acropora* remaining, so that this genus is threatened.

Several hypotheses have been put forward to explain the outbreaks of the Crown of Thorns on the Great Barrier Reef. According to Endean and Cameron, the most likely explanation is that humans have over collected its predators, among which the giant triton shell *Charonia tritonis*, very much sought by tourists. The decrease of tritons would thus have allowed multiplication of *Acanthaster* in heavily visited areas.

The same explanation has been suggested in Mauritius. If this is correct, it is one of the impacts of man on coral reefs, which will be examined more thoroughly in Chapter five. However, other authors, such as Frankel (1977), have expressed the opinion that the present infestations on the Great Barrier Reef are not abnormal, but are a phenomenon inherent to the ecology of the reef system.

IMPACT OF HURRICANES AND EARTHQUAKES

Hurricanes or typhoons (the second word being used in the China Seas) are extremely important because of the damage made in reefs by the winds and heavy seas which they generate. They are practically unknown in the equatorial one, and thus do not affect the reefs of Malaysia, Indonesia, Papua new Guinea, the Solomon Islands, or Kiribati the Phoenix Islands, the Line Islands; but Tuvalu (formerly Ellice) and the Marshall Islands are subject to hurricanes; in the Indian Ocean, Kenya, the Maldives and Northwest Australia are exempt, as also the Red Sea and Northeast Brazil. They strike the tropical areas *sensu stricto* in both hemispheres, especially on the west side of the oceans. In the Taumotu and Society Islands, Southeast Pacific, they are quite rare, but can occasionally be exceedingly strong, such as the six which occurred in five months during the southern summer 1982-83, thought to be related to an exceptional oscillation of the Niño warm current, usually restricted to Ecuador and Peru coasts from December to April. Severe storms, not true hurricanes, damage reefs in Hawaii.

Hurricanes include winds in excess of 120 kilometres per hour, with occasional gusts of over 300 kilometres per hour. Such winds produce extreme wave conditions, with a maximum height of 23.6 metres recorded during Hurricane Camille in the Caribbean in 1969, and associated storm surges to 5 metres above predicted tidal levels.

Although the tsunamis, or waves generated by earthquakes and travelling over large distances across the oceans are not often mentioned as a cause of destructions on reefs, they seem to be responsible for a part of the damage in the Tuamotu Islands.

The most extensive damage to corals is observed on the outer edges of reefs, which receive the direct impact of the waves. According to Laboute in Harmelin-Vivien and Stoddart (1985), the worst damage in Tikehau and Takapoto Atolls, Tuamotus, during the 1982-83 hurricanes occurred under water on the slopes where, on the exposed sides, avalanches of blocks destroyed from 50 to 100 per cent of the living communities down to 40 metres at least. The main destruction were on the steepest slopes, the avalanches being not so dramatic where the slopes were more gentle. On Rangiroa Atoll in the Tuamotus, Bourrouilh-Le Jan and Talandier (1985a) have shown reef-edge fractures, with radii of curvature measuring 100 to 200 metres, from which blocks were detached. Blocks which were thrown up the reef flat were up to 20 metres long, their sizes decreasing from the outer edge towards the lagoon where blocks still measuring 2 metres were precipitated. Most of the blocks visible in 1985 on Rangiroa derived from the 1983 hurricanes, but this atoll has previously been struck by hurricanes in 1906, 1878 and earlier, and it is a place where tsunamis, recorded in old (sixteenth century?) Native traditions, could have played a part. In the Caribbean, the observations made in Belize after Hurricane Hattie in 1961 indicate that 'immediately north of the storm track, up to 80 per cent of the reef corals disappeared on the barrier reef, presumably transported by wave action into deeper water'. In contrast with the Tuamotus, there was very little scattering of large blocks on the reef flats. In all these Polynesian and Caribbean sites, beach rock has been remarkably successful in resisting erosion.

The recovery of coral growth on a slope after a hurricane varies according to sites. Two years after hurricane damage in the Tuamotus, *Pocillopora*, *Acropora* and *Porites* were among the species that had recolonized large areas down to 15 metres, and *Acropora* colonies of up to 60 centimetres in size were observed at a depth of 20 metres; but it will probably take several decades before the vigour of these biotopes recovers completely. At Funafuti Atoll in Tuvalu, observations made 8½ months after Hurricane Bebe indicated that, beside regenerating segments of the few surviving massive coral colonies, young branched corals (*Acropora*, *pocillopora*) were much more numerous, and were expected to have a decisive influence on the future zonation of the reef edge: this fits more or less what has been observed in the Tuamotus, but it is necessary to mõnitor changes that occur over a longer period. In Belize, recovery from 1961 to 1965 was slight: recolonization was limited to a few scattered *Acropora palmata*; *A. cervicornis*, formerly widespread, was still rare, and sponges, and gorgonians had made a more rapid recovery than corals.

As a consequence of erosion, accretion of elements smaller than huge blocks occurs during hurricanes behind the reef edge. At Funafuti, Hurricane Bebe built on the southeastern side of the atoll a rubble rampart 19 kilometres long and up to 4 metres high, with a calculated volume of 1.4×10^6 m^3 and a mass of 2.8×10^6 tons. Before the building of this storm beach, a 'hurricane bank' had been formed behind, in contact with the shore of the vegetated island. As its name implies, it is attributed to former hurricane(s), but it was displaced by Hurricane Bebe in some areas. In Rangiroa, the former beaches retreated lagoon wards, but not far since they were anchored on old beach rock. The 1958 hurricane at Jaluit Atoll, Marshall Islands built gravel tracts, shore ridges or ramparts, and gravel sheets or blanket deposits, but these were much smaller than the huge rampart built at Funafuti.

In Belize, a number of sand cays disappeared completely during Hurricane Hattie, 1961. These cays were generally vegetated, most were long-lying, all but one were predominately sandy, and all were small. In many areas that coral sandy cays are often swept by hurricanes, another example being off Tuo in the New Caledonia barrier reef, were Ain Cay, which was wooded, was destroyed during a cyclone in 1932. In such cases, remnants of beach rock may persist to commemorate the outlines of the lost cay.

Even where the islands are not destroyed, their vegetation is damaged to various degrees. There are large differences between islands which have kept their natural cover, and islands with coconut plantations. In Belize, Stoddart (1964b, 1971a) found that on islands with natural littoral thicket, or with coconut keeping a dense undergrowth, Hurricane Hattie was unable to destroy more than an outer fringe of vegetation close to the coast made a remarkable recovery, but tree growth was slower. Nevertheless in the 1962-65 period newly planted coconuts had grown to heights of 1-3 metres. More rapid revival was observed in the Tuamotus: at Mataiva Atoll for example, two years after the 1983 cyclones. It is certain that islands with vegetation left in natural state resist hurricane damage better than those with plantations; but as coconut trees are the only possible commercial cultivation on atolls, they are necessary on many of the islands.

From observations made so far after hurricanes, it seems that the recovery of the destroyed coral communities is diverse, and that, in many cases, one kind of system can be replaced by another. For example, as Stoddart pointed out, 'in the absence of herbivores (which have disappeared in a hurricane) a coral-dominated system may be replaced by an algae-dominated system, for an indeterminate period'. A hurricane may thus be the origin of a discontinuity inn reef genesis.

Damage on coral reefs by earthquake was described near Madang, northeast New Guinea. Resulting from an earthquake of force 7.1 on the Richter scale, the damage varied considerably with species and growth form. It also varied over short distances, from less than 10 per cent to more than 90 per cent over 2 to 3 kilometres, and was greatest on reef slopes, especially in almost enclosed basins. It was generally less than that associated to major hurricanes, but, considering the type of destruction, it may be that recovery is more delayed. 'Such events must have a considerable influence on the structure and composition of reef communities ... Earthquakes may favour rapidly growing branching species, forming colonies which are repeatedly destroyed but which rapidly regenerate'. Countries such as North Melanesia which are among the most seismically active in the world, are not struck by hurricanes, so that these two different kinds of damage are, partly at least, distributed over two different kinds of areas.

It remains that other coral communities, such as those living in the Chagos, the Phoenix, or Brazil, escape the two types of events. So that, according to a remark made by Stoddart, a distinction should be made, more than usually, between reefs which are repeatedly devastated, and those which has escaped such disasters. Would reefs such as those existing in the Tuamotus, the Caribbean or New Guinea differ markedly in their physical characteristics, population structures, and biotic networks, from reefs outside the hurricane and earthquake zones? At first sight, the differences between two such classes are not evident, but a programme of investigations might reveal in the former more permanent geomorphological features than the short-term damage so far described on reef surfaces and bordering slopes.

Coral and Coral Reefs

Corals are a group of marine coelenterates which are mostly colonial but some are solitary and occur only in polyp stages. Most (true corals) of them belong to the order Madreporaria but some others belong to subclass Octocorallia.

Besides a group of Hydrozoa, the Millipora and its allies also are called corals due to their skeletal structure but they have no real similarity with true corals. Since the true corals are the builders of coral reefs and island so only the Madreporariana corals are considered here.

Definition of Coral Reef

Vaughan (1917) has defined a coral reef. "A coral reef is a ridge or mound of limestone, the upper surface of which is near the surface of the sea and which is formed of calcium carbonate by the action of organism chiefly corals". *Bayer* and *Owre* (1969) mention, "The uppermost layer is a living stratum composed of growing corals, alcyonarians and millepores, together with a virtually endless array of other organisms that live on or in this frame work, some acting to break down the coral skeletons, others serving to cement the resulting debris together into a conglomerated mass that can act as a foundation for further coral growth. Boring sponges molluscs, worms, and barnacles permeate the coral substance and open it to erosional forces that weaken it until it crumbles. Algae, sponges, hydroids, tunicates and other organisms, as well as chemical process, consolidate the fragments and provide a platform for new growth.

Requirements and Distribution

The reef-building corals require warm, clear, shallow water. Therefore, they are confined to continental and island shores in tropical regions (latitude 28°N-28°S). They flourish best at temperature between 22°C-28°C. As temperature falls with an increase in the depth of sea water, the reef building corals occur up to depth of 30 metres, or at the most 50 metre. They fail to grow in shaded areas, which may otherwise be suitable They inhabit water subject to strong wave action, because they cannot remove large amounts of sediment likely to accumulate on them in quiet waters. Excessive rains and fresh water are harmful for corals.

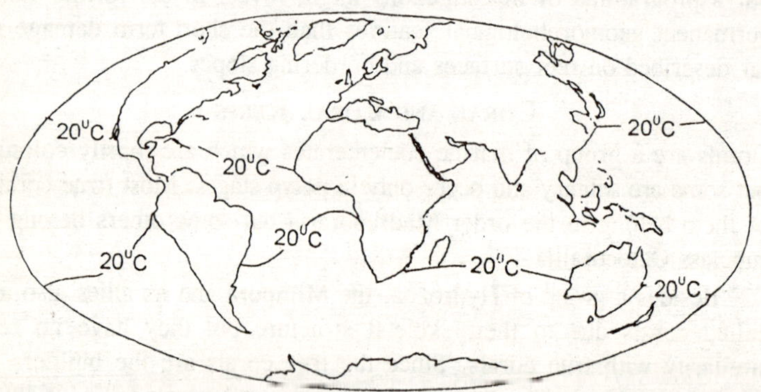

Fig. 12.3. Distribution of coral reefs today.

The reef building corals are found in to general regions: (i) the *Caribbean waters* including Florida, Bermuda, the Bahamas, and the West Indies; and (ii) the Indo-Pacific water from the east coast of Africa through the Indian Ocean and the Western Pacific as for as Hawaii. The second region specially abounds in coral reefs, and in fact the Pacific north-east of Australia is known as the coral sea.

Classification of Corals

The Madriporarian corals are divided into three groups:

1. *Imperforate* or *Aporous Corals*. They have a complete theta, compact sclerosepta and partitioned into loculae e.g., Astraeid corals, *Favia*, *Flebellum*, *Meandra* (= *Meandrina*, brain corals).
2. *Perforate corals*. In these corals the Corallum is extremely porous everywhere and is of loose construction e.g. *Porites*, *Acropora* (=*Madrepra*), *Montipora*.
3. *Fungid Corals*. They may be either perforate or imperforate e.g., *Fungia*.

Corals can also be classified as follows:

(a) *Hydrozoan corals*. Some of the animals such as *Millepora*, *Stylaster* etc. belonging to the order Hydrocorallina are colonial and are surrounded by calcareous exoskeleton. The skeleton is secreted by ectoderm. The individual has two types of polyps namely gastrozooids and branched dactylozooids loged within the exoskeleton. The dactylozooids are arranged around the central gastrozooid. They help in the formation of coral-reefs.

(b) *Octocorallian corals* include soft corals. The coral is formed of a colony of polyps with endoskeleton of separate calcareous spicules embedded in the massive mesoglaea. In the colonial coral, *Tubiprora* or organ-pipe coral, the skeleton is made of calcareous spicules consisting of vertical tubes connected together by lateral platforms. The vertical tubes are also partitioned by smaller cross plates. The tubes contain polyps. In *Haliopora* or blue coral, the calcareous spicules form a massive skeleton of corallium. In *Gorgonia* or sea fan, the colony branches in one plane and the axial skeleton is made up of horny material intermixed with calcareous spicules arranged around the polyps.

(c) *Hexacorallian corals*. These constitute the *stony corals* or *true corals*. They may be solitary or colonial and assume a great variety of forms. They are the main constituents of coral reefs.

Solitary Corals

Fungia, *Flabellum*, *Caryophylla*, etc., are the solitary corals or cup corals. The corallite is disc-like, cup-like or mushroom-shaped in form and measures 5 mm. to 25 cm. across. It is often without a theta.

Colonial Corals

Majority of the stony corals is colonial with plate-like, spherical cup-like or vase-shaped skeleton (corallium) The typical examples of colonial corals are *Acropora*, *Oculina*, *Favia*, *Madrepora*, *Meandrina*, etc. The colony is produced by asexual methods from a single sexually produced polyp. The polyps live at the surface of the calcareous skeleton. Depending on the various methods of asexual budding, varying forms of colonies are produced. Some of the colonies are branched. In *Acropora*, there is always a primary polyp at the top of the colony with lateral branches on either side. In some corals, like *Oculina*, the polyps remain widely separated, each occupying a separate theca. In others, like *Favia* and *Astraea*, the thecae are so close together as to have common walls. In the brain-coral, *Meandrina*, the polyps as well as the thecae become confluent, occupying valleys separated by ridges, on the surface of the corallium.

Abode of Animals

Coral reefs are the abode of several other animals as there are numerous interstices, cervices and other hiding places together with sheltering branches of the corals themselves. As such sponges, anemones, sea-urchins, starfish, crabs, tubicolous-annelids, holothurians, snails and bivalves all live in coral reefs.

Magnificent Colouration

Hyman (1940) writes that as coral beds consist of multitude of organisms of varied shapes and colours, viewed through the deep blue water of a lagoon, constitute one of the most beautiful sights in the world, excel the most splendid flower gardens. The coral polyps are yellow, green or brown due to their zooxanthellae, but almost any hue can be seen in them. The skeletons of corals are usually white but as they are sometimes permeated with red and green algae whose colours they take.

Coral Polyps

The coral organism is a small authorzoan polyp measuring about 1 cm. in length. It lacks a pedal disc and the oral disc bears tentacles in cycle of six. The pharynx is devoid of siphonoglophs. The mesenteries follow the hexamerous plan and are restricted to the upper part of the polyp. The muscle are poorly developed.

Structure of Coral Polyp

In structure the coral polyp is much like an Anemone except the skeletal portions. The structure of the soft part is like sea anemone but pedal disc is absent. Basal region or pedal disc is occupied by skeletal cup. There is no oral cone. Mouth is situated in the middle of the oral disc. A typical oral disc with tentacles on typical

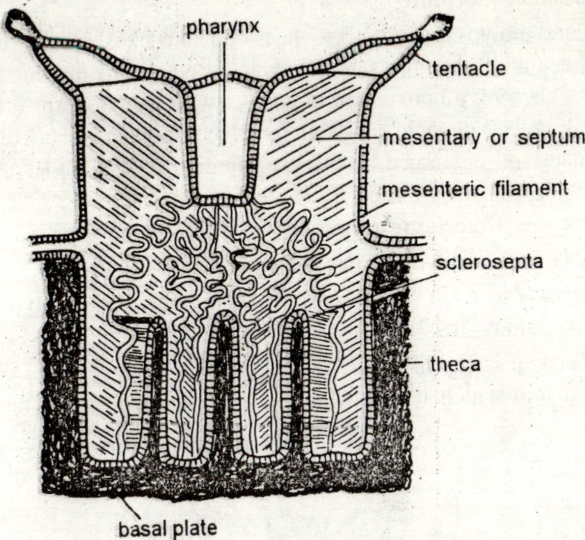

Fig. 12.4. V.S. through a coral in its theca.

hexamerous plan is present. Tentacles are simple and of moderate length ending in a terminal knob of nematocysts. A circular mouth surrounded by a flat peristomium leads into short stomodaeum or pharynx devoid of siphonoglyph. There are complete and incomplete septa or mesenteries arranged in cycles arranged on the typical hexamerous plan. Coral polyp being small does not possess more than three cycles of mesenteries. Usually two pairs of directives are present and there are ostia on the septa. The polyps of colonial corals are all interconnected but the attachment is lateral rather than aboral, as in hydroids. The column wall folds outward above the skeletal cup and connects with similar folds of adjacent polyps. Thus, all the members of the colony are connected by a horizontal sheet of tissue which represents folds body-wall and so it contains an extension of the gastrovascular cavity as well as an upper and lower layer of endoderm and ectoderm. The lower ectodermal layer secretes the part of the skeleton that is located between the cups in which the polyps lie. The living coral colony thus lies entirely above the skeleton and completely covers it.

Structure of Coral Skeleton

The skeleton of a colony is termed as *corallum* and that of coral polyp a *corallite*.

Structure of Corallite

Coral polyps remain fixed in a cup like exoskeleton, the *theca*. The theca and other parts of the skeleton wholly to the outside of the body of the polyp but remain in contact with its ectoderm throughout. According is *Voucoch* the skeleton of coral is made up of calcium carbonate and is framed by the precipitation of calcareous crystals in a colloidal matrix secreted by the ectodermal cells outside the body wall for the protection of the polyp.

A typical polyp coral has the following parts:

1. *Basal Plate*. A basal plate lies between the polyp and the substratum. It is the bottom of the cup.
2. *Theca*. It is cup like structure from which the polyp projects outside and into which it can be retracted.

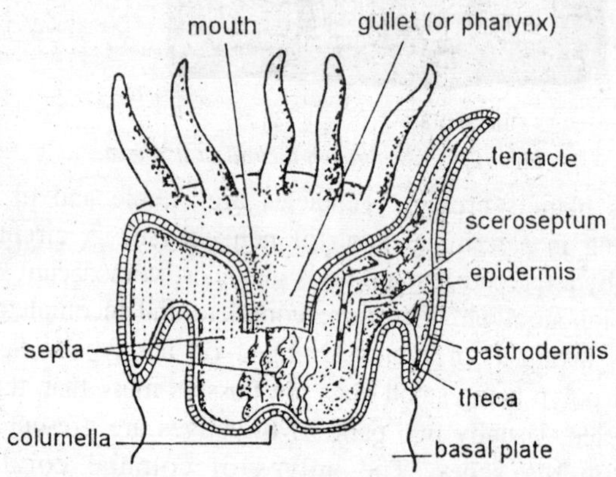

Fig. 12.5. One halt of a simple coral polyp showing relationship of soft parts of theca (corallite).

3. *Sclerosepta*. These are calcareous ridges or partitions arranged vertically or projecting radially inwards and are connected at the base by basal plate and at the side with cup like theca. They look like mesenteries but in fact lie between the mesenteries and thus lie in definite relation to them like the mesenteries, the sclero septa typically, occur in hexamerous cycles, 12 tertiaries, 24 quaternaries etc.; but others numbers are also met with. The sclero septa are usually endocoelic i.e. each skeletal ridge pushes up between the two septa of a pair. The original second cycle is usually, however, exocoelic. The formation of the sclerosepta usually precedes that of the corresponding septa (mesenteries) and thus there are often move sclerosepta than mesentery pairs. The

first cycle of sclero septa reach the columella, while the later one fall short of it and may fuse with adjacent larger sclerosepta. It may be emphasized that the corallite lies entirely outside the body of the polyp and the polyp base is pushed into ridges over the sclerosepta and goes down into blind pockets between sclerosepta and mesentery.

The sclerosepta are commonly sping thorny with jagged or toothed upper edges.

4. *Columella.* It is a pillar like irregular central skeletal mass which may be either an independent outgrowth from the basal plate or one formed by the union of the central ends of the sclerosepta, then called *pseudocolumella*. The columella may be solid or trabeculae.
5. *Epitheca.* It is a distinct calcareous layer which surrounds the base of the theca in a ring-like manner.
6. *Costae.* The space between the theca and epitheca is crossed by continuation of the sclerosepta called costae.
7. *Pali.* Small ridges between the columella and the main parts of the sclerosepta are termed as Pali.

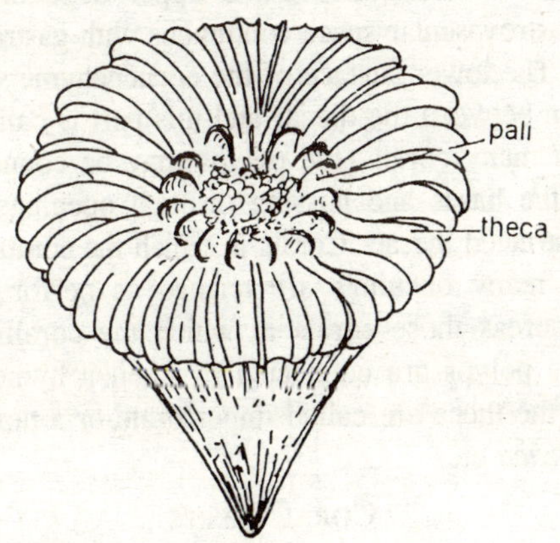

Fig. 12.6. Theca of a solitary coral showing pali.

8. *Synapticula.* These are the .skeletal bars connecting adjacent sclerosepta.
9. *Dissepiments.* Horizontal plates between sclerosepta are known as disepiments. These are of small extent.
10. *Trabeculae.* When the horizontal plates between sclerosepta are large and extend completely the corallite they are termed as trabeculae.

Formation of Coral

The coral polyp develops from a planula which settles down and begins to secrete a skeletal rudiment or *prototheca*. It is secreted by ectoderm first as a basal plate. Following it, the larva develops radial folds which secrete septa (sclerosepta) and at the same time a rim is built up as a thecal wall around the polyp, laying at the top. Meanwhile, further skeletal material is added into the gaps between the septa. The septa of the skeleton usually alternate with the mesenteries of a living coelenterate.

In living condition, the polyp fills the whole of the interior of the corallite and projects beyond its edge. The proximal portion of its body-wall is in contact with the theca which is a product of the epidermis. The free part of the body-wall of polyp is folded over the edge of the theca so as to cover its distal portion. Each skeletal septum is covered by an inturned portion of the body-wall. Thus the septa are actually external and are in contact with the epidermis throughout. The space between the theca of the coral colonies is occupied in life by an extension of the polyp walls, *coenenchyme* continuous with the latter above the upper edge of the theca and containing a gastrovascular space continuous with gastrovascular cavity of the polyps. The lower surface of the coenenchyme secretes the part of the corallum between the theca, and this part is called *coenosteum*. In addition, in many corals, the polyps may be connected by canals coming from the bases and passing through openings in the loosely connected constructed thecae. Corals in which the corallite is perforated like this with many openings are termed as perforate corals e.g., *Madrepora* whereas those corals in which the corallite are of solid texture and the polyps are connected by coenenchyme only over the upper edge of the theca are called imperforate or a porous corals e.g. *Flabellum*, *Astraea* etc.

Coral Reefs

A coral-reef is a ridge or mound of lime stone the upper surface of which is near the surface of the sea and which is formed of calcium carbonate by the action of organisms chiefly corals (*Vaughan*, 1917). Though the reefs are built by stony corals but other organisms such as foraminifera, millipora, tubipores, heliopores, the molluscs, echinoderms, some coralline, algae and sponges contequeute together in the formation of compact structure, the coral reefs. The coral reefs are formed by incrusting their skeletal parts on the deposited lime. Coral reefs composed of multiple of organisms vary in shape and colour. The

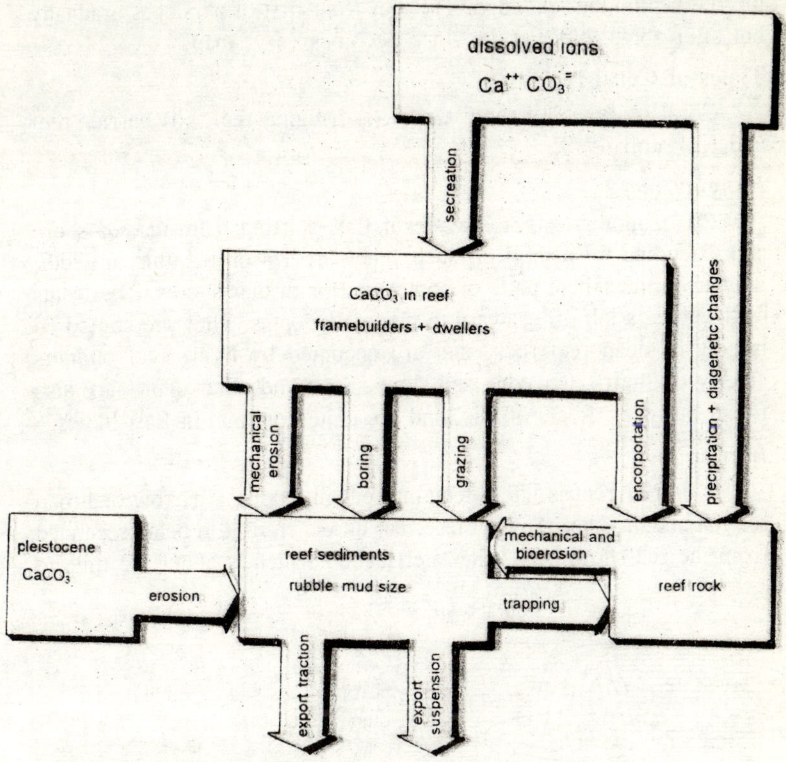

Fig. 12.7. Fate of calcium carbonate on a coral reef.

Zooxanthelae is responsible for the rich colouration of corals, which may be brown, yellow, or green. The reef building corals require warm, shallow waters, and consequently are limited to continental island shores in tropical and sub-tropical zones. They cannot tolerate temperatures below 18°C and they flourish nicely only above 22°C. Consequently their distribution is limited to the zone existing between about 28°C on either side of the equator. They rarely remain alive at a depth greater than 4 fathoms.

However, none have of the reef types are like man made continues wall but is broken up into many reefs and islands by passages, the larger of which may be drowned valley, sunk below the sea by land subsequence or rise in sea level. The lagoons usually contain inner island, or island, and reefs, etc. The flats are more or less exposed at the lowest tides. The reef front is exposed to forceful waves which knock off big and small fragments of the reef and heaves them up on the flat behind the reef edge. The reef front first of all slopes gradually but after about 200 feet the slope becomes very steep.

Types of Coral reefs

Coral reefs are of three types: (i) fringing reef, (ii) barrier reef and (iii) atoll

Fringing reef

The fringing reefs are sea-level flats starting from the sea-shore and extending for a short distance. They are 1/4 or 1/2 mile in width, built upon the salient parts of continental or insular shores. The fronts of these reefs fall off seaward to moderate depths. They are composed largely of dead reef rock, and are occupied by living reef builders chiefly on their outer edge and slope. Sand and other debris are also found on reefs. Reefs of this kind are quite common in East-Indies.

Barrier reef

A barrier reef is like a fringing reef in having a narrow or broad sea-flat and an outer growing face; but differs from it in being separated from the sea shore by a salt water lagoon which is about 1/2 mile to

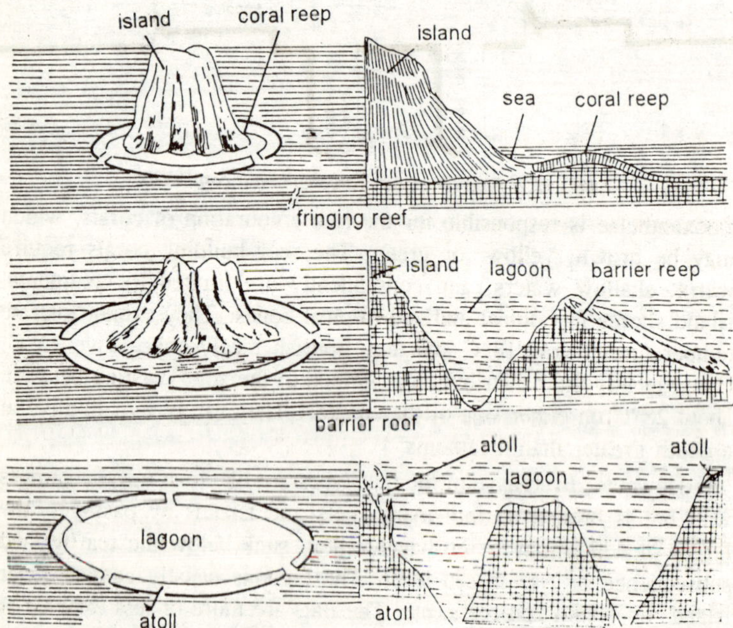

Fig. 12.8. Diagrams to represents different types of reefs, entire as well as in cross-section.

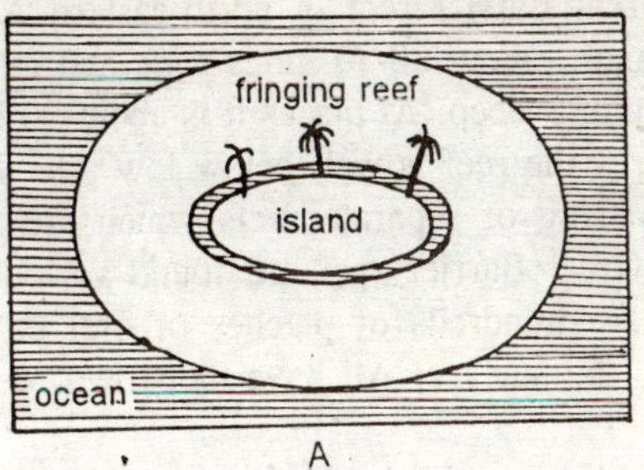

A

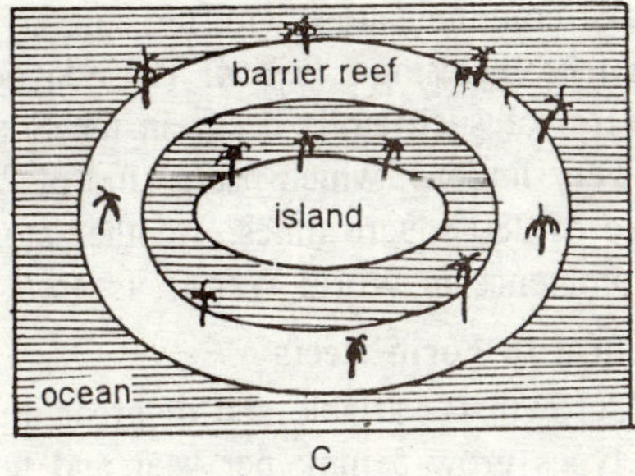

C

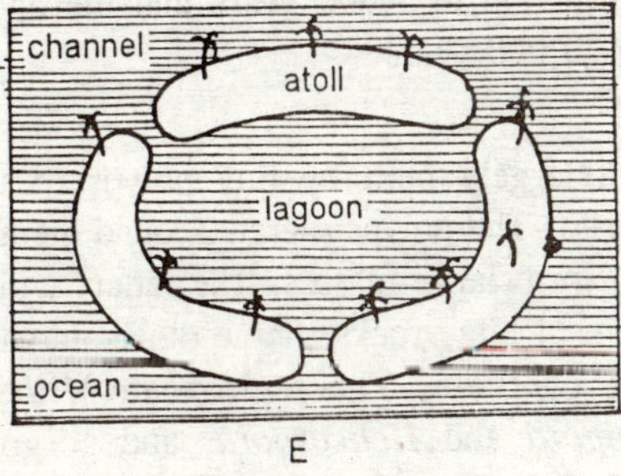

E

Fig. 12.9. Coral-reefs. A—Fringing reef. C—Barrier reef. E—Barrier reef in section.

scores of miles in width. The lagoon may be 20 to 40 fathoms or even more in depth. The inner shore is generally occupied by a fringing reef, the growth of which is less vigorous than that of the non-enclosed fringes. The outer or growing face of the barrier is continued in gentle slop 40 to 50 fathoms in depth follow a steep pitch to great depths.

Barrier reefs are frequently interrupted by "*passes*" or *passages* through which even the ships may enter. Barrier reefs may encircle whole islands. Sometimes small islands may make their appearance in the protected lagoons. These reefs are sometimes a great danger to shipping. The Great Barrier reef of north-eastern Australia is about 1350 miles in length, about 20-70 miles *wide* and *encloses* a channel from 10 to 25 fathoms deep. At places it is about 90 miles away from the shore. Much of the reef is well below low tide. It is not a single reef but a long string of separate reefs which are not always in a line. Inside the Great Barrier reef are found various types of coral growths. There are hundreds of patches of oral or elongated coral islands built upto the surface. All have their own reefs.

Atoll

It is a coral island and consists of a belt of coral reef having a central shallow lake communicating with the sea. It is a horse shoe-shaped or circular reef enclosing a *lagoon*. These lagoons are 40 or 50 miles across. Several of such atolls occur in the South Pacific. The *Atoll of Biknii* is very famous, which has a land of 287 square miles and a shallow lake of 280 square miles. Another atoll which is very famous due to prominence in World War II, is *Atoll of Tarawa*.

Growth of the Coral to Form Reefs

Rate of coral growth is variable and of great importance. Slow-growing massive types grow 5 mm. per year and fast-growing ones from 10 to 20 cm. per year. *Vaughan* estimated that a reef of 50 mm. deep is formed in 1,000 to 7,000 years and the age of the present reefs is 10,000 to 30,000 years.

Formation

The reefs are largely built by tiny polyps of the stony corals (order Madreporaria). The polyps secrete around them limestone cups, which coalesce to form large masses. The latter, with the passage of time, take the form of huge rocks. Some other animals also take part in the formation of reefs. The hydrozoan *Millepora*, calcified alcyonarians *Tubipora* and *Helicorpora*, and gorgonians add their skeletons to the reefs. The clams and tubeworms contribute their calcareous shell and tubes to the reefs. Lime secreting algae repair the windward edge of the reef by cementing loose pieces of coral rock into nearly solid ramparts. Sinking shells of foraminiferans fill the tiniest porse in the reefs. Whereas some organisms build up the reefs, others, like burrowing animals, break them down. This turns the reefs into labyrinths of ledges, crannies, grottoes crevices and tunnels.

The coral reefs grow very slowly. Most of them expand at the rate of 10-200 mm. per year. The existing reefs seem to have been formed in 15,000 to 30,000 years.

Theories to Explain Great Vertical Thickness of Coral Reef

Since the present reef building corals do not grow below 150 feet at the outside, and because geological evidence indicates that corals of the past ages were also littoral, in their habits, it becomes essential to explain the great vertical thickness often attained by the coral reefs. Many theories have been propounded, which the following four are the main ones:

Darwin's subsidence theory

Charles Darwin's theory explains the formation of reef. According to him the reefs begin as fringe around slowly sinking shores, which continue to grow upward and outward as the land sinks. Thus the fringing reef turns into a barrier reef when coast subsidies. The lagoon separating it from the reef becomes wider and wider. Islands surrounded by barrier reefs finally sink beneath the lagoon. Thus the encircling reef is left on which islands of wave tossed. Atolls are formed as accumulation of loose fragments of rocks.

Semper-murray solution theory

Sir John Murray was the chief biologist on the 'Challenger', the British ship that sailed into the oceans from 1874 to 1876 to explore extensively the conditions and life in the sea. *Murray* proposed that (i) Corals grow on high summits of the ocean bottom when these have been built up to the right level, (ii) the high summits are built by deposition of sediments and (iii) barrier reefs and atolls are formed due to better growth of corals at the reef edge and through solution of the inner coral rock. This theory is now not accepted at all.

Submerged bank theory

This theory has been supported by many recent studies. According to this theory the coral formation grow on flat, preexisting surface during or after the submergence of surfaces.

Daly-glacial-control theory

The main points of this theory are as follows: (i) During the last glacial period much water of the ocean turned into ice forming glaciers due to a very low temperature and thus the level of the ocean was lowered by 60 to 70 metres below the present level. (ii) Various terraces were then cut or islands levelled by wave action. (iii) Later, with the rising, temperatures, corals began to grow upon these platforms and kept pace with the rising sea-level as the ice melted.

This theory explains nicely about the uniform depth of coral lagoons, whose bottoms, below the debris since deposited, would consist of the platforms cut when the ocean was at its low level. Theories 3 and 4 supplement each other and at present are most favoured by the students of the problems, although Darwin's idea (Darwin-Dana subsidence theory) continues to find much support. The submergence theory agrees with Darwin's subsidence theory in that both consider the reef foundations to be now at great depths than they were when the coral growth started. But the submergence theory does not admit any relationship between that various kinds of reefs and postulates that the barrier reef and atoll have grown upon pre-existing flat platforms. The atolls are considered to have been shaped by winds, waves and currents.

Boring have been made to find out the truth of these theories.

(i) Boring of Funafuti atoll in the South Pacific north of Fiji was made in 1904 by an expedition of the Royal Society of London. The boring was 3 inches to 5 inches in diameter and went up to 1114 feet without reaching the reef base. Twenty eight genera of reef building corals were discovered, 2 of which are now living on the reef in that locality above 60 metres. The material obtained from the boring did not contain any of the deep water corals lives in that locality at the depth to which the boring went. This finding supports subsidence theory.

(ii) *Cary* (1931) made 3 boring at different distances from the shore into a reef a Samoa (Pacific island) and concluded that the reef rested on a level platform cut by the action of waves. This supports the glacial control theory.

(iii) The Great Barrier Reef Committee made two boring, one in 1928 and the other in 1938 on the Great Barrier Reef. Both boring gave the same result that the coral material extended out only to 400 to 450 feet and below this there was nothing but shore sand containing shells of various animals. There was no evidence of any underlying platform. Therefore this finding also supports Darwin's subsidence theory. Thus Darwin's subsidence theory applies of many reefs but some reefs may have been laid down on pre-existing platforms.

Some Important Facts about Reefs

Tropical storms can modify the reefs very much. Quarrying can also damage reefs as in India, where in 1971 *Pillai* has estimated that nearly 250 cubic metres of reefs material are removed per day for use in the production of cement, calcium carbide, calcium carbonate etc. Recently much concern has been expressed for the safety of some oceanic islands and their reef food chains because some reefs are being destroyed by ever increasing population of the Crown-of-thorns starfish.

Acanthaster planei. These starfishes feed upon the living coral and have caused much reef destruction in some parts.

Economic Importance of Coral Reefs

The coral reefs are of much importance to oil industry. They form highly favourable sites for the accumulation of petroleum deposits.

The coral reefs are of importance for *curio trade*. Many plant and animals like sponges, molluscs, fishes, echinoderms etc. grow on these reefs. Even some humans inhabit them.

Some corals form highly priced decorative pieces.

Corcllium rubrem is considered a precious stone. The red coral and organ pipe coral are used in medicine in South India. Skeleton of a few corals are used as building material.

Coral skeletons are also used in the formation of time, mortar and cement. The skeletons are also used in making ridges which act as natural barriers against sea erosion and cyclonic storms.

The coral reefs serve as good nursery grounds for commercially important fishes. They form more colourful and beautiful fishes.

13

Fossils

Corals, sea anemones, jellyfish and the small colonial hydroids are all representatives of Phylum Cnidaria. They were formerly grouped with the ctenophores, 'sea-gooseberries' or 'comb-jellies' in Phylum Coelenterata, but cnidarians and ctenophores are now regarded as separate though related phyla. Ctenophores are globular or elongated gelatinous organisms swimming by serried combs of fused cilia arranged in rows. They catch food with their long tentacles, which are armed with stinging cells. Until recently they were believed to be the only phylum with no fossil record, but a specimen found in the Lower Devonian of Hunsrückschiefer, and revealed by X-radiography, extends the known range far back in time.

Cnidarians are the simplest of all true metazoans, but they are of an evolutionary grade higher than sponges since their cells are properly organised in tissues which are normally constructed on a radial plan. The cell wall is *diploblastic*, having cells organised in two layers only. These are the outer *ectoderm* and inner *endoderm*, which have no body cavity between them but only a jelly-like structureless layer. In this *mesogloea* runs a simple nerve net. The mesogloea is sometimes invaded by cells from the two primary layers. The single body cavity (*enteron*) has only one opening, the mouth, which also serves as an anus and is normally surrounded by a ring of tentacles. It is lined by endoderm which is sometimes infolded to form radial partitions or mesenteries, increasing the area over which digestion may take place, for the primary function of the endoderm is digestive. The cells of the endoderm ingest food by means of amoeboid pseudopodia: alternatively they can extend flagella into the enteron to stir its contents.

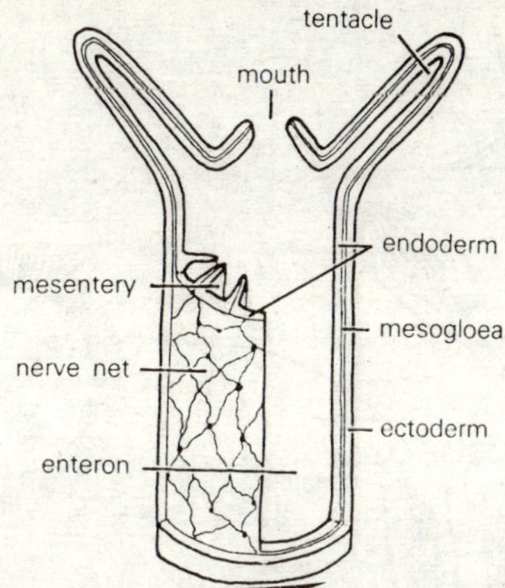

Fig. 13.1. Schematic diagram showing cnidarian hydroid cut longitudinally.

The cells of the ectoderm are more highly differentiated. Of these the principal types are the *musculo-epithelial cells*—columnar cells with contractile fibres and there are also sense cells leading to the nerve net below. In addition there are ectodermal stinging cells, the *nematocysts*. These are large cells with a sealed central cavity containing poisonous fluid, within which lies a tightly coiled elongated tubular thread carrying barbs and stylets inside it. A small sensory hair on the outside of the nematocyst, the *cnidocil*, is sensitive to vibrations in the water, and when a small organism passes close to the cnidarian it triggers the nematocyst to discharge. When this happens the central cavity is intensely compressed by strong muscle fibres and the seal breaks; the thread is shot out with great force, turning inside out as it extends so as to expose the spiral barbs, which penetrate the prey so that it is injected and killed by the poison. Batteries of nematocysts are found on the tentacles, which all discharge together. The captured prey, held fast by the threads, is conveyed to the mouth as the tentacle then bends into it. Further transport of food is by mucus strings.

Cnidaria are characterised by a life-cycle in which successive generations are of different kinds. This *alternation of generations* or *temporal polymorphism* is typical of the less specialised Cnidaria but may be suppressed entirely in the more 'advanced' kinds. Two types of individual, the fixed *polyp* and the free *medusa*, alternate successively so that the polyp gives rise asexually to medusae, which reproduce sexually so that their zygotes produce polyps, and so on. A polyp is a sedentary animal with a cylindrical body and an upward-facing mouth

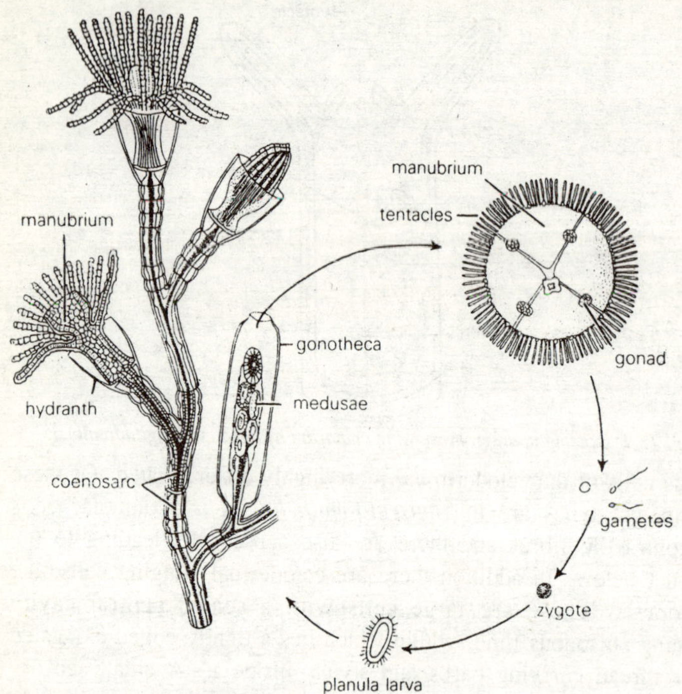

Fig. 13.2. Obelia—morphology and life-cycle.

surrounded by a ring of tentacles. The mesogloea is comparatively thin. Medusae are free-swimming inhabitants of the plankton, often very small; their orientation is inverted relative to that of polyps. The medusoid body is discoidal, with the mouth prolonged centrally into a downward-facing funnel or *manubrium*.

The mesogloea is very thick, and below it the enteron extends through four or more radial canals which join a peripheral circular canal, forming a kind of circulatory system for dissolved food material. The tentacles hang down freely beneath the outer rim, each usually carrying at its base an organ of balance or statocyst. Since medusae are the sexual as well as the dispersal phase of the cycle they carry the four gonads, one of which is located along each of the primary radial canals. When the gametes are ripe they are shed by rupture of the gonad wall, and fertilisation takes place in the sea water. The zygote then develops into a small ciliated *planula larva*. This has at first a solid core of endoderm, which eventually develops a central cavity, and settles down to become the next polyp generation.

Medusae move by rhythmic pulsation of the bell-like body through highly modified musculo-epithelial cells. They normally die after reproducing. Though alternation of generations seems to have been basic to the more generalised Cnidaria, such as most of Class Hydrozoa, it has been modified or suppressed in others. In Class Scyphozoa the medusoid phase is dominant and the polypoid phase very reduced, whereas in Class Anthozoa the medusoid has been eliminated entirely and the polypoid phase has become the sexual generation. A description of the major divisions of the Cnidaria (simplified) follows.

Major Characteristics and Classes of Phylum Cnidaria

The Cnidaria (Precam.–Rec.) are metazoans with radial or biradial symmetry. They have body walls of ectoderm and endoderm separated only by mesogloea; a single body cavity (enteron) with mouth but no separate anus; a simple nerve net; and no separate excretory or circulatory systems. Nematocysts are present. They may be solitary or colonial, and they are often polymorphic with alternate polyps and medusae. Many have calcareous or organic skeletons and they are frequently colonial. There are three classes.

Class 1. Hydrozoa (Precam.–Rec.)

Class 2. Scyphozoa (Precam.–Rec.): Most large jellyfish.

Class 3. Anthozoa (Precam.–Rec.): Corals, sea-anemones, gorgonians, sea-pens.

Class Hydrozoa

The Hydrozoa are usually polymorphic, with radial and often tetrameral symmetry. Several hydrozoan orders embrace a wide range of forms.

Order 1. Hydroida (Cam.–Rec.) Have colonial sessile polyps and planktonic medusae, e.g. *Obelia*.

Order 2. Trachylina (? Jur.–Rec.): Free medusae only, the polyp stage having been suppressed entirely.

Order 3. Siphonophora (Ord.–Rec.): Large and complex floating colonies, lacking hard parts, with some division of labour amongst the members of the colony.

Orders 4 and 5. Milleporina and Stylasterina (Tert.–Rec.): Have calcareous skeletons and are reef formers.

Order 6. Spongiomorphida (Trias.–Jur.): Form massive colonies with radial pillars united by horizontal bars, and structures resembling the astrohizae of stromatoporoids (q.v.).

Order Hydroida

Hydroids are rather generalised hydrozoans in which there is a chitinous external skeleton (*perisarc*). The order includes three suborders: Eleutheroblastina (free solitary naked forms, the pond *Hydra* being one), Calyptoblastina and Gymnoblastina. Representatives of the two latter suborders are occasionally found in fossil form. *Obelia* (Suborder Calyptoblastina), a standard example of a colonial hydroid, has a hollow root-like structure for attachment from which branching tubes arise, giving rise to polyps (*hydranths*) alternately on each side. The perisarc is annular in places, especially below the cups (*hydrothecae*) which surround each polyp. Though each polyp is an individual, the polyps are all connected together by a tubular system (*coenosarc*) through which the enteron continues. Each polyp has many tentacles and a prominent bulbous funnel or manubrium upon which the mouth is set.

Towards the base of the colony are cylindrical gonothecae: flask-shaped structure with a constricted aperture. Within each of these is a central stem arising from the coenosarc, and from this the medusae form, being budded off continually from the upper end of the stem as they mature. These escape through the aperture and thereafter swim freely and grow until they are old enough to reproduce sexually. Gonothecae are typical of the Calyptoblastina; in the Gymnoblastina the medusae form directly from the coenosarc without being enclosed, and there is no hydrotheca. Fossil chitinous hydroids are well known in the Ordovician and have been recorded from several localities.

The calcareous tubes of several Mesozoic and Tertiary serpulid tube-worms (e.g. *Parsimonia*) are often found infested by a colonial organism (*Protulophila*) which has been interpreted as a hydroid buried in the outer part of the calcareous tube. This has two components: polyp chambers, and stolonal networks which connect them in a regular scale-like pattern. The polyps grew around the rim and were probably incorporated into the tube as the serpulid grew, forming polyp chambers opening through semicircular apertures. This hydroid-serpulid association was probably symbiotic or commensal, in the same way that several species of the modern hydroid *Proboscidactyla* are symbiotic with tubular sabellid worms.

Protulophila would have been able to feed on some of the supply of food particles drawn in by the feeding arms of the worm, whereas the serpulid in turn probably derived some protection from the nematocyst batteries of the hydroid. It must have been a successful association for the degree of infestation is sometimes as high as 95 per cent and infested serpulids are found in beds ranging from the Middle Jurassic to the Pliocene, 170 million years in all.

Orders Milleporina and Stylasterina

The milleporines and stylasterines are hydrozoans which superficially resemble corals and may be quite important in some modern reefs. *Millepora* has a thick laminar calcareous skeleton of many vertical tubes with cross-partitions connected by thin horizontal ramifying canals. The soft tissues invest only the upper part of the skeleton degenerating in the older lower layers as the skeleton builds up. There are three types of vertical tube, each with its own zooid arising from the surface and capable of retracting into the tube the *gastrozooids*, which are purely manubrium-like mouths; the *dactylozooids*, which are elongated tentacles with batteries of nematocysts; and the *ampullae*, which produce medusae. All these are connected by soft tissues running through the horizontal canals. This kind of colony illustrates an effective 'division of labour' and consequent modification of the zooids. *Millepora* in its natural reef habitat is very variable in form. The millepores of Barbados (Stearn & Riding 1973) can be grouped in four forms:

(a) encrusting dead corals on gorgonians in thin sheets less than 1 cm thick.
(b) 'boxwork', i.e. forming thick near-vertical blades joined along their edges to form box-like, subangular, upward-opening cavities up to 30 cm. high.
(c) bladed, i.e. with erect blades buttressed by other blades, making colonies up to 50 cm high.
(d) branching, up to 20 cm high, with delicate or stubby branches.

It is possible that the boxwork, bladed and branching forms are distinct non-intergrading biospecies, but the encrusting form grades into the others and is probably an environmental growth form of the other species, being most common in shallow water or encrusting gorgonians. The habitat preferences of the other species seem to be governed by water turbulence, boxwork forms being the strongest and bladed the most delicate.

Class Scyphozoa

The Scyphozoa are free-swimming medusae, entirely marine, and usually with radial symmetry based on a tetramerous plan. The sessile polyp stage is called a *scyphistoma*, which buds by *strobilation*. This means that towards its upper end it is continually growing and being divided by transverse grooves, so that a pile of small medusae form one on top of the other and are released in turn to break free as *ephyra larvae*: the juveniles of the adult jellyfish. Few jellyfish are found in the fossil record. Of these, some are found as compressions in very fine sediment; others, including the many Ediacara species, are preserved as sand in fillings of the gut cavity.

Brooksella (M. Cam.) shows the original lobes of the gut, likewise preserved by sand in filling, but as these differ in form from those of modern jellyfish *Brooksella* is sometimes placed in a separate class, the Protomedusae. Some steeply pyramidal chitinophosphatic fossils, the Conulata (Cam.–Trias.) (conulariids), which have a quadrate cross section and often marked herringbone ridges down the sides, have also been considered by some authorities as belonging to Class Scyphozoa. This is based upon the reported presence of tentacles preserved below the pyramidal structure in rare instances and upon the resemblance in shape of one rather flattened genus (*Conchopeltis*) to a medusa. On the other hand, most conulariids do not look at all like jellyfish and may belong to an extinct phylum.

Class Anthozoa

The Anthozoa are polyps with no trace of a medusoid stage. Anthozoans (corals, sea-anemones, gorgonians and sea-pens) are solitary or colonial, entirely marine cnidarians. The polyps produce gametes which develop directly into planulae larve after fertilisation and thence to a new generation of polyps; the medusoid generation has been eliminated entirely. They resemble hydrozoan polyps in many respects though are often very large. They always have a tubular *gullet* or *stomodaeum* leading down into the enteron, which hydrozoan polyps do not have. The interior itself is divided by radial partitions (mesenteries) whose number and morphology is important in the subdivision of the class. Those that secrete hard parts, and especially the corals, are of great geological importance, often forming thick beds in the Palaeozoic and sometimes, in association with stromatoporoids, reef-like masses.

From Tertiary time onwards true coral reefs of vast thicknesses have formed, and corals living as reef formers are probably more important now than at any other time. Anthozoans are grouped in three subclasses. Ceriantipatharia. Octocorallia and Zoantharia. Only the last of these which includes the corals, will be considered in detail.

Subclass Ceriantipatharia

Ceriantipatharia are solitary or colonial polyps which for morphological reasons are placed apart from other groups. They are virtually unknown as fossils.

Subclass Octocorallia

Octocorallia (?Precam., Ord.–Rec.), poorly known as fossils, are especially represented by the gorgonians (Order Gorgonacea) common in many modern coral reefs. The colony forms a flat fan of anastomosing branches. The skeleton consists of horny branching tubes each of which may have a central calcified core. From the outer surface of the branching tubes projecting the many small polyps (*autozooids*), each of which has eight tentacles, characteristic of the subclass. These autozooids are set in a thick gelatinous outer coating over the branching tube, within which are set innumerable calcareous *spicules* which help to support the autozooids. These are the only part to preserve as fossils, and only from them can the former presence of gorgonians be inferred. The fossil record of octocorals is sparse, but recently organic skeletons believed to be of gorgonians have been found in rocks as early as the Lower Ordovician.

Furthermore, abundant spicules of Lower Silurian age, usually isolated, but sometimes found in nest-like aggregations, have been shown to belong to octocorals of Order Alcyonacea. These warty, spindle-shaped spicules of the genus *Atractosella*, once thought to pertain to sponges are virtually indistinguishable from those of modern alcyonaceans, hence this group has much more ancient origins than was formerly believed. *Atractosella* probably lived on a sand-gravel sea floor; it was not a reef former.

Heliopora (Order Coenothecalia) has a massive aragonitic skeleton, bright blue in colour. It is an important reef former in the Indo-Pacific province. Its similarity to the tabulate *Heliolites* has suggested that the latter is a Palaeozoic octocoral, but this resemblance is far more likely to result from convergent evolution.

The red octocoral *Tubipora*, the 'organ pipe' coral (Order Stolonifera), has polyps inhabiting long horny tubes and often forming large colonial masses. The sea-pens (Order Pennatulacea) are another kind of octocoral. Here the colony has the form of a feather, the base of which is set in mud while the feathery upper branches are lined with autozooids. The presence of *Rangea* in the late Precambrian of the Ediacara fauna, interpreted as a sea-pen, testifies to the success of the pennatulaceans in ancient times.

Subclass Zoantharia: Corals

The structure of the Zoantharia is exemplified by the Upper Jurassic to Recent solitary coral *Caryophyllia* (Order Scleractinia). Recent species are of cosmopolitan distribution, living on various substrates and at depths of 0–2750 m. *C. smithii* lives on cool, temperate waters off western Scotland. In regions of strong tidal currents it is found attached to pebbles or boulders, but in weaker currents on the outer shelf it can live in a variety of substrates; the most common form of attachment in sand patches is to calcareous tubes of the worm *Ditrupa*.

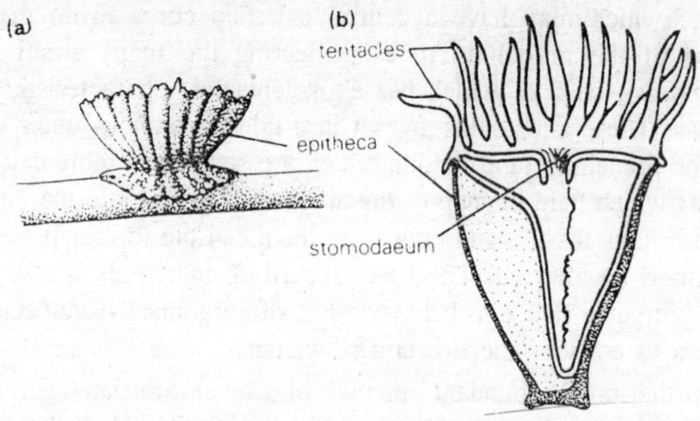

Fig. 13.3. (a) Young Caryophyllia, a Recent scleractinian coral attached to the tube of the worm Ditrupa; (b) Caryophyllia, vertical section.

The polyp of *Caryaphyllia*, which is some 2-3 cm across, is many-tentacled and has a stomodaeum leading down to the enteron, which is divided biradially by numerous mesenteries. Each of these has frilled free edges. The soft basal tissues secrete an aragonitic cup or corallum which is short and horn-shaped. It has an outer wall (*epitheca*) within which are numerous radially arranged *septa* between which lie the mesenteries. The first-formed *prosepta* are larger and more pronounced than the *metasepta* intercalated between them. When preserved as a fossil the aragonite is often altered to calcite or dissolved completely leaving a negative mould in the matrix. Aragonitic skeletons are known as far back as the Triassic, but not all Scleractinia are aragonitic.

In *Caryophyllia*, as in all Scleractinia, the septa are laid down in multiples of six. When the coral is very young the corallum is simply a *basal plate* with six small prosepta inserted between the mesenteries. Later the metasepta appear, and there may be two or three generations of these so that the whole becomes multiseptate. A slight shift of their spacing throughout growth as the corallum becomes elliptical imparts the characteristic biradial symmetry. Throughout growth new material is added to the exposed edges of the setpa and to the epitheca so that the coral expands as it grows. In broad and general terms the simple structure of *Caryophyllia* applies to most living and fossil corals, but there are major differences in skeleton structure, septal arrangement and mode of colony formation. Four orders have been defined:

Order 1. Rugosa (?Cam., Ord.-U. Perm)

Order 2. Tabulata (?Cam.-Perm)

Order 3. Scleractinia (Trias.-Rec.)

Order 4. Heterocorallia (U. Dev.-L. Carb.)

Order Rugosa

Morphology

The Rugosa are an almost exclusively Palaeozoic group of solitary and colonial corals. Primitively they show bilateral symmetry, arising because the numerous metasepta are inserted in four loci alone. Such bilaterality is clear in some of the Rugosa though in others it is obscured by the proliferation of septa, and in such genera as *Hexagonaria* no pattern at all is visible. Two examples are chosen to show the basic structure of rugose corals. The widespread Carboniferous *Amplexizaphrentis* (Superfamily Cyathaxoniicae) is a small solitary 'horn-coral', so-called because of its shape.

The outer horn-shaped part of the corallum is covered by a thin calcareous skin, the epitheca, which extends most of the way from the tip to the upper (distal) surface or *calice*, where the skeletal elements filling the inside of the corallum are exposed. These internal elements form the basis of classification of the Rugosa and are of two main types: the vertical elements (septa, *axial structure*), and the horizontal elements (*tabulae*), which will be considered separately, (*Dissepiments*, which are important horizontal elements in some Rugosa, are lacking in *Amplexizaphrentis*.)

The septa are thin vertical plates arranged in a characteristic biradial pattern, which develops to maturity throughout the ontogeny of the coral. Their manner of insertion can be studied by investigating the ontogeny of the coral from the early stages; this is usually done by making serial sections normal to the axis, from the tip to the calice, and arranging them in a successional series. This provides a full record of how the coral grows, since structures once formed by accretion are not resorbed but retain their original form through the life of the coral.

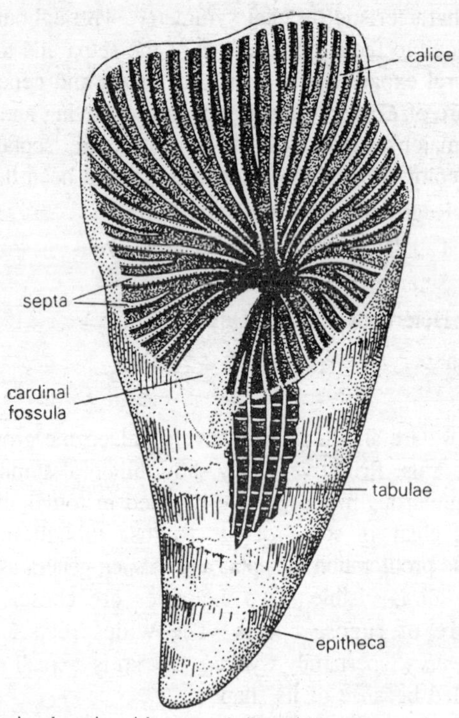

Fig. 13.4. Amplexizaphrentis with part of the epitheca removed, showing bilateral symmetry.

When *Amplexizaphrentis* is very young a single proseptum divides it, soon becoming separated into *cardinal* (C) and *counter* (K) prosepta. Two other pairs of prosepta follow: the alar (A) adjacent to the cardinal septum, and the *counter-lateral* (KL). Even at this stage some bilaterality is evident, which becomes more pronounced when the metasepta are inserted in four quadrants only, on the 'cardinal' side of the alar and counter-lateral septa. Through differential growth the counter-lateral and alar septa move towards the counter-septum to make room for new metasepta. Short minor septa (*second-order metasepta*) are laid down between the first-order metasepta. Around the cardinal septum there is a *cardinal fossula* where metasepta are not inserted, and especially during the intermediate growth stages there are lateral (*alar*) *fossulae* as well. These are visible in the adult *Amplexizaphrentis*, but in many large Rugosa, where the major and minor septa have proliferated greatly, the fossulae are so compressed as to be hard to distinguish on account of the increasing radial distribution of septa. Many Rugosa, e.g., *Philipsastraea*, have no trace at all of a fossula.

In *Amplexizaphrentis*, as in most other Rugosa, two other skeletal elements are (a) the *epitheca*, and (b) *tabulae*, which are flat horizontal plates forming a floor to the cavity in which the polyp resided. The major septa join together in a central boss formed of undomed tabulae.

Heliophyllum (Subfamily Zaphrenticae) a large North American rugosan, is sometimes found as poorly developed colonies In this coral there are so many septa that it is not easy to see the four main lines of emplacement. The spacing of the septa is more or less equidistant, and the symmetry is virtually radial. There are some structural elements not found in *Amplexizaphrentis*, especially the dissepiments, which are concentrated in a broad marginal zone or *dissepimentarium*. These are small curving plates located between the septa and set normal to them, inclined downwards at about 45° to the epitheca. The major septa may reach the centre, in which updomed tabulae are the most prominent components. Another characteristic feature is the presence of *carinae*: short 'yard-arm' bars projecting laterally from the septa. Though bilateral symmetry has been masked by radial structure the cardinal fossula can still be seen where the dissepimentarium narrows and the axial ends of neighbouring septa are shortened and curve round it.

Range of form and structure in Rugosa

Form, Type and Habit of Corallum

Rugosa are either solitary or colonial (*compound*), and either form may exist even within closely related genera or even species of the

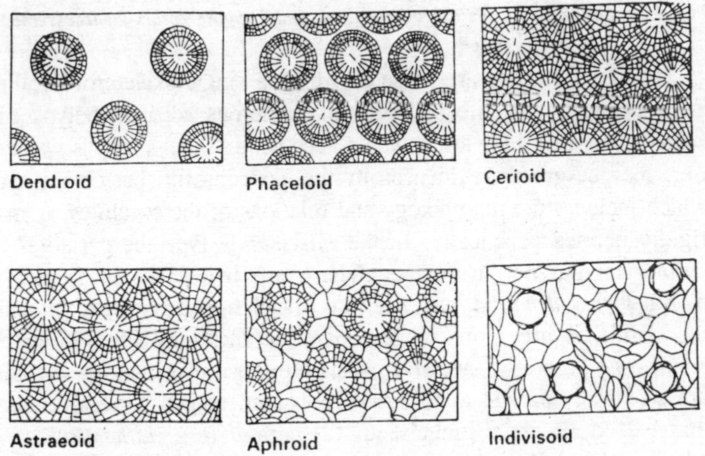

Fig. 13.5. Transverse sections through compound Rugose coral colonies, with terminology of different types.

same family. In solitary Rugosa the usual shape is a curved horn: the *ceratoid* shape of *Amplexizaphrentis*. If the 'cone' has expanded very fast, a flat *discoidal* shape may result, with progressively less abrupt expansion *patellate*, *turbinate*, trochoid and *ceratoid* forms will be the product. *Cylindrical* corals are virtually straight-sided, except for the first-formed part; scolecoid forms are irregularly twisted cylinders; pyramidal types have sharply angled sides; whereas calceoloid genera, oddest of all, possess a curved corallum with one flattened side and a hinged lid or *operculum*.

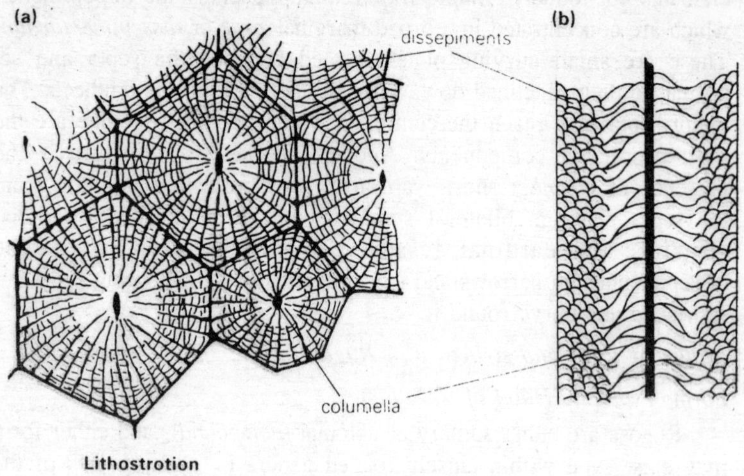

Fig. 13.6. Lithostrotion, a Carboniferous compound rugose coral: (a) transverse; (b) longitudinal section.

Colonial Rugosa are those in which a single skeleton (corallum) is produced by the life activities of numerous adjacent polyps each contribution a *corallite* to the whole. The habit of the colony is *phenetic*, i.e., may be greatly influenced by the environment, but colony type, which includes the morphology and relations of the corallites, is more rigidly defined genetically. In the *Fasciculate* type the corallites are cylindrical but not in contact. There are two kinds of fasciculate morphologies; *dendroid*, with irregular branches, and *phaceloid*, in which the corallites are more or less parallel and sometimes joined by connecting processes. *Massive* corals are those in which the corallites are so closely packed as to be polygonal in section. Several kinds of massive corals are distinguished: (a) *cerioid* (e.g. *Lithostrotion*), in which each corallite retains its wall; (b) *astraeoid* (e.g. *Phillipsastraea*), in which the corallite walls are wholly or partially lost but the septa

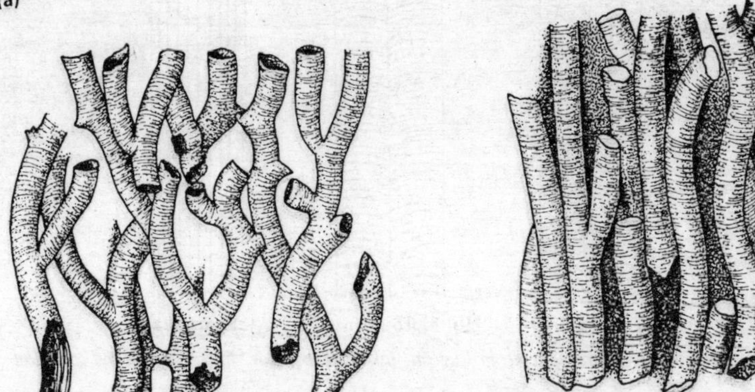

Fig. 13.7. *Form of corallum in Rugosa: (a) fasciculate dendroid Lithostrotion; (b) phaceloid Lithostrotion.*

stay unreduced; (c) *thamnasteroid* (e.g. *Orionastraea*), in which the septa of adjacent corallites are confluent and often sinuous or twisted; (d) *aphroid* (e.g. *Orionastraea*), where the septa are reduced at their outer ends so that neighbouring corallites are united by a zone of dissepiments alone; and (e) *indivisoid*, in which septa are absent and dissepimental material is dominant.

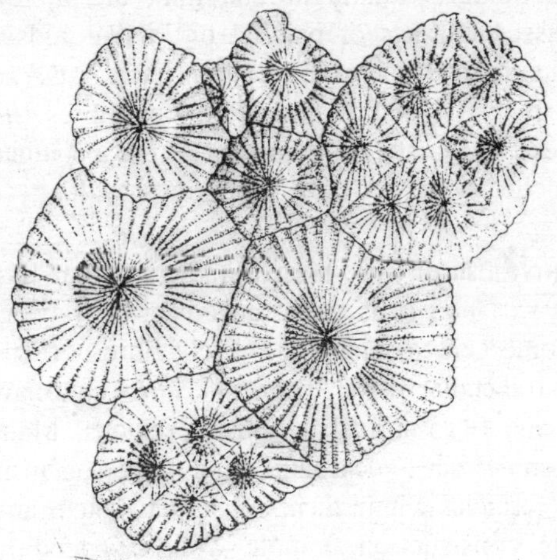

Fig. 13.8. Axial increase in ceriod Silurian Acervularia.

Fine Skeletal Structure of Septa

The basal ectoderm has numerous radial invaginations, each lying above a septum which it secretes by crystallisation. The skeletal

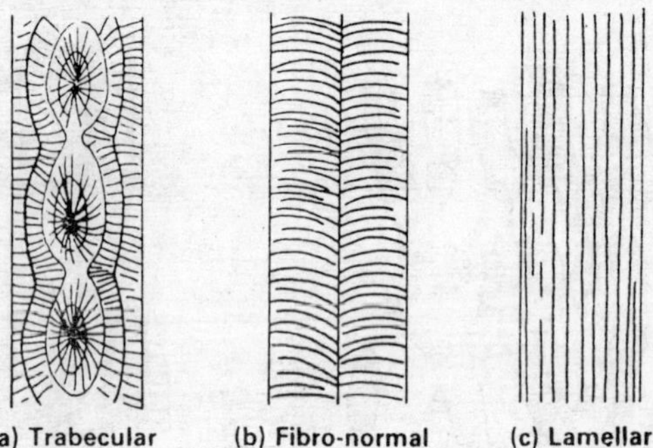

Fig. 13.9. Fine structure of septa in Rugosa: (a)-(c) trabecular, fibro-normal and lamellar tissue.

elements are formed of calcium carbonate, some or all of which may have been initially calcite. According to Wang (1950) and other authors, the process of crystallisation results in the production of three main components: *trabecular*, *fibro-normal* and *lamellar tissue*, the latter usually investing structures made of trabecular or fibro-normal tissue. When septa consisting of fibro-normal tissue are examined in section a median dark line is seen bisecting each septum, and projecting from it there are narrow parallel calcite fibres arranged normal to this median line, though radially around it at the tip of the septum. Trabecular tissue consists of parallel or fanlike rods arranged in a kind of palisade in the plane of the septum. Each *Trabecula* is composed of serially arranged radiating whorls of tiny fibres (*fibre fascicles*). These fascicles are secreted in linear series from a continually producing centre of calcification; the trabecula grows as long as calcification goes on.

The relative use of fibro-normal and trabecular tissue in rugosan septa is very variable, as is the arrangement of trabeculae within the septum. In some genera septa are entirely fibro-normal and in others there is only trabecular tissue, *semitrabecular* septa have a trabecular median part and fibro-normal peripheral regions. Many other septal types have been described, depending upon the arrangement of trabeculae and the fibre fascicles within them. Wang has noted, amongst types of fibre fascicle construction, simple widely spaced fascicles (e.g. *Keriophyllum*), grouped fascicles (e.g. *Palacosmilia*) or fascicles in continuous series (e.g. *Dinophyllum*). In *Tryplasma* the trabeculae are discontinuous as a result of intermittent calcification from the same centre. These are also many kinds of trabecular grouping and orientation in the septa. They are, however, generally directed upwards and inwards from the wall, subparallel and fan-like, but sometimes diverging from the wall in a complex and irregular manner. They commonly turn upwards axially and may be involved in the formation of axial structures.

Dissepiments, tubulae and epithecae are formed of fibro-normal tissue. Carinae are trabecular, formed either of single swollen trabeculae or of compound trabeculae, branched or in bundles.

Lamellar tissue, which usually surrounds the septa and other skeletal elements in some corals, was first described as a primary calcitic material. However, many recent workers have argued that lamellar tissue is probably secondary after aragonite and may result solely from the diagenetic enhancement of growth lines. Some authorities have even suggested that the whole rugose skeleton was originally aragonite and that trabeculae are diagenetic artefacts. Sandberg (1975), however, who studied ultrastructural and chemical changes in aragonitic Pleistocene to Recent bryozoans converting to calcite, has shown that the results of such diagenesis do not in any way resemble the forms taken by trabecular and fibro-normal tissue. Hence he has contended that all the preserved parts of rugose coral skeletons must originally have been calcite and that most of the fine skeletal structure actually seen is primary. For the moment, therefore, the issue is unresolved.

Since the gross morphology of corals is so highly variable and has been subject, in so many independent stocks, to repeated convergent evolution, there is a clear need in taxonomy to find more 'stable' characters to use as the basis of classification. Although micro-structure has proved useful here along with other characters, its interpretation is still the subject of much discussion and it is by no means a universal key. Microstructure has proved of some value in evolutionary studies, Kato (1963), for instance, suggesting a general stratigraphical change from lamellar to fibro-normal tissue.

Types of Septa

In the Rugosa septal microstructure and arrangement may vary greatly, and those differences are of high taxonomic value. Septa are normally laminar, but perforated septa do occur in some genera where the trabeculae are not closely joined. Usually they are straight, sometimes zigzag. Commonly septa are not of uniform thickness throughout their length. In *Kodonophyllum*, for instance, the outer parts of the septa are so greatly thickened that they are contiguous, forming a well marked *stereozone*. In *Aulophyllum* half of the coral only usually has medially thickened septa, so the sterozone is incomplete. The sides of rugosan septa are usually smooth but are often provided with *flanges* or small *denticulations* where the trabecular tips show through. In *Orionastraea* and some other aphroid genera the outer parts of the septa are replaced by dissepiments, whilst adjacent corallites of thamnasteroid types have confluent septa. *Amplexus* illustrates a type of coral in which septa are imperfectly developed, only being properly formed immediately above each 'calicular' surface.

Axial structure

Very many Rugosa have a central structure which may be one of three basic kinds. An *axial vortex* is formed by the joined axial ends of the major septa, slightly twisted as in *Ptychophyllum*. A *columella* is simply a dilated end of the counter-septum and is representative of *Lithostrotion*, *Lopho-phyllidium* and other similar genera. An *axial column*, as in *Aulophyllum*, is a broad axial zone formed of *tabella*, i.e. small incomplete tabulae within the axial complex. In *Lonsdaleia* and *Dibunophyllum* the axial column in section appears web-like and is likewise composed largely of tabulae with septa crossing it; in the latter genus it is divided by a central *septal plate* formed by the counter-septum. *Aulina* has a very simple axial column, the *aulos*, which is a vertical tube truncating the inner ends of the *septa* and traversed by horizontal tabulae.

Tabulae and dissepiments

Tabulae are transverse plates which may be flat, convex or concave. They usually occupy a central space or tabularium, and if there is an axial complex they joint with it. Tabulae may be replaced by a number of smaller plates called tabellae. Tabulae in the Cyathaxoniicae, the most primitive superfamily, generally extend right across the corallum, but in most Rugosa they terminate externally in a *marginarium*, which is either a septal stereozone or a dissepimentarium, and hence do not reach the edge.

Dissepiments, the small plates usually found towards the edge of the corallum, lie peripheral to the tabularium and like the tabulae are constructed of fibro-normal tissue. They are of various kinds. They may be simply inclined or slightly swollen plates dipping towards the axial region, but in *Phillipsastrea*, *Thamnophyllum* and relatives, for instance, there evolved an extraordinary range of dissepimental types, including horseshoe-shaped ones, which have proved useful in classification.

Dissepiments are not found in the earliest rugose genera. On their first appearance in late Ordovician streptelasmatinids and columnariids they are confined to a thin peripheral dissepimentarium. In later genera, however, they become more important structural components, most noticeably in types such as *Lonsdaleia*.

Rarely, as in *Cystiphyllum*, the dissepimentarium may become very wide and, together with tabellae with which dissepiments may integrade, make up virtually the whole skeleton, since the septa are reduced to separate trabeculae developed only on the upper surface of the tabellae and dissepiments.

Calice

The *calice* is that part of the coral which, in life, was in contact with the basal ectoderm of the polyp. It is in effect a mould of the secretory surface. It is not commonly seen in specimens preserved in limestones unless these have been weathered, but in individuals collected from shales calicular surfaces can often be clearly seen. Usually there is an annular calicular platform surrounding a deep central depression (calicular pit). Some genera with an axial complex have a marked calicular boss where it emerges to the surface, but this is not always present. The form of the calice is quite variable even within families, which illustrates various forms found in the Spongophyllidae in which there are conical, bell-shaped, saucer-shaped and everted types.

Asexual budding and the production of new corallites

When a compound coral grows the first-formed protocorallite gives rise asexually to offsets, which may form in a number of ways. The term *budding* is used for the soft parts, *increase* for the skeleton. Detailed studies have been made in the Lithostrotionidae and many other groups of the three main methods of increase and budding:

(a) *Axial increase*. In this mode the corallite splits by fission into two or more daughter polyps within the calice. Such budding is inevitably parricidal (i.e. destroys the parent). The upper surfaces of many compound corals can often be seen with several corallites undergoing axial increase at approximately the same time. As the colony continues to grow the parent corallite is buried and the daughter corallites grow to full size. This is the least common of modes of increase in the Rugosa, being very rare in *Lithostrotion* but normal in the Silurian *Acervularia*.

(b) *Peripheral increase*. Here small offsets arise round the parent corallite. This may or may not be a parricidal system; usually the daughter arises from the parent's side without killing it. It is quite common in fasciculate genera and has been reported in a few *Lithostrotion* species.

(c) *Lateral increase.* This is the most common method of increase in the Rugosa and is not parricidal. In *Lithostrotion* it has been well investigated but has proved to be so variable as to be of limited taxonomic value at the generic level. Three distinct types of lateral increase have been distinguished in *Lithostrotion* using serial sections. In one type (e.g. *L. junceum*) the lateral bud arises fully external to the parent calice and is initially aseptate; septa only form later. In a second type,. typical of species with a narrow dissepimentarium (e.g. *L. arundineum*), the new corallite arises from near the periphery and, though septate from the start, does not inherit septa from the parent. In a third type, in species such as *L. martini* with a wide dissepimentarium, the daughter arises from well within the parent calice and inherits its initial septa, especially those of the outer wall, from the ends of the parent septa. Cerioid species of *Lithostrotion* increase in a manner analogous to the third type, but the daughter remains attached to the parent and is polygonal from the start. Nudds (1981) has extensively reviewed colony form in *Lithostrotion*.

There are some rugose corals (notably cerioid *Lonsdaleia*) which show a curious feature associated with offset production. For there may be seen a pair of narrow tubular ducts extending from the base of the offset obliquely upwards through the dissepimentarium of the parent. The ducts may then merge at higher levels in the parent corallite. They can only have served to prolong gastric and nervous communication between the parent corallite and the daughter until the latter was well established. Oliver (1976) has reviewed many aspects of colony formation in Rugosa and other corals.

Rejuvenescence

In solitary cylindrical forms the corallite is often found to be constricted at irregular intervals, leaving a broad or narrow shelf where the septate older calice is exposed. Above this it may expand again, before once more being constricted. These constricted bands probably represent periods of famine during which the polyp resorbed its own tissues in order to stay alive and shrank away from the edges, becoming smaller while retaining its form. The next period of increased food supply permitted growth to begin once more; this is rejuvenescence. Starvation and regrowth, however, are only one explanation. In some species rejuvenescene may be especially strong, so that sharp rims are formed with deep and almost slit-like contractions between. In some instances small buds are found on the upper surface of a calice.

These probably result from a late rejuvenescence of a coral which almost died, from small pieces of still living tissue.

Evolution in the Rugosa

The Rugosa have relatively few phylogenetically useful characters. Furthermore, they were very plastic in their evolutionary potential and notoriously subject to homeomorphic trends in separate stocks. Yet however common such iterative trends may have been, the guiding principle controlling their evolution was always towards a strong and firm skeleton. Thus many of the observed structures can be interpreted functionally; for instance, the complication of the trabeculae increased skeletal strength, the carinae held the polyp more firmly and the axial column gave greater support in the central region.

There is thus some functional meaning in all the trends that have been observed, though it is not always clear what the function was.

Overall pattern of evolution

The first known corals from the Middle Cambrian of New South Wales, Australia, include the genus *Cothonion* which has a conical septate corallum and a biradial operculum. This has been tentatively referre to the Rugosa. Undoubted Rugosa do not appear until the Black-riveran (M. Ord.) in North America. They spread rapidly and evolved in a series of five major episodes. During each of these one particular faunal group was dominant, and subsidiary elements at the time were: (a) small persistent bradytelic stocks; (b) early members of a later dominant stock; and (c) remnants of a formerly dominant stock. The main phases are as follows:

(a) In the Middle Ordovician to Lower Liandovery period small solitary corals dominated, mainly without dissepiments and with only feeble development of trabecular tissue. The Streptelasmatina and Columnariina were dominant.

(b) The Upper Llandovery to Lower Devonian period saw the first really prolific burst of Rugosa and the exploitation of the reef habitat, by both solitary and colonial corals, though the Rugosa were subsidiary components of most reefs relative to stromatoporoids and tabulate corals. Even so, at generic level the Rugosa outnumbered the Tabulata in the Middle Silurian and thereafter kept ahead.

A dissepimentarium was independently evolved in all three suborders, and seems to have been associated in some way with the great success of the Rugosa at this time. The Columnariina were dominant, but other groups also flourished including a curious short-lived family Calostylidae, which had perforate septa. (Perforate septa are also known in the Scleractinia, but the Calostylidae were not ancestral to them.)

(c) There were relatively few Lower Devonian corals, and not many of the Silurian genera lasted into the Devonian. A new burst of evolutionary diversity in both simple and compound corals took place in Middle to Upper Devonian times, with digonophyllids (Suborder Cystiphyllina) and the colonial phillipsastraeids being numerically important. These are all large forms with a wide dissepimentarium and long septa with trabecular fans. But by the late Devonian coral faunas had become impoverished and the dominant Phillipsastraeidae and the last Cystiphyllina had died out.

(d) The Lower and Middle Carboniferous saw the climax of rugosan development with Superfamily Zaphrenticae dominant (Family Aulophyllidae, Family Lithostrotionida) and a whole variety of solitary and colonial forms evolving rapidly. The independent development of an axial structure was characteristic of virtually all groups, and the microstructure became very complex. Some primitive Cyathaxoniicae still continued. By this time the Tabulata were declining whereas the Rugosa were relatively more successful. Late Middle Carboniferous and Upper Carboniferous forms were few in number, and most families had died out by the Artinskian.

(e) The final radiation of the Rugosa was in the permian, based mainly on the surviving Cyathaxoniicae, in which the protosepta were strongly developed. Such genera as *Plerophyllum* and the columnariid *Waagenophyllum* were dominant. The last known Rugosa are known from the uppermost Permian. The early Scleractinia may have been derived from one of these later Cyathaxoniicae; on the other hand, many authorities consider that the Scleractinia derived from a soft anemone stock and are not descended from the Rugosa.

Evolutionary trends

Rugose corals had only a limited evolutionary potential; hence it was inherently likely that the same kind of structure would be evolved several times over. According to Lang (1923), who first tried to establish such repetitive 'trends', in Carboniferous corals, 'each character follows in varying degrees and at varying rates one of a comparatively few possible developments pointing to the existence…of limited tendencies… repeated in each lineage'. In taxonomy, which becomes confused by such trends, the only way to separate and identify the genera is by very detailed study of morphology, fine structure and ontogeny. In general terms the development of a marginarium (stereozone or dissepimentarium) is perhaps the most persistent and universal of all trends in the Rugosa, and since this happened in all three suborders independently in the later Llandovery, just prior to the first success of the Rugosa as reef builders, it seems to have been important and fundamental, though it is not entirely clear how. A number of other trends fundamental to all the Rugosa have also been distinguished.

The phylogeny of the Rugosa, however, is still too poorly known to allow trends to be properly documented except in a few specific cases. Evolutionary studies of trends within Family Lithostrotionidae have shown that the parallels are so close in several descendent stocks that species originally lumped in the same genera were actually derived polyphyletically. Thus 'Diphyphyllum' is a composite genus of which several species were originally derived from an ancestral *Lithostrotion* stock. In all of these the central tabulae have become so shaped that the outer down-turned edges have formed a single central tube, the advantage being that a stronger axial structure results. Similarly '*Orionastraea*', which has a weak or absent columella and a tendency for septa to be withdrawn from both the axis and the periphery, is evidently polyphyletic and has had to be split since it included morphs that are of similar form but derived from different ancestral species.

Apart from these trends, there is a cerioid trend in some Lithostrotionidae and in addition a particularly interesting trend which involves the loss of dissepiments through time. In the early forms (e.g. *Lithostrotion praenuntius*) there are six circlets of dissepiments, reducing in the descendent *L. martini* to four and then to two. In addition the septa generally become fewer and the tabulae more widely spaced. The evolutionary principle governing the origin of the new types is probably paedomorphic since the descendent adults tend to resemble ancestral juveniles. Such morphology would seem to be especially appropriate for an environment in which rapid growth with minimal metabolic expenditure on skeletal elements was desirable.

Likewise, the advantage of food sharing between the linked polyps in the 'orionastraeoid' aphroid morphologies, and the fact that in these all the available surface space is covered by polyps, would lead to greater efficiency of the corallum as an integrated unit. Each of these trends therefore, expressed polyphyletically, has its own advantage suitable for particular ecological conditions.

Microevolution

There are relatively few well documented accounts of small-scale evolutionary change in the Rugosa. One of the best known is that of Carruthers (1910), who worked on the *Amplexizaphrentis delanouei* group in Scotland and England. The species group actually ranges as far as Belgium. His collections were made from thin limestones in an alternating limestone–shale–sandstone–coal sequence, all well located stratigraphically. In studying the populations in ascending sequence he noted that most characters (shape, size, tabular spacing, etc.) tended to remain constant but that there seemed to be successive changes in the shape of the cardinal fossula and the length of the major septa. The oldest beds contain *A. delanoeui* ssp., which has septa meeting in the centre and a large cardinal fossula expanded axially. A variant of this, *A. parallela*, with a parallel-sided fossula, occurs in the same beds. It was probably derived paedomorphically.

A short-lived side branch led to the small *A. lawstonenis*. In younger beds *A. delaneoui* and *A. parallela* are rare and replaced largely by *A. constricta* in which the axial end of the cardinal fossula is constricted to a keyhole shape. In the youngest lime-stones there occurs. *A. disjuncta*; this form passes through a *constricta*-type morphology in its juvenile development, but thereafter septa retreat away from the centre.

Though this has been regarded as a classic study it is severely limited in two respects. One is that since the corals occur only at certain horizons within a highly variable sedimentary sequence, it was not possible to establish continuity of variation. Secondly, the study was done within a limited area alone, and it is not clear how far the sequence can be substantiated in other regions. And since different 'species' coexisted in time (*A. delanouei* and *A. parallela*) and there is some evidence of the same species being found at different horizons in different areas, the pattern of evolution was probably very complex and is perhaps best regarded as the expression of a trend within a species group rather than as a matter of linear descent involving separate species. In functional terms the morphology of *A. constricta* makes for a stronger axial region than that of *A. delanouei*, but that of *A. disjuncta* is clearly weaker. Perhaps the need for rapid growth, as would be allowed by septal reduction, may in this case have offset the axial weakness

So many factors seem to have been involved in the evolution of the *A delanouei* group—including trending, response to local environmental conditions, paedomorphosis, allopatry and polyphyly—

that the evolutionary pattern now seems to be so complex as to elude analysis until better data are available.

Classification of the Rugosa

The following classification, erected upon the various structural features already described, is based upon that in the *Treatise on invertebrate paleontology*. It is largely founded on megascopic features; knowledge of microstructure is not yet advanced enough to be of much use. The inevitable effect of parallel trends and convergent evolution, however, creates taxonomic problems which if resolved may lead to a radical re-arrangement of the group.

For the purposes of this textbook the simple classification given in the first edition of the '*Treatise*' is retained, being more familiar and easier to use.

Order Rugosa (?Cam./Ord.-U.Perm.): Solitary or colonial corals with the major septa inserted in four quadrants only relative to the protosepta. Epitheca normally present, and up to three orders of septa. Dissepiments occur in more advanced groups; tabulae normally present. First- and second-order septa well developed; third-order septa very rare.

Suborder 1. Streptelasmatina (Ord.-L. Trias.): Solitary or colonial. Marginarium either a septal stereozone or a dissepimentarium constructed of small globose interseptal dissepiments. Tabulae domed.

Superfamily 1. Cyathaxonicae (Ord. -1. Trias.): Small and solitary. Usually without dissepiments, with only a narrow septal stereozone and with outwardly declined tabulae. Example include *Cyathaxonia*, *Syringaxon*, *Amplexizaphrentis*.

Superfamily 2. Zaphrenticae (Ord. Perm.): Solitary or colonial. Usually with a dissepimentarium of stereozone and with conical or domed tabulae. Examples include *Heliophyllum*, *Dibunophyllum*, *Lithostrotion*, *Phillipsastrea*, *Streptelasma*.

Suborder 2. Columnaruna (Ord.-Perm.): Usually colonial, rarely solitary. Marginarium variable (absent, septal stereozone, or with lonsdaleoid dissepiments). Septa thin, Tabulae flat, depressed or sagging axially, sometimes domed. Examples include *Lonsdaleia*, *Spongophyllum*, *Waagenophyllum*.

Suborder 3. Cystiphyllina (Ord.-Dev.): Solitary or colonial, Septa with large and often complex trabecular structure, or sometimes absent. Marginarium a stereozone or wide dissepimentarium with small globose dissepiments. Tabulae present. Examples include *Cystiphyllum*, *Calceola*, *Goniophyllum*.

The 1981 '*Treatise*' classification is more complex: it elevates Rugosa to a subclass with two orders: (a) *Cystiphylliida*, which include mainly the Lower Palaeozoic solitary corals and cystiphyllines, and (b) *Stauriida*—the solitary and compound corals dominant in the Upper Palaeozoic and grouped in some 18 suborders. This classification will probably be more generally used in future.

Ecology of the Rugosa

Solitary Rugosa had no really effective means of anchorage on the sea floor. They did not normally cement themselves to the substratum though some have cicatrices of attachment or root-like 'talons' which may have helped to fix them. They seem to have preferred soft substrates, but since relatively few 'hardgrounds' are found preserved it is not known how far they could clonise such hard substrates. The horn-like shape would give a reasonable degree of support if the coral were to be half sunk in mud, convex side down. This would allow the polyp to be exposed above the sea floor and would enable it to stay in the same general attitude whilst the coral grew. There is some evidence that Rugosa such as *Aulophyllum* could grow like this, orientated with respect to currents. Commonly, however, in any population some or all of the coralla are twisted; they must have toppled over and then grown up again.

The solitary *Calceola* lay on the sea floor, convex side down with its operculum open to 90°. Presumably it could shut it in case of emergency.

Compound Rugosa did not normally fix themselves to the sea floor either, though rare examples of hardgrounds encrusted by Rugosa are known. In softer substrates the weight of the colony in the mud would no doubt give some stability. Since they were uncemented they were environmentally restricted and could not build proper reefs. This seems to have been their most severe evolutionary limitation, from which, in spite of their 230 million year history, they were never able to escape. Thus they are very rarely found within the limits of strong wave action and so were never really involved as frame builders in the algal and coral-stromatoporoid reefs of the Palaeozoic.

Rugose coral faunas were very sensitive to environmental conditions, and the form of both solitary and colonial species was strongly influenced by ecological factors.

Separate facies faunas can be distinguished, adapted for particular circumstances and normally found with characteristic associations of brachiopods and tabulates. Three such facies faunas were noted by

Hill (1937) working mainly with Scottish Carboniferous successions. These are as follows:

(a) *The Cyathaxonia fauna.* Usually found in black or greenish calcareous shales, though sometimes in bryozoan reefs also, these corals are all small and solitary with dissepiments poorly developed. This facies fauna is very long-ranged and certainly antedates the Carboniferous; indeed the earliest of all rugose coral faunas seem to have been of this type, though occurring in mudstones. It is usually found with tabulates, small brachiopods and trilobites. As represented typically in the Carboniferous the sea floor was generally muddy, with decaying organic matter, but must have been fairly well oxygenated.

(b) *The Caninia—clisiophyllid fauna.* This fauna, found first in the Tournaisian but probably older, consists of large solitary genera with dissepiments. Clisiophyllids gradually took over from caniniids with time. These funas were adapted to shallow limestone seas, well lighted and oxygenated but with little terrigenous material.

(c) *Colonial coral faunas.* Compound rugose corals with dissepiments may occur in small and scattered 'reefs'; they are represented in bedded limestones as well. These were first considered to be reef faunas analogous to modern reef corals, but since the Rugosa were not really reef formers they are best considered as just a group of colonial corals. These colonial corals evolved from the *Caninia*-clisiophyllid fauna.

Hill's three facies associations have generally been well substantiated, though the *Cyathaxonia* fauna is very much separate whereas the other two are some-times integrading and ecologically less well defined. Thus it is not uncommon to see rapid alterations of faunas (b) and (c) in vertical sequence, as in the Carboniferous of the west of Ireland where, in addition, thickets of fasciculate lithostrotionids are found with the large solitary. *Aulophyllum* growing within the thickets.

More recent work has concentrated upon the distribution of rugose and other corals in Palaeozoic reefs. Klovan (1964) and Krebs (1974) have charted the facies associations of corals and other reef organisms in Devonian reefs of western Canada and central Europe respectively. In these reefs colonial Rugosa tend to be confined to the quieter fore-reef facies around the flanks of rigid frame-reefs; they do not normally occur within the reef itself nor, except in localised patches, within the back-reef facies. Solitary Rugosa are likewise confined but are numerically insignificant.

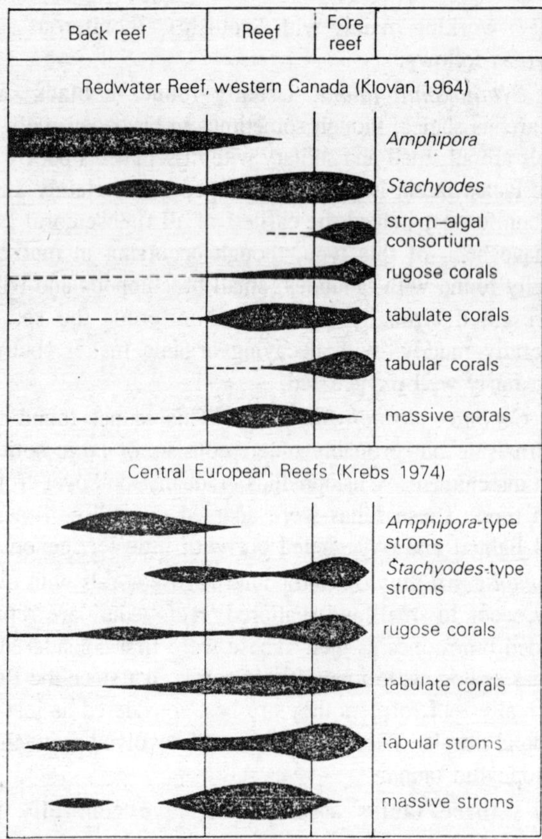

Fig. 13.10. Distribution of rugose corals, stro-matoporoids and other reef-associated organisms in Devonian reef facies.

Tabulate corals, especially those with a creeping or foliaceous habit, were somewhat less restricted but could not compete with stromatoporoids and algae in the turbulent zone of most active reef growth. The hard skeletons of both rugose and tabulate corals, however, seem to have contributed to establishing a firm foundation for later reef growth of algae and stromatoporoids. Although the broad pattern of distribution of rugose and tabulate corals with respect to ancient reefs has been generally confirmed, a more detailed picture of the various coral-stromatoporoid facies associations is emerging. Scrutton (1977), for example, who worked upon an evolving reef complex in southwestern England, has shown how the genera in stromatoporoid reef limestones differ from those in biohermal, bioclastic and bedded dark limestones.

Copper (1974) has described a generalised palaeoecological succession of a Devonian reef from pioneer through intermediate to a mature reef system. Devonian reef systems were evidently more extensive and elaborate than those of the Silurian but by the end of the Devonian had collapsed. A fuller treatment of the theme has been given by Walker and Alberstadt (1975). On a global scale, evidence is also accumulating on the biogeographical distribution of rugose corals

Additional Common Genera

Solitary Rugosa

Calceola. 'Slipper-shaped', semicircular in transverse section; calice deep, closed in life by thick lid marked by striae on inner surface; septa from fine ridges on thickened wall.

L-M Devonian

Dokophyllum. Conical to top-shaped with deep calice; septa long, becoming discontinuous at higher levels; tubulae flat in axial region and replaced by dissepiments towards periphery; root-like outgrowth from base.

M-U Silurian

Caninia. Horn-shaped, becoming cylindrical, long; major septa initially long, later withdraw from axis becoming short; cardinal fossula small with short cardinal septum; minor septa very short, may be discontinuous; dissepiments irregular, arranged around periphery; tabulae flattish in axial region and turned down at margins.

L-U Carboniferous

Aulophyllum. Cylindrical with well-defined axial structure consisting of a haphazard arrangement of the radial and concentric plates; seen in transverse section the axial structure has a sharp protection towards the cardinal fossula; numerous major septa extend to axial structure; minor septa half as long; wide zone of numerous small dissepiments; tabulae irregularly domed.

L-U Carboniferous

Dibunophyllum. Cylindrical, with axial structure showing spider's-web pattern of concentric and radially arranged plates; prominent median plate aligned towards cardinal fossula; numerous major septa extend to axial structure; minor septa irregular, short: wide zone of closely spaced dissepiments.

L-U Carboniferous

Palaeosmilia. Cylindrical, long; numcrous radial septa extend to axial region where their ends may fuse; minor septa extend half-way

to centre; cardinal fossula narrow, not easy to distinguish: wide zone of abundant dissepiments: tabulae small, incomplete. *Palastraea* is similar but compound, cerioid.
L-U Carboniferous

Compound Rugosa

Acervularia. Massive, cerioid; corallite wall hexagonal in transverse section, well-defined inner circular wall formed by thickening of septa: longer septa extend to centre where their ends may fuse; peripheral zone of 'bubbly' dissepiments; incomplete tabulae; small domed plates in axial region.
M-U Silurian

Phillipsastrea. Similar to *Acervularia* but astraeoid: inner wall encloses smaller area.
M-U Devonian

Lonsdaleia. Fasciculate, phaceloid; axial structure similar to that of *Dibunophyllum* with median plate; major and minor septa confined to space between axial structure and peripheral zone of large dissepiments, tabulae flat or gently declined; *Actinocyathus* (L Carboniferous) is similar but cerioid.
L-U Carboniferous

Geological History

M Ordovician-U Permian

The Rugosa appeared in mid-Ordovician times and by the Silurian were common and widespread. They were, however, subordinate to the tabulate corals and stromatoporoids in the clear-water calcareous facies, e.g. the Wenlock Limestone (M Silurian). Here forms like *Dokophyllum* and *Acervularia* are common examples. There was a further expansion in numbers and diversity in the Devonian when new families appeared including both solitary (e.g. *Calceola*) and compound (e.g. *Phillipsastrea*, forms.

In the later Devonian numbers plummeted and many families became extinct. More disappeared at the close of the Devonian, few surviving into the Carboniferous. There was, however, a dramatic recovery in the lower Carboniferous with the arrival of new families, some showing new features, e.g. complex axial structures, not seen in earlier forms. During this time rugose corals were at their peak of abundance. Compound forms such as *Lithostrotion*, *Lonsdaleia* and their relatives occur extensively at some levels in the later part of the lower Carboniferous.

Rugose corals are not found after the lower Carboniferous in Britain, suitable marine rocks being absent. Elsewhere, for instance in Europe, Asia, North America and Australia, they persisted, though with reduced diversity, until the end of the Permian.

Order Tabulata

The tabulate corals are entirely Palaeozoic, and though they appear a little earlier than the Rugosa they have an otherwise similar time range. They are always colonial, never solitary, and usually their corallites are small. Invariably they have prominent tabulae, but other skeletal elements, in particular the septa, are reduced or absent.

Having relatively few structural elements, the Tabulata are of comparatively simple construction. *Favosites* (Sil.-Dev.), for example, has elongated, thin walled, prismatic, polygonal corallites with horizontal tabulae extending right across the corallum. The septa are reduced to short and somewhat irregular spines. All the walls are perforated by numerous *mural pores*, connecting the corallites. In *Favosites*, as in *tabulates* generally, elements of 'skeletal microstructure' (tabulae and walls) are fibro-normal. Trabeculae are scattered in tabulates. In *Favosites* they are normal to the surface of the tabulae but inclined upwards and outwards along the axis of the septa. In other genera septal spines are not necessarily trabeculate. Some genera (e.g. *Trabeculites*) have trabeculate walls, and *Halysites* has trabeculae in the walls.

Range in Form and Structure of the Tabulata

Form of Corallum

The corallum (colonial skeleton) is built up by individual polyps which may or may not be directly connected to each other. *Cerioid* forms (e.g. the Favositina) have polygonal corallites all in contact. *Cateniform* colonies (e.g. the Halysitina) have elongated corallites joined end to end in wandering palisades. Fasciculate tabulates (e.g. *Syringopora*) have cylindrical corallites which may be dendroid or phaceloid and may be provided with connecting tubules. *Auloporoid genera* have a branching (*ramose*) tubular structure and often an encrusting creeping (*reptant*) habit; others are erect. Finally, *coenenchymal* types, characteristic of the more advanced tabulates (e.g. the Heliolitina), have no dividing walls between the corallites but instead a common mass of complex tissue, the coenenchyme, originating as a marginarium and forming a dense calcareous mass in which the corallites are embedded. As in the Rugosa, colony form, which refers to the relations of the corallites, must be distinguished from colony habit; thus *Thamnopora* is a ramose but cerioid favositid. Other tabulates may be *massive*, *foliaceous* (in which habit the coral forms thin overlapping laminar sheets) or *creeping*.

Skeletal Elements

Tabulae, the most important skeletal elements, are always present and commonly traverse the corallite horizontally. Sometimes they are replaced by smaller tabellae. In the fasiculate genus *Syringopora* the tabulae are funnel-shaped (*infundibuli-form*) and run through the tubules that connect the branches. Septa in tabulates are rarely more than short spines, commonly 12 in number, but in some cases they do reach the centre of the corallite. In *Caliapora* and other genera of the Favositina they may be replaced by small shelves (*squamulae*) which protrude inwards at intervals from the corallite wall.

Where there is a marginal zone, as in the more advanced tabulates, it can be of two kinds. It may be only a thickened zone of annular lamellae as in *Straitopora*, constricting the aperture. In the other kind the growth of the marginarium is accompanied by the loss of the corallite walls, resulting in a coenenchyme. This can be constructed in various ways and evidently evolved independently in several stocks; tabulae, trabeculae borne on dissepiments and trabeculae closely packed or sometimes organised into tubes may all have a part in it. The production of a coenenchyme is one of the more important evolutionary trends in the Tabulata, another being the development of mural pores or connecting tubules for interconnection of the corallites.

Asexual Budding and Growth of the Corallites

The offsets may grow in various ways: (a) axially (parricidaly) through the growth of dividing walls; (b) peripherally, i.e. erupting in the coenenchyme; (c) laterally, in which each bud originates by outgrowth from the parent and thereafter grows intermurally.

Evolution and Ecology of the Tabulata

Possible tabulates are recorded from the Cambrian, though the earliest undoubted *Tabulata* are known in the Chazyan of North America, somewhat antedating the arrival of the first Rugosa. By Trenton times they had spread to many parts of the World and the first Heliolitina and Halysitina had appeared, while the Sarcinulina and Licenariina were already on the wane. The Favositina from their first appearance in the Middle Ordovician became a very dominant group throughout the Silurian and Devonian, though the Heliolitina were also important.

With the Carboniferous the Favositina, which had undergone many changes, lost their dominance and became subordinate to the Auloporina and Syringoporina. These remained the important groups until the end of the Permian.

Larger tabulates are found in coral-stromatoporoid reefs and were relatively important, though they were not really frame-builders since they had no proper means of attachment. Smaller tabulates tended to occur in deeper waters, and fasciculate genera usually lie in quieter environments. Ordovician and Lower Silurian tabulates of small size are often found with early solitary Rugosa, forming a characteristic association in calcareous mudstones. In later rocks also the two are frequently associated, and evidently the Rugosa and Tabulata had similar habitat preferences. Generally they lived in similar conditions to those of modern corals, but they are not normally found in very high energy environments.

Tabulates are not of great stratigraphical value, but they do sometimes occur in useful marker bands. The Lower Devonian *Pleurodictyum*, a small domed tabulate always found with a commensal tubiculous worm in the centre of the colony, is characteristic of sandy facies, by contrast with other corals. Along with other fossils of the same facies it has been used successfully to tell directions of funal migration across Europe.

Classification of the Tabulata

Order Tabulata is divided into seven or eight suborders which are quite clearly circumscribed and easily distinguished. The following classification is much abbreviated and the 1981 '*Treatise*' classification is somewhat different:

Order Tabulata (ord.-perm): Invariably colonial. Generally composed of small corallites with tabulae: septa (usually 12 in number) reduced to vertical rows of spines of absent.

Suborder 1. Licenariina (L. Ord.-L. Sil.): Massive colonies, Corallites prismatic with sixteen or more septa and with mural pores in horizontal rows. Examples include *Lichenaria*.

Suborder 2. Sarcinulina (U. Ord. M. Sil.): Massive coenenchymal colonies, with coenenchyme enclosing horizontal spaces and formed largely of tabular and septal extensions. Up to 24 septa. Examples include *Sarcinula*.

Suborder 3. Favositina (M. Ord.-Perm.): Variform colonies with slender corallites having mural pores and short spinose septa. Examples include *Favosites*, *Alveolites*, *Michelinia*, *Pleurodictyam*.

Suborder 4. Auloporina (Ord.-Perm.): Corals with creeping or erect habit, fasciculate, with tabulae widely spaced or absent. Examples include *Aulopora*.

Suborder 5. Syringoporina (U. Ord.-Perm): Large erect coralla, dendroid or phaceloid, with cylindrical corallites connected by horizontal tubules. Septa often absent; tabulae horizontal or funnel-shaped. Examples include *Syringopora*.

Suborder 6. Halysitina (M. Ord.-L. Dev): Cateniform colonies of imperforate coralla having in some cases small vertical tabulate tubules between the large tubes. Examples include *Halysites*.

Suborder 7. Heliolitina (M. Ord-M. Dev.): Massive colonies with 12 septate corallites embedded in coenenchyme. (These are sometimes classified apart and have some similarity to certain octocorals.) Examples include *Heliolites*.

Order Scleractinia

All post-Lower Triassic corals are included in Order Scleractinia. Most of them, like *Caryophyllia*, secrete an aragonitic exoskeleton in which the septa are inserted between the mesenteries in multiples of six. After the first six prosepta grow, successive cycles of 6, 12 and 24 metasepta are inserted in all six quadrants. Through their pattern of septal insertion the Scleractinia are immediately distinguished from the Rugosa. Each of the earlier cycles is complete before the next cycle grows, though in some cases this system breaks down in the higher cycles.

The skeleton originates as thin basal plate from which the septa arise vertically. As this skeleton grows upwards the lower rim of the poly phangs over it as an edge zone. Within the cup there may be dissepiments as well as septa. In general, the structure of both simple and compound scleractinians is light and porous, rather than solid as in the Rugosa. In several respects other than the primary septal plan the Scleractinia differ from Palaeozoic groups, as mentioned briefly below: this has suggested to some authors (Oliver 1980), that Scleractinia arose from a group of sea-anemones rather than from any Palaeozoic coral.

Range in Form and Structure of Scleractinians

Type and Habit

Scleractinians may be solitary or compound. In solitary scleractinians the form of the corallum depends on the relative rates of vertical and peripheral growth once that basal plate has been secreted.

Where peripheral growth has been dominant a discoidal form results, often with everted septa in cases where these have grown rapidly. Perhaps the commonest kinds are those with turbinate or conical coralla in which the axis is straight, though horn-shaped and cylindrical types are common.

In colonial Scleractinia, as with other colonies, the corallites are interconnected as a result of repeated asexual division by the polyps. Colony type is genetically defined and refers mainly to the relationships of the corallites. As with the Rugosa there are dendroid, phaceloid, thamnasteroid and, rarely, aphroid types. In cerioid scleractinians the thecae (i.e. walls) of adjacent polygonal corallites are closely united and of dissepimental or septal origin, by contrast with the walls of cerioid rugosans, which are epithecal. *Plocoid* types of corallites have separated walls but united by dissepiments. Ramose types of creeping habit, paralleling the tabulate *Aulopora*, are not uncommon.

In addition there are two other types which do not correspond to any rugosan type: (a) the *meandroid* type in which corallites are arranged in linear series with the cross-walls absent, and confined within the lateral walls which run irregularly over the surface like the convolutions of a human brain; (b) the *hydnophoroid* type in which the centres of the corallites are arranged around little hillocks or *monticules*.

In habit scleractinian colonies may be *branching* (ramose), massive, encrusting and creeping (reptant) or foliaceous. Sometimes adjacent colonies of the same species fuse to form a single colony.

Additional Common Genera

Michelinia. Corallum small, sometimes with root-like processes at base; corallites relatively large, tick-walled; tabulae numerous, vesicular. Distinguished from *Favosites* by larger corallites.

L. Devonian-U. Permian

Pleurodictyum. Similar to *Michelinia*, with large corallites (about 5 mm across) but walls thick with many pores; tabulae may be absent. A curved worm tube often present in centre: the association was probably commensal, i.e. the worm probably benefited (e.g. by gaining shelter) without incommoding the coral.

U. Silurian-M Devonian

Vaughania. Like *Michelinia* but smaller, discoid, with shallow corallites; walls thickened; no tabulae.

L. Carboniferous

Alveolites. Like *Favosites* but encrusting with inclined corallites opening obliquely at surface, rounded polygonal in transverse section.

M. Silurian-U. Devonian

Heliolites. Massive, with cylindrical corallites separated by coenenchyme of narrow polygonal tubes; with or without 12 short septa; tabulae flat, regular.

M. Ordovician-M. Devonian

Catenipora. Phaceloid; corallites oval in transverse section, arranged in single series which join at intervals to enclose irregular spaces; spinose septa in 12 vertical rows; tabulae regular.

M. Ordovician-U. Silurian

Halysites. Similar to *Catenipora* but has coenenchyme of a single tube separating adjacent corallites, each tube with closely spaced transverse plates.

M. Ordovician-Silurian

Syringopora. Phaceloid; corallites cylindrical, thick-walled, united at intervals by transverse tubes; tabulae unmerous, more or less funnel-shaped.

U. Ordovician-L. Carboniferous

Geological History

L. Ordovician-U. Permian

Tabulate corals appeared in the early Ordovician and became varied and widespread in the later part of the period. They occur most commonly in shallow-water limestones and calcareous shales and they played a significant role in reef formation in the Silurian and Devonian. They outnumbered rugose corals in reefs of that time. Heliolitids and favositids were the dominant forms. Halysitids were also common and widespread but did not survive the Silurian. Heliolitids and many other tabulates disappeared when numbers dropped abruptly in the late Devonian. There was a modest recovery in numbers in the lower Carboniferous, the most important examples being forms like *Syringopora* and *Michelinia*. Later tabulates are not found in Britain. Elsewhere, a small number of genera, including *Michelinia*. occur in the late Carboniferous and Permian, finally disappearing at the close of the permian period.

Other Primary Structures

The thin basal plate is semitransparent and firmly adherent to the substratum; it may later be thickened by secondary deposits. The epitheca is not developed in many scleractinians but if present may consist of chevron-like crystallites of aragonite. Dissepiments, like those of the Rugosa, are secreted by the base of the polyp. As the latter grows it moves upwards, detaching a small part of the base from the calice at a time and forming a dissepiment to enclose each space. Where the fine skeletal structure of dissepiments have been studied (Sorauf 1972) dissepiments have been shown to have two layers: a first-formed primary layer, overlain by vertically aligned clustes or aragonitic spherulites. Endothecal dissepiments are confined within the corallite, but in some colonial scleratinians the corallites are united by a common spongy tissue, the coenosteum, which may be formed partially of exothecal dissepiments and partially by rods and pillars (costae), as in *Galaxea*.

Tabular dissepiments are flattish plates extending across the whole width of the *corallite* or confined only to its axial part (thus resembling tabulae in rugosans). *Vesicular dissepiments* are small arched plates, convex upwards and overlapping, and usually inclined downwards and inwards from the edge of the corallite.

Secondary Structures

Sterecome is an adherent layer of secondary tissue which may cover the septal surface. It is composed of transverse bundles of aragonitic needles. Sterecome normally thickens the epitheca internally as well, but in some cases its function is taken over by primary thickening of the septa, by dissepiments or by synapticulae.

In compound scleractinians individual corallites are usually separated by a complex perforated tissue: the *coenosteum*. This may consist entirely of exothecal dissepiments (tabular) or other material, but it provides a support for numerous canals linking the individual corallites and binding the living tissues together in a functionally cohesive mass.

Ascxual Rcproduction and Colony Formation

The form of the colony in the Scleractinia is largely determined by the mode of budding combined with relative growth rates. The following types have been distinguished:

(a) *Intratentacular budding*. Here the polyp divides by simple fission across the stomodaeum, each daughter (of which there may be two, three, four or more) retaining part of the original stomodaeum regenerating the rest. The products of such *poly-stomodaeal* budding remain linked and except in very rare instances never separate as individuals. Meandroid corals are clearly the results of this kind of budding.

(b) *Extratentacular budding*. In this type new stomodaea are produced outside the tentacular ring of the parent. These extratentacular buds soon separate and do not remain linked. The colony that results from such budding (whether branching or massive) thus consists simply of numerous separate individuals and is not integrated functionally in the same way as are the products of intratentacular budding. There may, however, be some nervous or chemical linkage if the soft tissues are in contact, even if the enterons are not united.

Evolution in Scleractinians

The oldest known scleractinians are Middle Triassic in age. They may have been derived from a group of soft anemones or from a group of late Palaeozoic cyathax-onians. On their first appearance they are already differentiated into a number of important families. *Hermatypic* in type, they formed small patch reefs, may have possessed zooxanthellae and were confined to warm, shallow, well lighted waters. By the late Triassic they were expanding fast in many parts of the world, and many new families arose. This pattern persisted through the Lower Jurassic but with the first true deeper-water *ahermatypic* corals (Suborder Caryophylliina) emerged at about that time, though these were of limited importance. By the Middle Jurassic scleractinians were on the increase everywhere, and patch reefs erupted wherever conditions were suitable. In the Tethyan Ocean the first large reefs developed, persisting until a general setback in the early Cretaceous. Later the large reefs and coral banks developed, and although some of the older families died out others arose. The success of reef corals fluctuated in the Cretaceous, and the ahermatypes became really important for the first time.

By the late Cretaceous both hermatypic and ahermatypic corals were becoming of distinctly modern type, though like many other groups they suffered a major extinction even at the end of the Cretaceous. By the end of the Eocene the dominant groups were much like those of the present since the formerly important Mesozoiic families (e.g. the Montlivalitiidae) had disappeared. During the late Tertiary the two main coral faunal provinces of the present—the Indo-Pacific and Western North Atlantic provinces—had become distinct, and reef-building corals flourished on an unprecedented scale. Many reefs were killed by the lowering of sea level in the Pleistocene, but their remains have provided stable surfaces for the regrowth of coral since their submergence. Comparative few general were affected by the conditions of the Pleistocene, so that Pliocene coral faunas are effectively the same as those of the present.

Mode of Life of Rugose Corals

The clues to the ecology of rugose corals lie in the type of rock in which they occur, the associated fossils and, of course, in analogy with modern corals. Some Rugosa occur mainly in shallow-water limestones and calcareous shales. *Lithostrotion* and *Lonsdaleia*, for instance, often form extensive sheets which can be traced for many miles in the Lower Carboniferous of the north of England. The enclosing rock is a light-coloured limestone in which brachiopods are next in abundance to corals; and it formed, perhaps, in shallow and relatively warm waters. Some of the small horn-shaped solitary corals like *Zaphrentiles* occur in dark limestones and calcareous shales which were formed in deeper, quieter, conditions; they represent an environment from which the compound rugosa are conspicuously absent.

Additional Common Genera

Montlivaltia. Solitary; corallum cup-shaped to cylindrical; many septa project above shallow calice; no columella; many arched dissepiments. Probably zooxanthellate.

L. Jurassic-Cretaceous

Thecosmilia. Like *Montlivaltia* but forms small colonies. Dendroid to phaceloid.

M. Jurassic-Cretaceous

Thecosmilia. Like *Montlivaltia* but forms small colonies. Dendroid to phaceloid.

M. Jurassic-Cretaceous

Isastraea. Colonial; massive, cerioid corallum: corallites hexagonal to pentagonal in transverse section; about four cycles of septa; thin dissepiments. Probably zooxanthellate.

M. Jurassic-Cretaceous

Parasmilia. Solitary cup-coral fixed at base; elongate becoming cylindrical in mature stage; surface of calice ridged by many septa arranged in four cycles; prominent spongy columella; non-zooxanthellate; common in the Chalk (U Cretaceous).

L. Cretaceous-Recent

Geological History

M Trias-Recent

By late Trias times, scleractinians were widespread and varied. They continued to expand in the Jurassic and Cretaceous, reaching peak numbers in the late Cretaceous. In Britain they are represented by only a few genera in the earlier Jurassic, usually of solitary forms like *Montlivaltia*. The upper Jurassic saw the first real development of coral reefs, especially in the Tethys area and extending to about 54° N, present day latitude. This is about 20° further north than today's reef belt. Reef-like structures with *Isastraea* occur, for instance, in the Corallian of southern Britain. There was little reef formation in the early Cretaceous, but in mid and (more so) in late Cretaceous, reefs were again extensive. In Britain, however, corals were not an important component of the fauna, mainly being represented by solitary cold-water-cup-corals like *Parasmilia*. Numbers fell sharply at the end of the Cretaceous and did not begin to revive until the Eocene. There was then a renewed expansion of coral reefs, reaching its peak in the Miocene in substantially the areas where the reefs are best developed today. These are the tropical and subtropical regions around oceanic islands and along the east coasts of larger land masses, e.g. the barrier reefs of Australia.

Apart from small specimens of *Goniopora*, a reef-building form occurring in the Eocene (Bracklesham Beds only), only cold-water cup-corals are found in later deposits in Britain, e.g. *Balanophyllia* in the Pleistocene Red Crag.

Most families of scleractinians show evolutionary trends, especially from solitary to compound coralla and, in colonial types, from phaceloid through plocoid and cerioid to meandroid form. This probably increased the degree of colonial integration.

Ecology of Scleractinians

Scleractinians fall into two main ecological categories: hermatypic and ahermatypic.

Hermatypic corals are those in which the endodermal cells are replete with symbiotic algae (dinoflagellates or zooxanthellae). These algae exist in enormous numbers, and they are so essential to the metabolism of the coral supplying it with nutrients and oxygen that hermatypes can only flourish in the photic zone, at depths normally less than 50 m (rarely 90 m). Since the corals have firm anchorage they grow successfully and most abundantly within the zone of wave action, at depths less than 20 m.

Hermatypes generally need a minimum water temperature of 18°C and flourish best between 25 and 29°C and at normal salinity. Hence most reef corals, which are hermatypic, are restricted to shallow, well lighted, warm water. In addition they need a good supply of oxygen and generally prefer a firm non-muddy substrate for otherwise the planulae will not settle. There are, however, hermatypes living and building reefs in muddy regions. Light is perhaps the most important of all the limiting factors, though a reef can be killed by tidal emergence during heavy rain, during a hurricane for instance, and may take 20 years to re-establish itself.

Ahermatypic corals have no zooxanthellae and do not have the same environmental restrictions. They are found at depths up to 6000 m but are most abundant down to 500 m, and though they can survive temperatures below 0°C they flourish best between 5 and 10°C. Furthermore, they can live in total darkness. Over two-thirds of these 'deep sea corals' are solitary and do not form reefs, though the colonial dendroid caryophyllid *Lophelia* forms deep water banks or thickets at the edge of the western European continental shelf.

The earliest (Triassic) scleratinians were evidently hermatypes, though perhaps not so adapted for the reef habitat as those of the present since they formed only small reefs. Ahermatypic corals developed from many hermatypic stocks by the slow spread of certain types into colder deeper waters. The first known ahermatypes were the Jurassic caryophyllids whose descendants retained this habit, but representatives of many other groups also become ahermatypic, so that the hermatype/ahermatype division today cuts across many families and even genera.

It is possible that the great expansion of the hermatypic reef-builders was directly linked with their adoption of zooxanthellae. Such symbiotic algae are found in other organisms too. The ecology of modern corals has been much studied; useful references are Yonge (1968), Muscatine and Lenhoff (1974), Lewis and Price (1975), Gill and Coates (1977), Goreau *et al.* (1979) and Vosburgh (1982).

Classification of the Scleratinians

In classifying the Scleractinia, as with the Rugosa and Tabulata, it has been necessary to find stable taxonomic characters. Septal structure can be used for defining subordinal categories whilst mode of colony formation is of value only at the familial level. Habitat and dimensions at any level of classification are of little value.

Order Scleractinia (M. Trias.-Rec.): Solitary or colonial corals with mesenteries paired and the septa arranged in multiples of six. Basal plate gives rise to septa and epitheca.

Suborder 1. Astrocoeniina (M. Trias.-Rec.): Normally colonial hermatypes with small corallites. Septa of up to eight trabeculae, spinose to laminar. Examples include *Acropora*, *Thamnasteria*.

Suborder 2. Fungiina (M. Trias.-Rec.): Solitary or colonial with perforate septa linked by synapticulae. Corallites usually large. Mainly hermatypic. Examples include *Fungia*, *Cylolites*, *Isastraea*.

Suborder 3. Faviina (M. Trias.–Rec.): Laminar septa with fan-like trabeculae. Dissepiments present but synapticulae rare. Mainly hermatypic. Examples include *Favites*, *Montlivaltia*, *Thecosmilia*.

Suborder 4. Caryophyllina (Jur.-Rec.): Normally solitary and ahermatypic. Septa laminar with simple trabeculae in fan system; rare synapticulae. Examples include *Parasmilia*, *Caryophyllia*.

Suborder 5. Dendrophyllina (U. Cret.-Rec.): Solitary or colonial. Septa laminar but irregularly perforate. Wall of swollen synapticulae. Mainly ahermatypic. Examples include *Dendrophyllia*.

CORAL REEFS

Recent coral reefs are confined to tropical seas and are best developed in the Indo-Pacific area and in the western North Atlantic. The Indo-Pacific reefs, where maximally developed (Melanesia to South-East Asia), have 92 recorded genera and subgenera and over 700 species, while the coral fauna of the North Atlantic is smaller with only 34 genera and 62 species. The North Atlantic fauna seems to have been derived initially from a Pacific source since Oligocene reefs of the West Indies have many of the same corals, but it has since developed in isolation, particularly since the Americas were linked by the Panama isthmus in the Pleistocene.

In *Structural* reefs the scleractinians have built up massive wave-resistant structures, often to great thicknesses, where the bulk of the material is contributed by corals, even though calcareous algae may play an important part in reef structure. Such a large rigid structure built up by generations of corals is the habitat of a host of diverse organisms, the whole forming one of the most complex ecosystems known. In coral reefs productivity, calcium metabolism and carbonate fixation (3500 g m^{-2} a^{-1}) are very high. The range of habitats is such that there is a high degree of specialisation in the associated fauna, especially fishes, molluscs and arthropods; these are found in zoned and localised niches in various parts of the reef.

In reef communities, however, though equally confined to tropical seas the corals merely live together without building up a rigid structure.

Types and Genesis of Coral Reefs

Structural coral reefs today belong to three major types: *fringing reefs*, *barrier reefs* and *atolls*. Fringing reefs develop along shorelines, especially those of volcanic islands. Barrier reefs are formed some distance from the shore. Atolls are circular or horseshoe-shaped and usually orientated with respect to the prevailing wind, with a central shallow lagoon in which small patch reefs may develop. The Great Barrier Reef off eastern Australia is a special case, being a linear feature parallel with the fault-bounded margin of the continent, where coral growth has initiated on the edges of an upraised fault.

It is now well known that the three major types represent successive stages in the growth of coral around a sinking volcanic island. The growth of corals has kept pace with subsidence. While broken reef talus has formed cascading fans going down to depth on the outside. This theory, originally suggested by Darwin, has been confirmed by deep borings into atolls. At Eniwetok atoll, bored in 1951, basalt was encountered at about 1.25 km below the surface. The limestone overlying the basalt was Eocene in age. Thus Eniwetok is a limestone pillar 1.25 km thick standing on a deeply drowned volcanic island some 3.2 km high on the ocean floor. The maximum rate of subsidence has been about 50 cm per year. Confirmatory seismic and drilling evidence from other atolls suggests that this is the common pattern. The now sunken volcanic islands arose at 'hot spots' on a submarine ridge crest and have been carried away into deeper waters (as on a conveyor belt) as the ocean floor spread away from the ridge.

Darwin's theory of atoll formation, though now vindicated, did not take into account Pleistocene changes of sea level, which were far reaching since the reefs were exposed and the corals killed off. The exposed surfaces were eroded to some extent and often fretted into a karst topography. Hence the veneer of corals which has grown since the end of the Pleistocene has re-established itself on an anomalous topographic surface, and it is because of this that the form of a modern reef is not simply a reflection of constructional activity.

Growth and Zonation of Corals

Corals on reefs have few natural enemies other than coral-eating fishes and the starfish *Acanthaster*. The latter has recently wreaked havoc on some Pacific reefs, for after the depredations of the starfish the dead coral becomes covered with an algal film which prevents coral planulae from settling.

Because of the steep environmental gradients within the reef complex the corals themselves are ecologically zoned with specific reference to the shoreline. Each species occupies a particular habitat and is not found elsewhere. Within particular habitats there is a kind of 'pecking order' in corals; dominant species may actually eat other species of the same genus by mesentery extrusion and digestion. Different species inhabit the windward and leeward sides of the island. There is, furthermore, a great change both in species and in their growth forms, from the upper part of the fore-reef slope across the reef flats to the central lagoon with its shell sand floor and patch reefs. In some cases massive rounded corals such as *Diploria* tend to inhabit the surf zone, while the stout branching *Acropora* (the 'stag's horn' coral) lives at greater depths out of the range of surf action. In other areas massive corals are commoner in quiet waters, and *Acropora* occurs in high energy zones. The apparent structural strength of corals in therefore only one factor amongst many that control ecological zonation; ability to compete for space and to cope with sediment influx may likewise be important.

Zonation varies from reef to reef depending on various factors. The presence of resistant algal ridges, for instance, can greatly affect the pattern of zonation. Not only the scleractinians but also the gorgonians (in the Caribbean reefs) and octocorals (in the Indo-Pacific reefs) are ecologically zoned, and so are the various bivalves and other molluscs that form specialised communities in the very varied reef habitats.

The growth rate of corals varies greatly with the environment and the season but may reach 10 cm per year in branching corals and 1 cm per year in the more massive colonies. Reef growth as a whole may reach a maximum of 2.5 cm per year, but in the past it has generally been less.

Within reefs there is not only continual construction by the scleractinians and encrusting algae, but also a constant attrition through wave action and, equally important, destruction by boring organisms and coral-eating fishes. These produce a great quantity of fine communited coral sand,which fills up the lagoon and is transported to deeper waters.

Deep-water Coral Banks

Deeper water corals (ahermatypes such as *Lophelia* and *Dendrophyllia*) may form structures differing from those of their shallow water counterparts. Some of these deep water corals are solitary, others are colonial and dendriform.

The development of these structures takes place in four stages. At first there is a single *colony*. When this colony is joined by others it becomes a *thicket* a few metres across, in which the members may be all of the same species or of different species. This newly developed environment attracts other organisms, e.g. fishes, molluscs and crustaceans, giving it a new ecological character. As the thicket spreads and matures, skeletal debris accumulates from broken and bored coral, providing a new substratum which again attracts other animals. This stage is now a *coppice* and may be several metres across. Eventually, with the accumulation of more debris and further coral growth, a *bank* develops: a topographic entity with a core of solid skeletal debris, a covering mat of more open debris and a capping of live coral. The proportion of living to dead coral decreases with time.

The exposed sequence studied by Squires were Tertiary in age, but similar banks are forming today. The study of fossil thickets, coppices and banks can yield valuable palaeocological data.

Geological Uses of Corals

Corals as Stratigraphical Indicators

Corals are generally too long-ranged for use as zone fossils. Nevertheless, they have been used where no shorter-ranged fossils are available, especially in the Carboniferous. In 1905 Vaughan, working on the Lower Carboniferous of the Bristol region, England, erected a zonal scheme based upon the first appearance of corals and brachiopod. This scheme, though crude in terms of time range, has proved to be effective and to be applicable, with minor local modifications, to other parts of Britain and Europe.

Vaughan believed that the faunal sequence he erected reflected evolutionary lineages, but in fact the corals really occur as assemblage biozones with each of the main cycles bringing in a new migrant fauna. The correlations arc somewhat inexact and are being supplemented or even replaced where possible by a stratigraphy based on goniatites, conodonts and for-aminiferids. New stages based upon transgressive cycles in the Lower Carboniferous are now standard, but much refinement is still needed. Coral zonation in the European Devonian, once elaborated in great detail by Wedekind, has proved to be over optimistic.

Corals as Geochronometers

In well preserved specimens of many rugose and scleractinian corals the epithecal surface shows fine growth ridges, some 200 per centimetre. Each of these represents a growth increment: the former position of the rim of the calice. These fine growth ridges are often grouped in prominent bands or annulations between which the epitheca

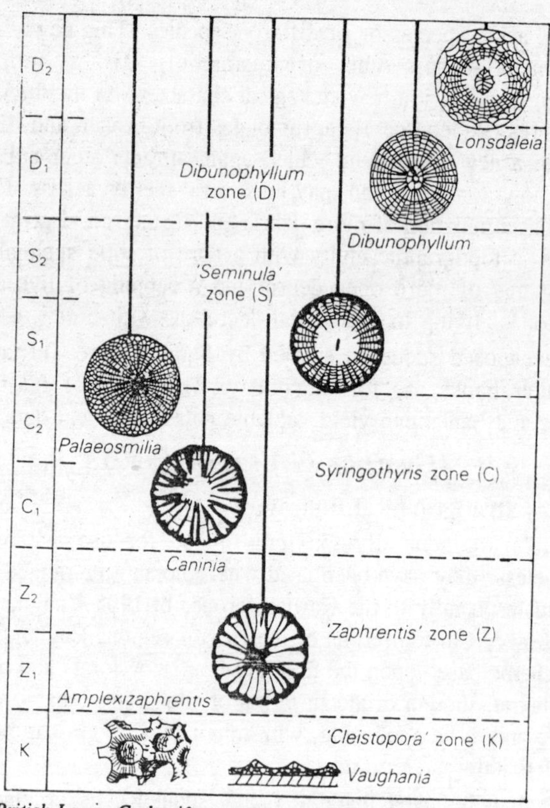

Fig. 13.11. ***British Lower Carboniferous coral-brachiopod zones, based on the original work of Vaughan and showing the time ranges of various coral genera. The Seminula zone is based on the eponymous brachiopod.***

is constricted. It is normally recognised that the fine growth ridges represent daily growth increments, whereas the banding is monthly and the broader and more widely spaced annulations are yearly. Wells (1963) first established in the scleractinian *Manicina* that the growth ridges were diurnal, and since the Recent *Lophelia pertusa*, from Norwegian fjords, has 28 growth ridges (presumably diurnal) per band, each band corresponding to a lunar cycle, the monthly banding hypothesis seems at least possible.

Wells (1963), working on Devonian corals with annual rather than monthly annulations, counted the number of growth ridges per annulation in a number of species. Though the results were limited by the preservation of the corals, he concluded, mainly from observations on *Heliophyllum*, *Eridophyllum* and *Favosites*, that there were an average of 400 days in the Devonian year. This figure corresponded well to astronomical estimates that the Earth's rotation has been slowing down through tidal friction by about 2 seconds per 100 000 years. Assuming that the Earth's annual circuit round the Sun was the same length, which seems to be fairly well substantiated, there must have been more days in the Devonian year, but they were shorter.

Annual banding is not often found in corals, having only been recorded in specimens that originally lived in waters where there were seasonal fluctuations. It is normally easier to work with the Rugosa that show the growth ridges grouped in monthly bands. These are more abundant and do not have to have a complete epitheca. Many Recent corals have monthly breeding cycles during which carbonate deposition is inhibited, and this may have been the case with the Rugosa.

The monthly banded *Middle* Devonian corals studied by Scrutton (1965) had an average of 30.6 growth ridges per monthly band, and if the figure of 399 (the astronomical estimate for length of the Devonian year) is divided by the average growth-ridge count per band, the result is consistently 13. Hence it seems that 13 bands each of about 30.6 growth ridges were laid down by these corals in the Middle Devonian year.

The 'coral clock' provides a consistent check on estimates of the slowing of the Earth's rotation based upon a biological rather than a physical system. It has, however, become increasingly clear that corals are not ideal tools for this, for while diurnal banding is clear enough on all unworn corals, clear monthly and annual cycles are harder to detect and may not be present in many corals. Random events may in any case have left a masking record in the growth of the skeleton. Periodic growth features in bivalves are now better understood and may prove to be more suitable for geochronometry.

Corals as Colonies: the Limits of Zoantharian Evolution

Colonies have been defined as 'groups of individuals structurally bound together in varying degrees of skeletal and physiological integration; all genetically linked by descent from a single founding individual'.

Such colonies (e.g. cnidarians, bryozoans, graptolites) can exhibit a wide range of organisation and integration. At one extreme the zooids budded off from the parent may be completely independent, whereas at the other end of the scale the individuals of the colony may become linked and co-ordinated so that the whole colony can function as a single unit. In some cases zooids may become highly specialised, promoting division of labour amongs the members of the colony.

Advantages of the colonial state presumably are mainly connected with greater efficiency in protection and stability as well as in reproduction, feeding and respiration. Thus a colony of asexually reproducing individuals can soon develop a large and effective biomass, with a strong and stable skeleton in which there is the potential for integration and co-operation between zooids.

In the octocorals there are two distinct functional types: (a) those which secrete a massive scleractinian-like skeleton with very little integration; and (b) those (the vast majority), including pennatulaceans, in which integration varies from the very simple, through numerous intermediate stages, to very complex colonies with the various zooids specialised for different functions. In the most extreme cases not only are interzooidal canals linking the enterons retained, so that food captured by a few zooids can be shared by the rest of the colony, but also there are specialised 'siphonozooids' organised for pumping water through the coenenchymal canals, to ensure that adequate supplies of oxygenated water for respiration reach all members of the colony. Such *polymorphic* zooids function together but cannot function apart, and the colony acts as a 'superindividual'. Despite the fact that this level of integration in the octocorals evidently originated very early in their history (if the interpretation of *Rangea* as a Precambrian pennatulacena is correct), the zoantharians have never fully reached the same levels of colonial organisation since their zooids are not polymorphic.

Throughout the Phanerozoic colonial corals have outnumbered solitary types, though in the case of the Rugosa solitary corals were more abundant than colonial ones. Colonial Rugosa vary in level of integration between (a) those phaceloid colonies in which individuals are functionally unconnected and (b) aphroid colonies where the corallite walls have gone and zooids were probably connected together by linked enterons. This is the highest level attained in the Rugosa.

Tabulate corals likewise did not achieve a high degree of colonial integration, though mural pores in *Favosites* suggest enter on connections. The heliolitids, however, had a coenosteal system paralleling that of coenosteal scleractinians, but for some reason which is not clear they were not especially successful. They too perhaps were limited in growth rate because the skeleton was massive rather than light and porous like that of scleractinians.

Although cerioid and phaceloid forms are known amongst scleractinians, a generally higher level of integration is apparent in the preponderance of meandroid and coenosteoid types. Meandroid genera have a *polystome* system in which hundreds of mouths with their adjacent tentacles connect in each 'valley' with a single elongated enteron. In coenosteal Scleractinia the enterons of adjacent polyps are also all connected, and skeleton building has become a cooperative venture. But even the most advanced scleractinians, with the possible exception of a very few rare genera, do not exhibit polymorphism.

Despite the fact that rugose and tabulate corals were able to make some headway towards an integrated colonial system, they were generally less successful in rocks of Permianin the Scleractinia. But their comparative failure in the reef-builders was probably due not so much to this and to certain other features of their organisation. The Rugosa and to some extent the Tabulata had slow-growing and rather solid and heavy skeletons in which the great deal of calcium carbonate was deposited in internal structures. They lacked an edge zone and could not attach themselves firmly to the substratum. It is not known whether Palaeozoic corals had zooxan thellae in their tissue; if they did not, one of the great advantages possessed by reef-building scleractinians was denied them, and this would have affected their growth rate, ability to get rid of waste material and other factors.

But the strong, yet light and porous, stable, fast-growing and at the same time increasingly integrated skeletons of the scleractinian permitted the development of the coral reef environment, so laying down the framework for some of the most complex and productive of all ecosystems that have ever evolved.

Minor Cnidarians

Subclass Octocorallia

?Precambrian L. Jurassic-Recent

These are marine, sessile, colonial forms with polyps projecting from shared soft tissue. Polyps have eight feathery tentacles and eight mesenteries. They may secrete a horny skeleton, sometimes with calcareous spicules, or be calcified.

Octocorallins include sea-fans, e.g. *Gorgonia*, and sea-feathers, which are important members of recent coral reefs; and sea-pens, the Pennatulacea. In the latter a primary polyp produces a long stem from which numerous secondary polyps are budded, often forming lateral leaf-like structures. It also forms a down-growing stalk which fixes the colony in soft sediments. Octocorallians are rare as fossils. Possible sea-pens have been identified from the late Precambrian, e.g. *Charniodiscus*.

Class Hydrozoa

U. Precambrian-Recent

Mainly marine, sessile or free-swimming; usually colonial with tiny polyps sharing common tissue. Typically their life history involves an alternation of generations between the polyp phase and free-swimming medusae (jellyfish). The skeleton, if any, is horny (chitinous) or occasionally calcareous.

Marine hydrozoans range down into deep water, about 8000 m. Some, with a massive or encrusting calcareous skeleton, like *Millepora* (Tertiary-Recent) are important reef-builders.

14

Degradation of Reefs

The traditional division of reefs into three main types is retained in this book. Fringing reefs, barrier reefs, and atolls will thus be examined successively, with examples and variations of each type. However, bank reefs and patch reefs are sometimes considered as a special type, and will be dealt with here just after fringing reefs. Ridge reefs, so far neglected, will be introduced: they occur mainly but perhaps not exclusively in the Red Sea. Another section will deal with special cases which do not fit this classification. Finally, the contrasts between Indo-Pacific and Atlantic reefs, including those of Brazil, will be reviewed.

Fringing Reefs

Fringing Reefs without Boat Channel

It is unusual to find fringing reefs without any boat channel, since there is generally some sediment arriving from the adjacent land which prevents good coral growth close to the shore. Such fringing reefs are found, however, particularly on the Red Sea coasts where, owing to the arid climate, the supply of sediment is frequently meagre. A good example, described with precision by Mergner and Schuhmacher (1974), occurs 6 kilometres south of Aqaba harbour in Jordan. This coastal reef has been built off a narrow beach of sand and pebbles, with rock slabs. The reef flat is 60 metres wide, with rich coral growth underlain by reef rock without large voids.

The reef flat is about 0.80 metre below sea level (tidal range is significant, but the level can be lowered by meteorological and seasonal variations). The outer edge is normally steep, with horizontally growing superimposed coral colonies. Seaward of shallow (4 to 5 metres) sandy

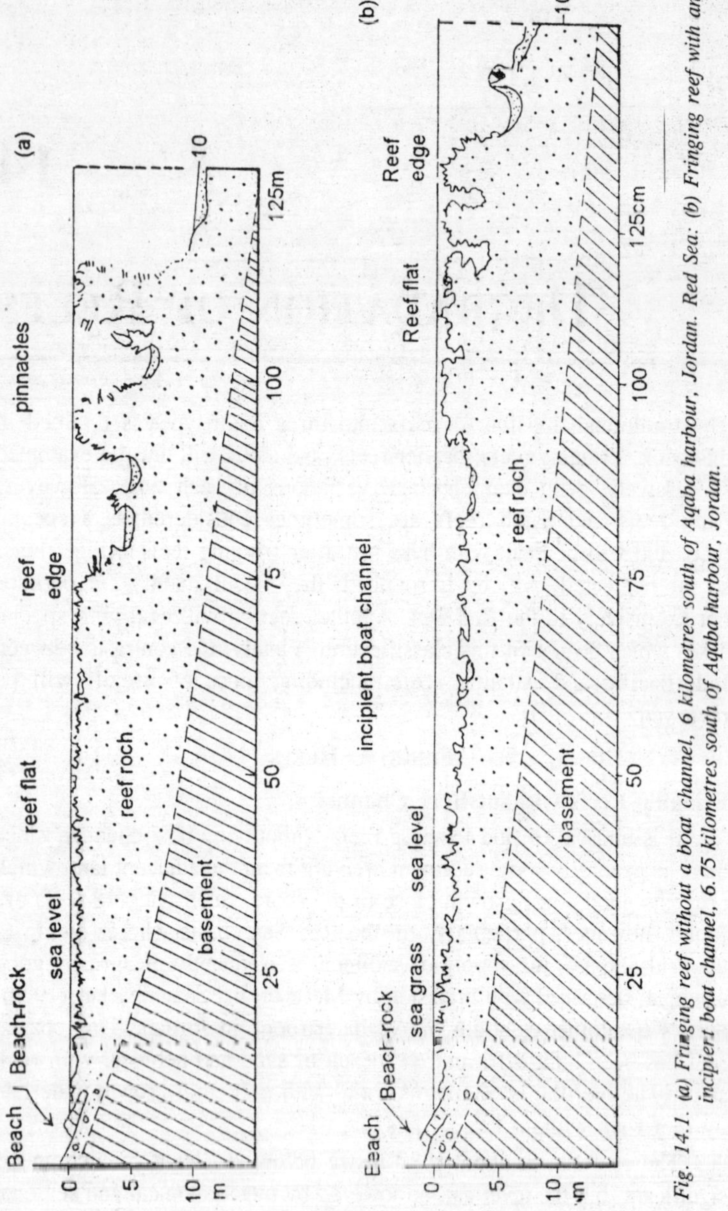

Fig. 14.1. (a) Fringing reef without a boat channel, 6 kilometres south of Aqaba harbour, Jordan. Red Sea: (b) Fringing reef with an incipient boat channel, 6.75 kilometres south of Aqaba harbour, Jordan.

or muddy depressions, pinnacles with rich coral growth and steep or overhanging sides rise to the same level as the reef flat. Another fringing reef without a boat channel has been described from the Sarso

Islands in the southern Red Sea. These islands are emerged reefs, probably originating as ridge reefs, common in the Red Sea.

The fringing reef is very narrow, below a coastal corrosion notch in the emerged fossil reef, and the beach is restricted to a thin sheet of sand and silt on the lower part of which sea grasses grow. Alternations of coral heads and small hollows with a shallow sand fill extend as far as the outer edge, but there is no boat channel. In rare instances, reefs growing offshore in coastal waters can become attached to the mainland by progradation of the coastline behind them. This has happened near Yule Point, North Queensland, where, under a humid tropical climate, rivers discharge large quantities of muddy sediment which is the cause of the progradation. In such conditions, the absence of boat channel occurs in a climatic environment just opposite to the above mentioned Red Sea reefs. The existence of such reefs is obviously threatened by excess of sedimentation.

Fringing Reefs with an Incipient Boat Channel

This subtype is represented 6.75 kilometres south of Aqaba harbour, i.e. 750 metres south of the reef. According to Mergner and Schuhmacher, the only difference is the presence here of a very shallow depression, 1.50 metres below the general surface of the reef flat. This 'lagoon' does not contain much sediment, but the coral growth is different in that it consists of numerous scattered colonies. The difference could be due to local minor variations in sediment supply. On the other side of the same gulf near Elat, Israel, an incipient boat channel is also found in the fringing reef. The beach is fed by granitic sand and stones, mixed with calcareous sand and including beach rock slabs. It is bordered by a boat channel, not more than 2 metres deep, and between 10 and 40 metres wide, with a sandy and muddy bottom from which grow coral heads of various shapes and sizes. The reef flat is 15 to 30 metres wide and 0.3 to 0.8 metre deep, with a complicated pattern of deeper holes partly filled by coral detritus.

Larger coral colonies occur on the outer edge with the usual horizontal growth. Detrital material appears against the foot of the slope, but it is largely covered by living corals as the depth is still very small. Knolls exist in front. Such reefs on both sides of the innermost parts of the Gulf of Aqaba or Elat are obviously young, being built over the apron of apparently Pleistocene material which covers the crystalline basement. Farther south in the Gulf of Elat along the Sinai Peninsula, fringing reefs display more complicated patterns and have had a longer history. They include remnants of sharms or guilles cut during the Pleistocene regressions and are backed by sebkhas and mangrove swamps. Relics of former fringing reefs outcrop locally at a similar level. Mahe Island in the Seychelles has many fine fringing reefs, with shallow boat channels containing rippled sands. Behind these, meadows of marine grasses such as *Thalassia* and *Cymodocea* are exposed, a feature which is frequent in the Western India Ocean (e.g. in Madagascar and Kenya).

The outer parts of these fringing reefs include ridges formed by dense growths of brown algae, which are replaced by branching corals on the outer sides of windward reefs. Fringing reefs with discontinuous boat channels and fringing reefs without boat channels alternate along the southwestern coast of Sri Lanka at Hikkaduwa near Galle, where they have been illustrated by Mergner and Scheer (1974).

Fringing Reefs with a Well-developed Boat Channel

Lord Howe Island in the Tasman Sea between Australia and New Caledonia bears the southern most reefs in the world, and provides a good example of a fringing reef with a distinct boat channel. The fringing reef extends over 6 kilometres along the west coast of the island. It is divided into several segments by passages, which are shallow except for Erscott's Passage (15 to 19 metres). The outer slope does not descend immediately to great depths (Lord Howe lies on an insular shelf. 30 × 18 kilometres wide at a depth of 50 metres), but displays well-developed spurs and grooves. The boat channel, locally known as the lagoon, is shallower than one metre at low spring tide, but distinctly below the level of the reef which is fully exposed under these conditions. The reef includes many living colonies, especially *Acropora* spp., but also a number of soft corals (*Alcyonaria* spp.), and algae such as *Caulerpa* and *Padina* spp. In the north part the reef continues into the boat channel as a series of arms which have grown in alignments recurved by the refracted south-westerly oceanic swell, and consisting of dead coral blocks of medium and small size. The boat channel ends landward in a beach of calcareous sand, with coral and shell debris, backed by recent dunes and consolidated calcareous aeolianites.

The climate is subtropial with Mediterranean affinities in the monthly distribution of rainfall. The aeolianites behind the beach have been cut back to form a shore platform in front of an overhanging cliff, with mushroom-shaped stacks rising from the platform. In the boat channel the sediment is of variable thickness, and fills buried valleys that have been discovered beneath its floor. It seems that the present (Holocene) fringing reef rests upon a Pleistocene one, which has been eroded during the low sea-level phase(s). The aeolianites probably formed during the last marine regression as they continue below present sea-level. In the Caribbean, a much narrower fringing reef with a boat channel may be mentioned at Jacmel on the south coast of Haiti. This reef is backed by a series of elevated coral terraces.

Seaward from the coastline is an inner platform with big blocks that have fallen from the overhanging cliff cut into the previous reef; a narrow but well-shaped boat channel with phanerogams growing on a sandy bottom; and an outer reef bearing many sargassoes. Scattered path reefs grow in front, as in the examples quoted from the Red Sea.

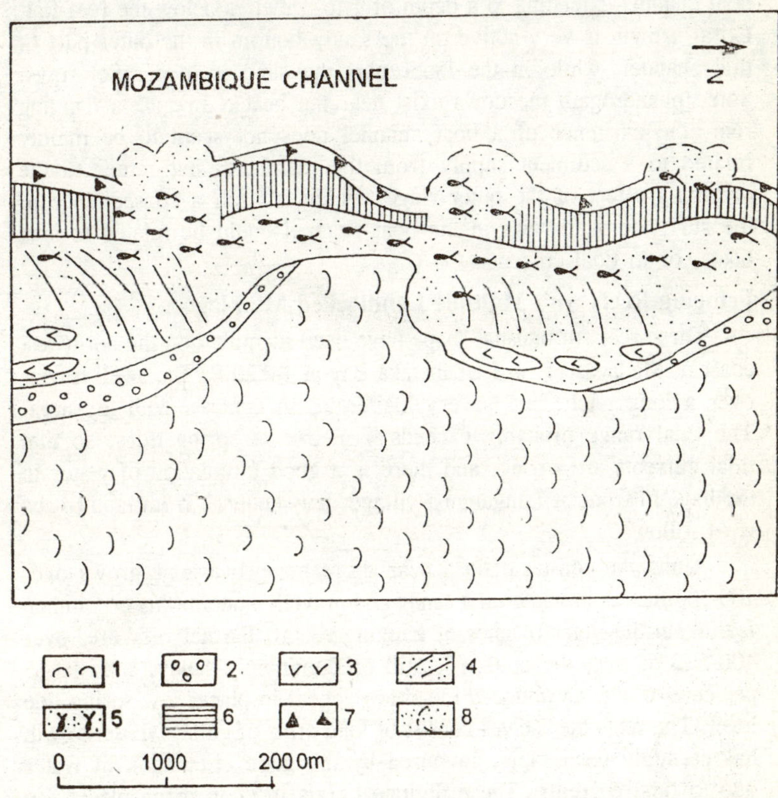

Fig. 14.2. Fringing reef on the southwestern coast of Madagascar (lat. 24°S). 1, Pleistocene and Holocene dunes; 2, beach; 3, Phnerogam meadows in inner boat channel; 4, boat channel with striations in sand; 5, scattered coral colonies; 6, reef flat; 7, detrital outer ridge with big blocks; 8, submarine sandy delta.

The fringing reef on the southwest coat of Madagascar, described by Batistini (1964) and Pichon (1972), has grown in a much higher wave energy environment since this coast is exposed to a southwesterly swell generated in the Southern Ocean. It is advisable here to proceed from sea to land in the description. On its outer side the reef is capped by a detrital outer ridge including big blocks (the so-called negro heads). This ridge is interrupted by gaps filled by submarine sandy deltas generated by strong ebb currents, the tidal range being 3.3 metres at spring tides.

The detrial accumulations in incisions in the reef are thus in a reverse direction to those in Lord Howe fringing reef. Inside the outer ridge the reef flat is exposed at low spring tides; it is backed by a boat channel extending to a depth of 1 to 3 metres below the reef flat. Coral growth is very active on the sandy bottom in the outer part of this channel, while in the inner part the sand is in parallel rows. Some phanerogam meadows exist near the beach. In such a fringing reef, the existence of a boat channel does not seem to be mainly related to a sediment supply from the mainland, but rather to the damming effect of the outer ridge, which is itself a consequence of the surf; the sand rows in the boat channel could be related to surf beats, or to tidal currents.

Fringing Reefs with Multiple Landlocked Mini-lagoons

These reefs of unusual shape have been identified on the northwest coast of Madagascar, in Ramanetaka Bay at 14°20'S. The swell travels over a long fetch, but is very moderate, in contrast with the area. The tidal range probably exceeds 4 metres at spring tides, so that tidal currents are strong, and there is a good circulation of water in the bay. In front of Langamena village, the sequence from land to sea is as follows.

Along the coastal plain a near shore mangrove zone grows over 100-200 metres in width on a sandy and gravelly bottom (this is common feature in the inner reaches of fringing reefs). Farther offshore, over 400-500 metres, the bottom is still predominantly sandy, with 10-15 per cent of silt. *Cymodocea* meadows appear in places low spring tide level. The reefs themselves consist of long arms of coral, whose growth has certainly been rapid, favoured by the good circulation of water due to tidal currents. These digitate corals meet in many places, so that they enclose parts of the sea floor, which in this area is 8 to 12 metres deep. The result is that inside the coral ridges are small enclosed lagoons, 5 to 10 metres deep at low spring tide, completely sealed but connected with the open sea by their surface except at low spring tides. The reefs are surrounded by rims of *Porites* and *Astraea*, their inner parts bearing sand, gravel, phanerogam meadows, and small coral colonies. The sediment in the mini-lagoons is finer, including 50 per cent of silt and clay, because of their sheltered situation.

Still farther offshore, the reef growth continues in the shape of coral knolls or patch reefs. Outside the area these coral knolls are extremely numerous. In some places several kilometres offshore the knolls have also enclosed mini-lagoons. Summarizing the facts, the Ramanetaka case seems to owe its features to a very unusual environment: water stirred by strong tidal currents; moderate surf, so that waves do not damage the reefs; moderate depth (10-18 metres) of the bay, so that reefs soon grow to the surface; and absence of pollution, since the mainland is very sparsely populated: Langamena is but a tiny village.

Reefs Intermediate Between Fringing and Barrier

These reefs may be exemplified by some of the raised reefs described by Chappell (1974) on the Huon Peninsula in New Guinea. Some of these emerged reefs had their outer part separated from the land (i.e. from the earlier reef) in the course of the general uplift, by a stretch of water deeper than the usual boat channel (10 metres). In such a situation the term lagoon seems more appropriate than the expression boat channel. The former 'lagoon' zone is easily recognized from its sediments, different in texture from those of the built-up reef. Patch reefs can exist in such a lagoon, with a greater height than the coral heads scattered in boat channels. This is a fossil reef, but such an intermediate subtype is also represented in living reefs. Along the Kenya coast, the reef belongs really to the fringing type is the surroundings of Mombasa.

Farther north, however, near Malindi, it passes laterally into a more complex reef, 3 kilometres wide, including a landward zone of phanerogam meadows an coral colonies on sandy bottom, two or more areas several metres deep with interspersed shallows, and finally a very shallow outer reef. Near Port Sudan in the Red Sea, the fringing reef includes discontinuous depressions, 20 to 25 metres deep, which could have been produce by karstification during a low sea level phase: in which case the fringing reef rests at depth on a dissected Pleistocene reef. In the Caroline Islands and the Marianas, several examples of intermediate situations have been described by Tayama (1952) at Kusiae Island, at Tinian Island, at Merizo, Guam Island. He called all these

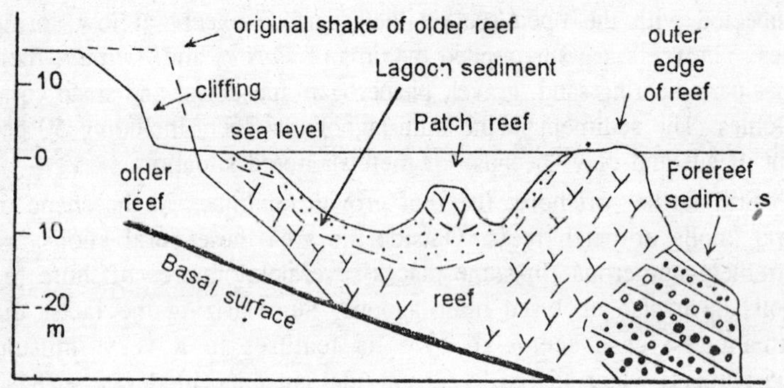

Fig. 14.3. Fossil raised reef of the Huon Peninsula, Papua New Guinea, intermediate between fringing and barrier reef.

structures almost barrier reefs. At Kusaie and Merizo there are elongated sand cays on the reef. Such transitional forms indicate that the distinction between fringing and barrier reefs is by no means absolute, although in the majority of sites discrimination is easy.

Asymmetrical Fringing Reefs

On islands where the direction of wave attack is well defined, with much stronger wave action on the exposed shore than on the sheltered side, it happens, when a small rocky island stand as offshore with moderate depths around it, that a fringing reef grows on the exposed side only, whereas on the sheltered side a sandy tail is built. The swell comes from the northwest as in Ramanetaka Bay where this subtype is also represented on four other small islands. The dynamics of such frameworks have some resemblance to the glacial forms known as crag and tails in Scotland. In the case of coral islands such as Nosy Iranja, the absence of a fringing reef on the leeward side seems due to the arrival o an excess of sediment supplied by erosion on the windward side. An emerged sand cay with beach rock indicates that Nosy Iranja belongs to a somewhat mature stage. Comparison should be made with the sand cays associated with isolated reef patches in large lagoons as behind the Great Barrier Reef and in New Caledonia, but without emerged rocky islet to support the structure.

Bank Reefs

Under this name, a number of reefs may be grouped. The smalles are often called *patch reefs*. They rise from various depths, and lie at various distances from the land. Their shapes can be quite different. Off the coasts of Cuba, numerous small bank reefs have been carefully described by Kuhlmann (1970, 1971, 1974). At Cayo Macao Reef in

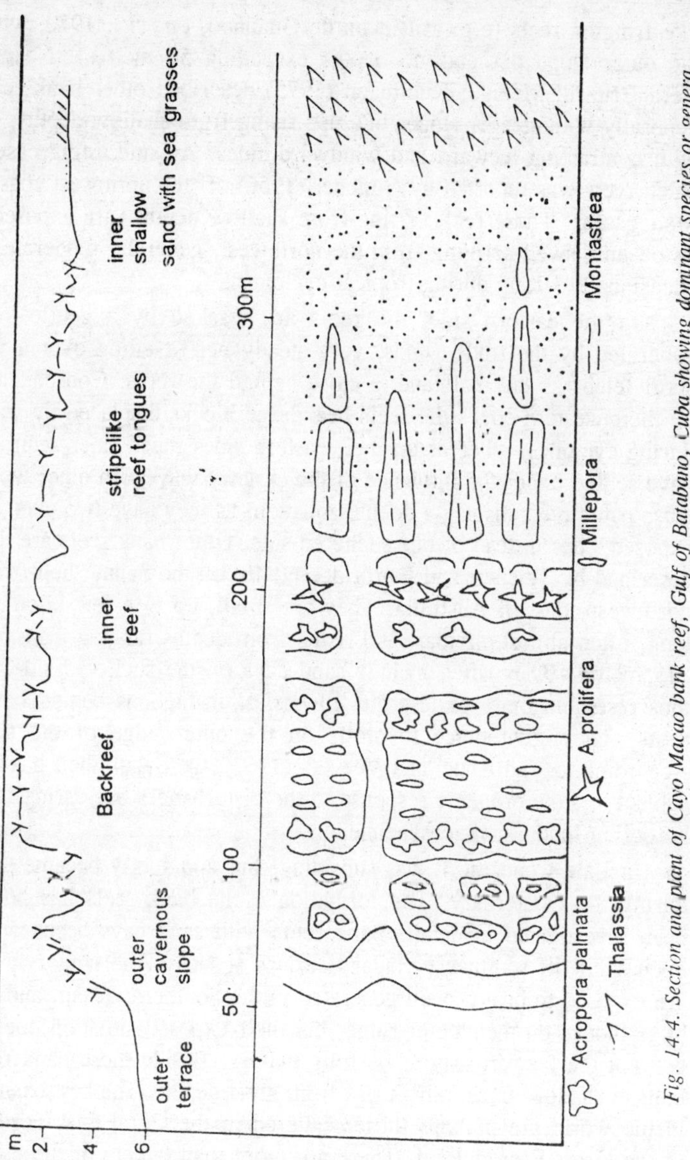

Fig. 14.4. Section and plant of Cayo Macao bank reef, Gulf of Batabano, Cuba showing dominant species or genera.

the Gulf of Batabano for example the usual zonation from exposed to sheltered side appears clearly. The depth from which the reefs grow is very small. The other examples described by Kuhlmann from various coasts of Cuba are not essentially different, although some are more like fringing reefs (e.g. at Rincon de Guanabo, *op. cit.*, 1970, where the outer slope descends to depths exceeding 50 metres). Near La Vera Cruz in Mexico, Kuhlmann (1975) described other bank reefs, generally with steeper slopes but also rising from shallow depths, and with contrasting leeward and windward sides. An interesting case of patch reefs was reported by Guilcher (1956) off the northwest coast of Madagascar. These reefs, rising from shallow depths, are exposed to an oceanic swell arriving from the northwest, generally moderate but occasionally strong during tropical cyclones.

On the eastern side, the reefs are washed by a gentle swell generated by the trade winds, very steady but ravelling over a very short fetch, as the mainland is close behind the reefs. Consequently, on their western edges the reefs bear large blocks displaced by waves during cyclones, whereas on their eastern sides much finer sediments have settled under the influence of the easterly waves. In other words, these patch reefs display a double zonation, as they have two unequally exposed sides instead of one sheltered side. Other bank reefs are those described by Teichert and Fairbridge (1948) on the Sahul Shelf off the northwest coast of Australia. They are small, up to a few kilometres long, often almost circular, and rising from depths ranging from 50 to 115 metres. They often include sand cays on the sheltered side. and thus resemble more or less the 'low isles' in lagoons behind barrier reefs. They give place to atolls on the outer edge of the shelf. According to Van Andel and Veevers (1967), the Sahul Shelf has been subject to slow orogenic response to the disturbances occ-urring in the bordering Indonesian geosyncline.

In Late Cenozoic it was subsiding, but apparently became stable during the Quaternary. The formation of its bank reefs has not yet been investigated in detail. Bank reefs with sand cays occur on the open shelf off southwest Madagascar, as at Nosy Andriangory. They are exposed to heavy swell generated in the Southern Ocean, and bear huge blocks on their outer ridge, but their exposed situation does not prevent the leeward cay to be fully mature. Beside these bank reefs, built in shallow areas, others rise from great depths. The best examples in the world are perhaps those scattered in the Coral Sea, northeast of the Great Barrier Reef. There are more than twenty of these coral banks, of different size. The largest one, Tregrose Bank, is 95 *g* 50 kilometres, being thus larger than all atolls except perhaps Kwajalein Atoll in the Marshall Islands (120 *g* 32 kilometres). Several others exceed 30 kilometres. They rise steeply from a wide plateau, 600-

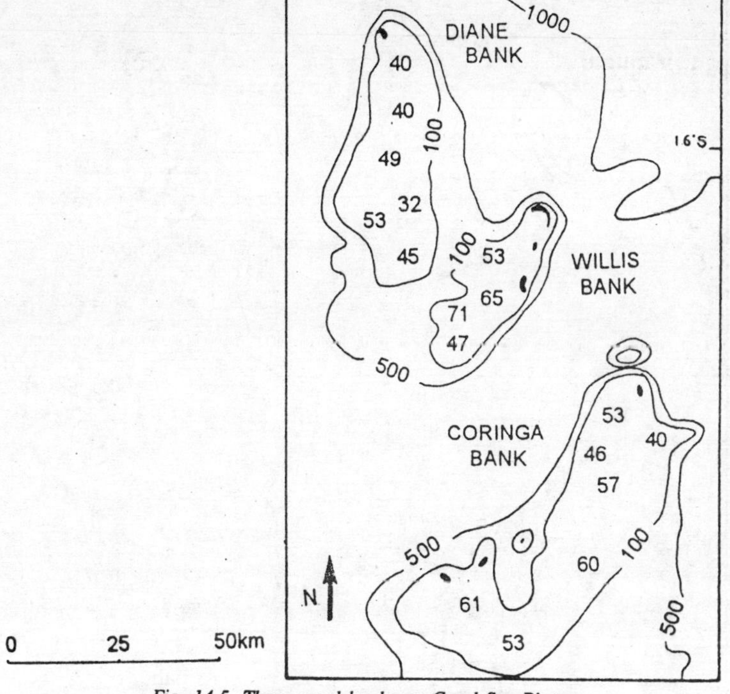

Fig. 14.5. Three coral banks on Coral Sea Plateau.

2000 metres deep, tilted from southeast to northwest, which forms an intermediate step between the Great Barrier Reef in the southwest and the Coral Sea Basin in the northeast where there are depths of more than 4500 metres. The relatively flat surface of these banks is at depths of some tens of metres (usually 50-60 metres), the viable reefs being shallower and mainly found near the bank margins. A few reefs bear cays and islets, both vegetated and unvegetaed, so that some banks such as Tregrosse, Coringa, and Willis more or less resemble huge drowned atolls. The plateau on which they stand is thought to be a result of a Tertiary differential subsidence, which separated it from the Queensland continental shelf upon which the Great Barrier Reef is built.

The continental nature of the crust here has been verified by seismic reflection. The formation of bank reefs on this plateau needs still further research. The large Chagos Bank in the Indian Ocean, even larger than those of the Coral Sea, has been mentioned. Several

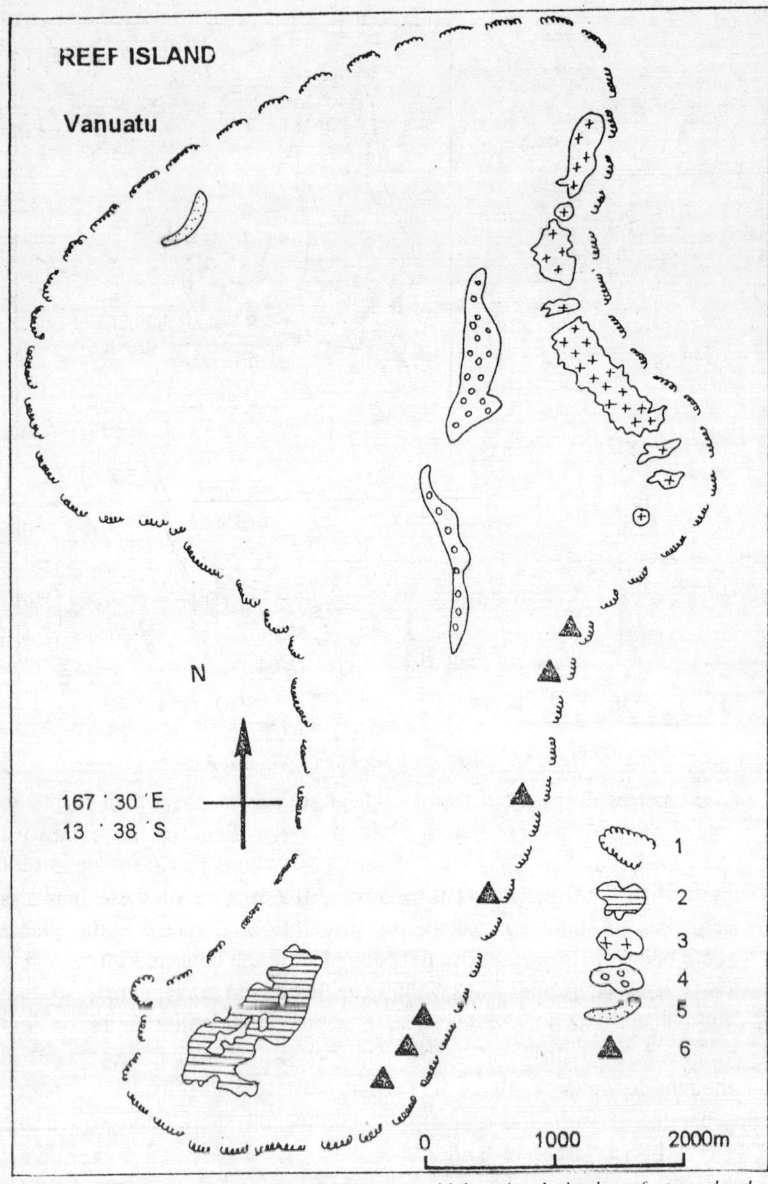

Fig. 14.6. A bank reef: Reef Island, Vanuatu, Melanesia. 1, bank reef at sea-level: basement probably Pleistocene with thin sheet of lose sediment; 2, karstic hole with pool, carved into 1; 3, islets (rampart) of Middle Holocene age; 4, sandy windward islets; 5, leeward sand cay; 6, big detrital blocks thrown by hurricanes onto windward edge.

smaller bank reefs in the same area, as Coetivy, Cargados Carajos, and Tromelin, still await investigation. Another small (8.5 × 5.2 kilometres) bank reef, called Reef Island, and located in the Banks Islands, northern Vanuatu, is an example of the diversification of landforms which can occur in this type of structures. It consists of a coral platform at sea-level, probably Pleistocene, and karstified during the Wisconsinian regression, hence the hollow 7 metres deep in the south. It bears three types of low islets: from east to west, a windward series, also facing the trade words; and a leeward sand cay.

Many small bank or table reefs, rising to the sea surface from great depths, occur in the Caroline and Mariana Islands, as Losiep Merir, and gaferut. In the same area, Tatsumi Bank is a table reef drowned under 7 metres of water in its shallowest part. At Saipan and Guam, tops of hills or mountains have been interpreted by Tayama as raised fossil table reef. In the Caribbean off the shelf of Nicaragua, San Andres Island is a raised tank reef, uplifted in Post-Miocene times, which is thought to have a deep-stated volcanic basement.

Barrier Reefs and Associated Features

Various Barrier Frameworks

Barrier reefs can be continuous, with only rare passages or fragmented into small units. Examples of continuous linear reefs occur atfulear in Madagascar and in the northern section of the Mayotte barrier for 18 kilometres without any break, with fine differences, however, between the oceanward and the lagoonward edges. The western and southern barriers of New Caledonia, and the Cook Reef at the northern end of the same island, provide many example of rectilinear reefs, as does the barrier encircling Raiatea and Tahaa Islands in the Society group, where almost all the existing passages are bordered on either side by twin sand cays: this feature exist in other coral seas, and has been explained by Steers (1929) as the result of refraction of the swell on either side of the passage, thus supplying sediment. The Bora Bora barrier provides another example. The long barrier reef bordering the southern side of Tagula Island, at the east of New gunea, belongs to the same subtype, with only three passages within a distance of more than 120 kilometres.

Linear reefs from barriers in the Fiji Islands especially in the Lau Group; in the Exploring Isles, the American Passage across the barrier is of qure exceptional depth, exceeding 100 metres. A different pattern occurs where barrier reefs are divided into many segments, each comparatively small and recurved inward at both ends, by the

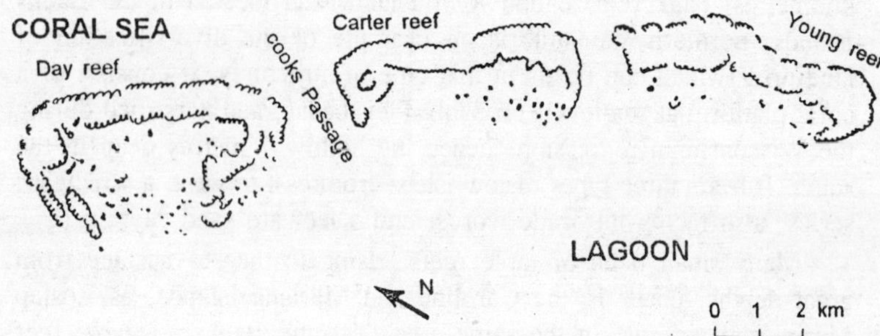

Fig. 14.7. Arcuate reefs forming elements of the outer edge of the Great Barrier Reef at latitude 14°30′S, where the lagoon is about 65 kilometres wide.

same process of wave refraction as in the twin sand cays must mentioned. This subtype is well illustrated in the Great Barrier Reef off Lizard Island (14°30'S) and off Cooktown (15°10'-15°30'). The subtype, which has been carefully analysed and pictured by Fairbridge (1950) and called by him and others ribbon reef, although this name would perhaps better fit the first subtype. The inner side of these reefs forms a sheltered area, in which a large number of tiny coral knolls grow.

At Day Reef, the recurves partially enclose a small lagoon. Other farolike features of this shape, more advanced in development, have been described from several parts of the Mayotte barrier. The structure of barrier reefs can also differ. In the barrier reef which griddles Martinique Island in the Caribbean, more than 50 kilometres long on the eastern side, exposed to strong swell generated by easterly winds, corals are replaced on the crest by a carbonate pavement covered with dense stands of fleshy algae, especially *Sargassum*, not only on the inner slope, but sometimes also on the upper part of the outer slope. Broad phanerogam meadows grow in the inner part of the lagoon, where fringing reefs are also common. Several barrier reefs in the Society Islands have an inner slope invaded by sediments washed over the reef flat by the surf and entering the lagoon as protruding fans or aprons.

The largest of these occurs at Maupiti; others are visible on the north side of Tahaa barrier reef, and there are several on Bora Bora. In the north of Bora Bora there is a combination of recurved segments of the barrier, convex outwards, which have emerged as wooded islands, and shallow fans, convex inwards and encroaching into the lagoon.

The fans are fed with sediment washed through shallow channels called *hoas* during periods of heavy swell. These hoas are described in more detail later in the Tuamotu atolls. Palau Island in the Western Pacific has fringing reefs in the northeast and east and barrier reefs elsewhere. The same change occurs around Tagula Island near New Guinea, and Rambi Island in the Fiji Islands. Similarly, Guam Island in the Marianas has fringing reefs along its coasts, except for the southernmost tip, where a barrier reef is backed by a lagoon, this structure being some kilometres wide.

Sediments in Lagoons Behind Barrier Reefs

The nature of sediments in lagoons behind barrier reefs indicates the ratio of supply from the central island or mainland and the supply derived from corals and associated marine organisms. This has been investigated in several lagoons in the Indian and Pacific Oceans and in the Caribbean, by means of the proportion of $CaCO_3$ in bottom samples, an adequate criterion where the central island is siliceous. The conclusion was that, in the humid tropical environments where all these elagoons lie, the percentage of terrigenous and skeletal particles depends mainly on the relationship between the areas of the central island and the bordering lagoon. There were significant contrasts between Bora Bora and Tahiti, on the one hand, and between the two lagoons surveyed around New Caledonia on the other.

The general distribution in Tuo lagoon, New Caledonia is similar to that in the lagoon behind the Great Barrier Reef at latitude 14°30'-15°S', where terrigenous sedimentation from the Australian continent and high islands dominate the inner part, while the influence of *Halimeda* gravels and sands is dominant in the outer part. However, such distributions have been recently modified by man, in places where the supply of sediment from the land has been increased by the cultivation of steep slopes, leading to an increase of lagoon silting by rivers, and a decrease of the $CaCO_3$ proportion in the lagoon. This has happened at Mayotte and is part of the problem of reef pollution which will be discussed.

Drowned Barriers

At Madagascar, Guilcher (1954) has described a drowned barrier which extends for hundreds of kilometres along the northwestern coast. At the northernmost tip of Madagascar it continues as an emerged Pleistocene reef. In its submerged part it bears only a few coral colonies, although the depth does not exceed 4 to 10 metres, and is thus quite suitable for coral growth. The 'lagoon' between the barrier

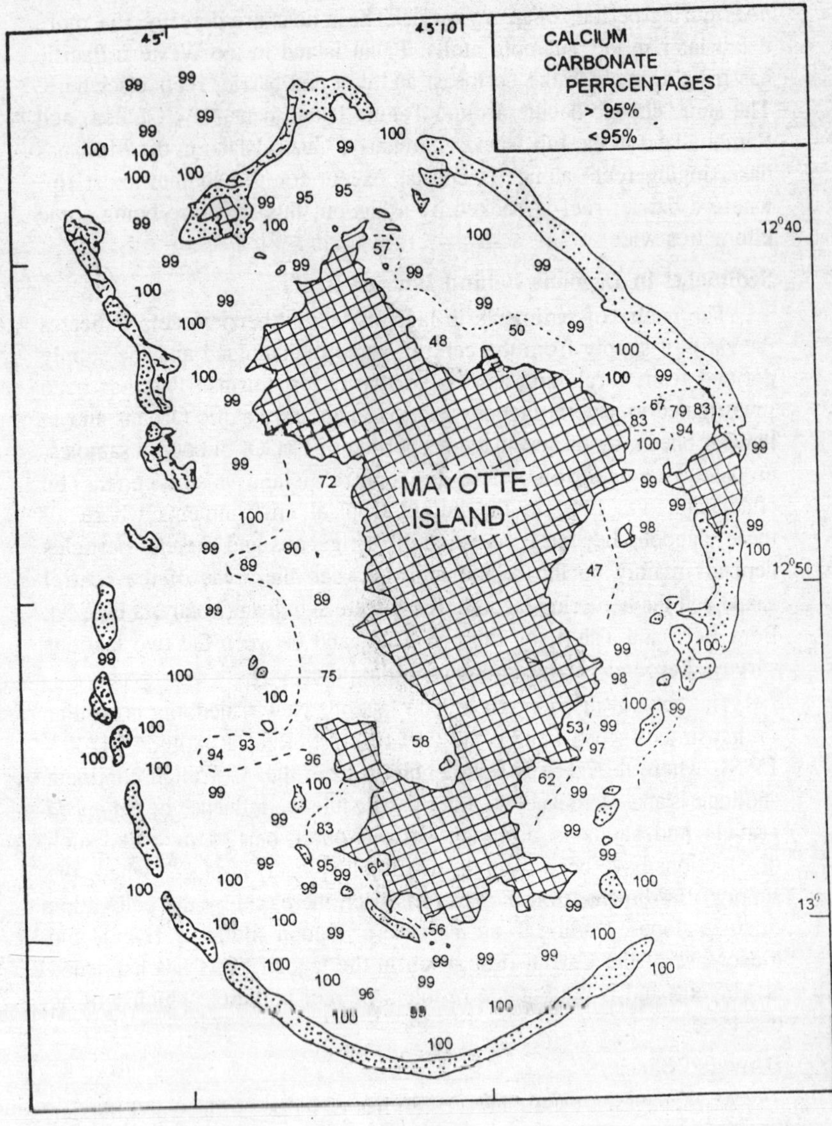

Fig. 14.8. Calcium carbonate percentages in Mayotte lagoon, Indian Ocean, in 1959. 1, volcanic island; 2, outer barrier; 3, inner barrier; 4, parts of barrier drowned under several metres of water.

and the mainland is much deeper, generally between 20 and 40 metres. The barrier is breached by deep passages off the main rivers, as in the case of true barrier reefs. The problems related to this drowned barrier (if it is really one, and not a structural feature) remain open so far. Also in the Indian Ocean, a partly drowned barrier reef lies in the Gulf of Marmar between India and Sri Lanka. Another site where a drowned barrier reef exists as such is Tahiti, where the barrier is submerged in three sections along the northeastern, eastern and southeastern coasts of the island, whereas it rises to sea-level in its other parts. Such features are understandable as adjustments of an island that has recently formed over a hot spot.

Another submerged barrier reef, mentioned by Davis (1928.) and shown on marine charts, extends along the southern coast of easternmost New Guinea for at least 160 kilometres. It is under 7 to 15 metres of water and is separated from the land by a lagoon 45 to 70 metres deep. It is a westward continuation of the ribbonlike barrier reef off Tagula Island, which is not submerged. Two or three atolls extend in between. This area and the nearby Louisiade archipelago are places where further reef studies are needed and will certainly be most fascinating. Still other submerged barriers are known in the Fiji Islands, an archipelago where they are adjacent to features described sometimes as raised barriers, but thought by Ladd and Hoffmeister (1945) and Purdy (1974) to be subaerial karstic features. The general tectonic context makes these features quite understandable.

Almost-atolls

The finest almost-atoll in the world is Truk in the East Caroline Islands. The reader should refer to the handsome chart published by Shepard. It is more or less circular or pentagonal, with a diameter of approximately 50 kilometres, and a circumference of 200 kilometres. About 20 high volcanic islands or islets rise above the lagoon surface, mostly on a central ridge running WSW-ENE. The largest of these islands, Tol, Moen and Fafan, are 7 to 9 kilometres long; the highest is 480 metres high. All have fringing reefs, some have mangroves on the inner reaches of their shores. The outer barrier reef is cut by about 50 passages; it is more continuous in the east than elsewhere; it is absent or drowned for only 5 kilometres in the southwestern corner. It bears many low cays in the southeast; others are scattered in the south, east, and north, and only a very small one in the west. These islets lie thus on the windward rather than the leeward side, a distribution similar to that observed in the Society Islands at Bora Bora, Mopelia and scilly. The submarien topography of Truk lagoon is irregular, with a series of depressions exceeding 40 fathoms (73 metres), generally not connected with passages across the barrier. Many coral knolls rise from the bottom. The interpretation of these features by Shepard (1970) is that the enclosed depressions result from solution of limestone during low sea-level phases, the knolls being either remnants of such karstification, or due to coral growth during the Holocene marine transgression, or both. These conclusions are similar to those from other coral areas as the Bahamas, Mayotte, Clipperton, etc

Not far from Truk, Pohnpei Island and its barrier reef stand in striking contrast, with a large central island and a narrow lagoon with extensive mangroves along the volcanic shores. There are depressions in the lagoon which have clear connections with submarien valleys and passages. The almost-atoll of Gambier has not been surveyed as closely as Truk: but an essential difference is that its barrier reef is drowned for at least half its length. Several possible explanations have been examined by Brousse *et al.*, but none of them is definite. The first idea is that there has been tilting, so that the barrier reef subsided in the south, but this does not explain the drowned northwestern section, so that several local sinkings would perhaps be a better hypothesis.

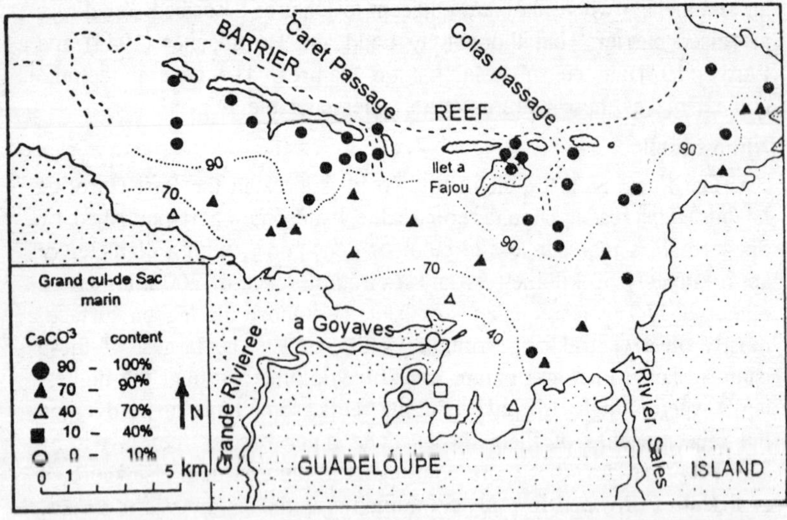

Fig. 14.9. $CaCO_3$ content of sediments in Grand Cul de Sac Marin, a lagoon behind the barrier reef of Guadaloupe Island.

The volcano upon which the Gambier framework rests is dated at only 5 million years, a surprisingly young age for such sinkings which normally would be expected later in the history of this structure. Another curious fact is the existence of the small Temoe Atoll 40

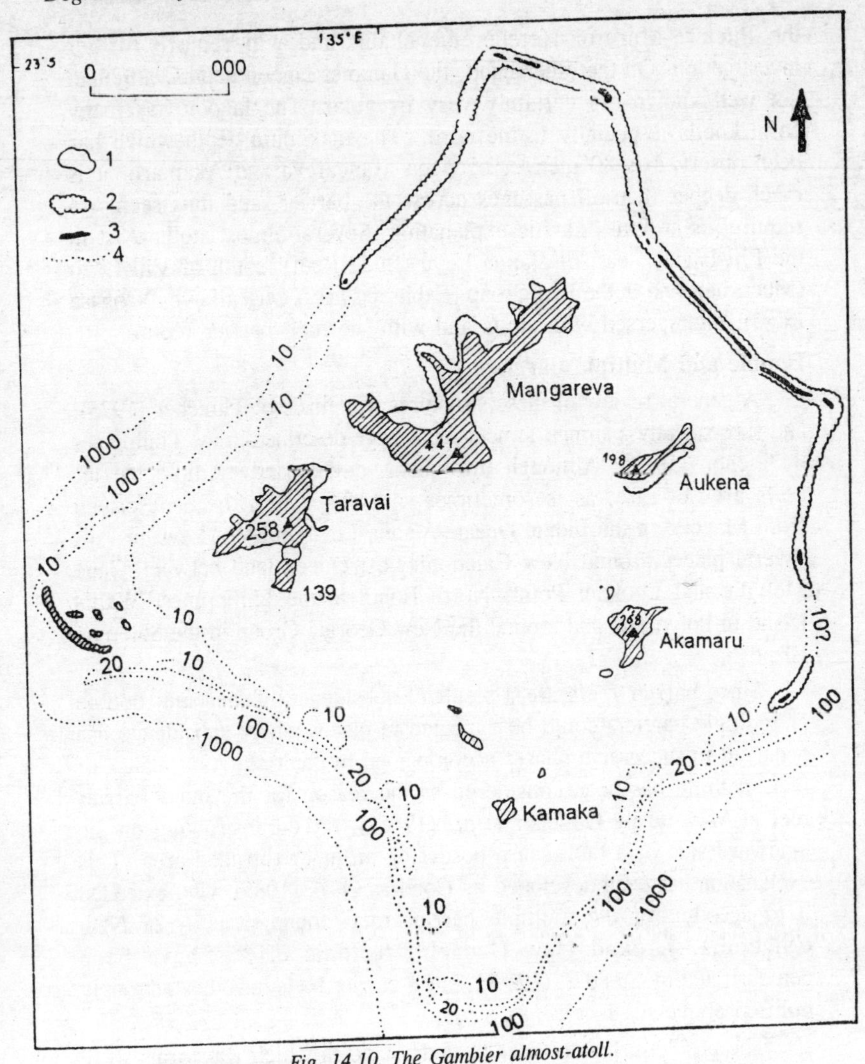

Fig. 14.10. The Gambier almost-atoll.

kilometres southeast of Gambier, and Portland Bank, 30 kilometres farther in the same direction, which was thought to be a shallow submarine bank, but remote sensing has indicated that it is really another submerged atoll, under 12-13 metres of water. Atolls were not expected here, because in this direction the type of reefs should have been younger, not older than the Gambier almost-atoll. It thus seems that events have occurred here, which do not fit the general history of the Pitcairn-Mururoa-Hereheretue chain, and will require further investigation. On the other hand, the Gambier lagoon depths, although not well known are certainly very irregular. The lagoon has many coral knolls, especially in the north. The maximum depth which has been observed is 80 metres, between Mangareva and Akamaru: it is much deeper than all passages across the barrier, and thus seems to require, as at Truk, karstic explanation. Several almost-atolls exist in the Fiji Islands, east of Vanua Levu (Budd Reef), south of Viti Levu (Mbengha), and in the Lau Group (Exploring Isles, Ongea Levu, Yangasa Levu), interspersed with atolls and with 'normal' barrier reefs.

Double and Multiple Barriers

A general review of these structures was made by Guilcher (1976), and subsequently a similar structure has been described in the Philippines by Pichon (1977). Although infrequent, double and multiple barrier reefs are not rare, as is sometimes said. They have been described from Mayotte in the Indian Ocean, Vanua Levu and Viti Levu in Fiji, several places around New Caledonia, off Queensland between Cape Melville and Lookout Point, North Bohol in the Philippines, Wallis Island in Polynesia, and around the New Georgia Group in the Solomon Islands.

Since barrier reefs are a result of subsidence of the land, double or multiple barriers could be expected in places where subsidence has occurred in successive phases accompanied by faulting. As a matter of fact, faulting has been considered as the reason for the inner barrier reef at Mayotte by Guilcher *et al.* (1965a), the coral growing on the landward side of a fault scarp bordering an inner subsided area. This explanation has been developed by Coudray *et al.* (1985), who extended it to account for the multiple barrier reefs found in parts of New Caledonia. Around New Georgia, tectonic effects have been demonstrated by Stoddart (1969c), since barrier reefs have been strongly uplifted above sea level.

The problem, however, is to understand in which waycorals were able to build barrier reefs inside a lagoon, where, as Davis (1928) pointed out, the ecological conditions of growth are much poorer than along the outer barrier reef, washed by the oceanic swell. It is true that fringing reefs are common along inner shores of lagoons behind barriers; but these are only thin coatings, not massive as barrier reefs.

Guilcher (1976) has suggested to consider five factors able to permit unusually favourable conditions of coral growth in these lagoons: firstly, a large tidal range allowing oceanic swell to penetrate freely over the outer barrier at high spring tides;secondly, many wide passages across the outer barrier, which have the same result;thirdly, a large distance between the outer barrier and the mainland, resulting in a long fetch which favours strong waves generated by local winds and thus good water circulation; fourthly, a lagoon stretching in the direction of the dominant wind, again permitting strong wave action and water stirrings; and fifthly, strong currents entering the lagoon through passages, and favouring the growth of coral ridges.

The Mayotte double barrier reef can probably be explained largely by the first factor, because the tidal range reaches 3.50 metres at spring tides. In New Caledonia, the second and the fourth factors seem to be dominant, especially the fourth one, since the prevailing wind, blowing from the southeast along the axis of the lagoon, produces large waves. At North Bohol, Pichon (1977) suggested that strong tidal currents (i.e. the fifth factor) were the main cause. Off Queensland, where the tidal range is large and the lagoon wide, the first and third factors are relevant; at Wallis Island, the fifth and perhaps the first, and at Vanua Levu, the fifth.

Conversely, the absence of double barriers in the Society Islands seems to be due to a combination of a very small tidal range, narrow lagoons, and lack of wide passages on the windward side, except at Tahiti. None of the suggested possible explanations, however, appears to apply to Viti Levu; and around New Georgia the barrier pattern remains difficult to understand in its details; but the intense tectonic activity may have modified the ecological conditions for coral growth in that area since the reefs began to be built.

Sand Cay Reefs and Low Isles in Lagoons

This subtype of coral reefs has been described more frequently and accurately than any other one perhaps. Fairbridge (1950), Stoddart and Steers (1977), Stoddart *et al.* (1978), Hopley (1982), all concerning the Great Barrier Reef, may be selected among references since they refer to previous papers an give general views. As Steers (1929) pointed out, 'the term Great Barrier Reef is, some ways, extremely misleading', since this huge area 'really consists of a number of reefs and reef patches cut by many channels' among which reefs with sand cays are widespread. Similar features have been described from Jamaica, Indonesia off the north-western and southwestern coasts of Madagascar off eastern Guadalcanal etc. Sand cays are found at various stages of growth in lagoons behind barrier reefs, as in Queensland, and around New Caledonia where they have not been described so frequently as in Australia although they are quite numerous and fine. They occur on reef patches in shallow, wide lagoons where there is a long fetch in the direction of dominant winds, resulting in strong wave action. Substantial amounts of reef debris are thus produced, and swept up on to the reef flat to form surface accumulations.

In most cases, these accumulations include ramparts of coarse material on the windward side, and islets (cays) of sandy sediments, more or less vegetated, on the leeward side. A 'lagoon', generally extremely shallow, occurs in between, sometimes with mangroves. Beach rock is frequent in the leeward cays. On different stages in sand cay development may be observed on New Caledonian reefs, with a simple tongue of loose sand on the reef in the foreground, two well-developed structures in the middle and far distances, and another incipient, circular structure in between. The so-called Low Isles in the Great Barrier Reef environment are a variable structure, with modifications that have been investigated several times since it was

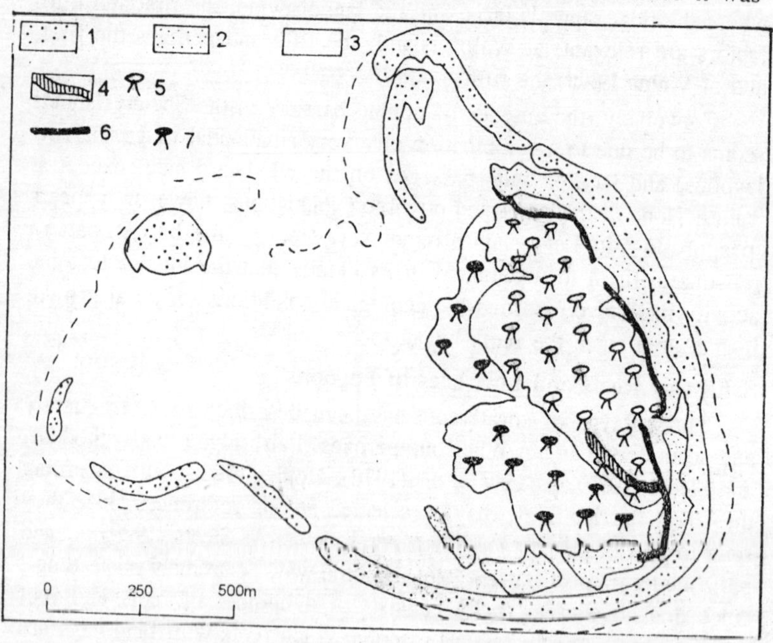

Fig. 14.11. Low Isles, Great Barrier Reef near Cairns at 16°S simplified. 1, sand cay; 2, outer shingle rampart; 3, inner shingle rampart; 4, older shingle ridge; 5, mangroves; 6, white shingle ridge in 1973, to which the structure has retreated, leaving only a few eastward remnants; 7, westward expansion of mangroves by 1973.

first surveyed in 1929. Then there were three ramparts of coars material on the windward side, extensive mangroves in the eastern part of the 'lagoon' (normally very shallow here), and a leeward sand cay with some beach rock. Subsequent surveys indicated that one or two other ramparts were added before 1945, but surveys in 1954 and 1973 showed the destruction of considerable parts of these outer framework, and progradation to the west of their remnants, presumably as the results of recurrent tropical cyclones. Yet the mangrove area has enlarged, and is thought to have impeded the migration of ramparts.

The sand cay has not moved much, probably because it is partly anchored on beach rock, but the exposure of layers of beach rock on its southeastern shore testifies to slight reshaping. Evidence from the Low Isles thus indicates that these structures do not necessarily show continuous growth, but can be diminished in area or change in shape during stormy phases. Also in the Great Barrier Reef area, Lady Musgrave Reef is a fully mature structure which, in fact, is not wholly

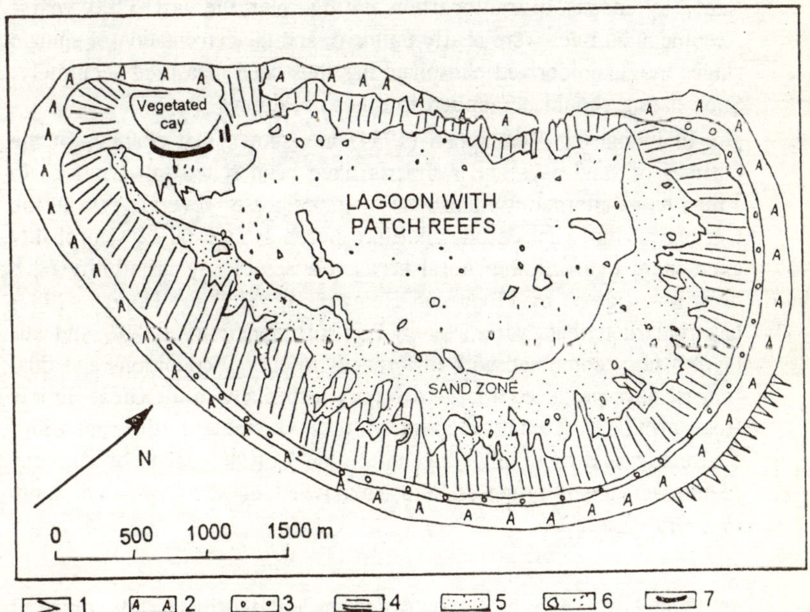

Fig. 14.12. Lady Musgrave Reef, at the southern end of the Great Barrier Reef (152°23′E, 23°54′S) a mature, large cay reef. 1, spurs and grooves on windward side; 2, algal rim; 3, coral rubble; 4, radial lineations; 5, sand zone; 6, lagoon with many patch reefs; 7, beach rock along vegetated cay, which consists of sand and shingle.

typical since it contains a real lagoon a few metres deep, that is, a little deeper than usual: Lady Musgrave Island has been sometimes considered as an atoll. But this reef displays an excellent zonation, with, from outside, spurs and grooves, a calcareous algal rim, a narrow crown of coarse sediment, a beautifully striated zone, a sand zone, a lagoon with small coral heads, and, on the leeward side, a mature, vegetated cay.

The Hiengu sand cay on a patch reef near Tuo, in the eastern lagoon of New Caledonia, is another example o a mature structure. It is covered by a dense broadleaf woodland, and occupies a larger part of the reef surface than is general. The windward side of the reef normally bears a recurved ridge of small blocks. The 'lagoon' between this and the sand cay is almot dry at lowest sprign tides. Slabs of beach rock exposed on the windward shore of the cay indicate that a leeward migration has occurred, shown also by the bare and tail prograded on the other shore. Chivas *et al.* (1986) investigated teh development of a coral cay, Lady Elliot Island, on the Great Barrier Reef, by means of radiocarbon dating over the last 3200 years. Accumulation rates were nearly uniform, and the cementation of shingle ridges by guano-derived phosphate has thus been measured accurately. Such dating should be applied to other coral cays.

In Indonesia, Verstappen (1954) has shown that changes in the outlines of sand cays and ramparts have been rlated to variations in strengths of alternating monsoons over periods of several yers. In the Caribbean, hurricanes erode and may sweep away sand cays, probably even more than in other coral areas. Causes of cay instaibility have been investigated by Flood and Heatwole (1986), and Hopley (1982) has remarked that 'variations in reef-top ges, morphology, and sea level history combined with differences in energy conditions and tidal ranges' have produced a large diversity, especially in the Great Barrier Reef province. Taking account of these factors of differentiation, destruction and rebuilding, a general model of the growth of sand cay reefs is tentatively proposed here, acknowledging that it may not apply to every case.

ATOLLS

There are 425 atolls recorded in the world; a few more if submerged atolls and some of the bank atolls are taken into account. The archipelago with largest number of atolls is the Tuamotus (76). The larges atol is Kwaajalein in the Marshali Islands (120 × 32 kilometres), followed by Rangiroa inthe Tuamotus (79 × 34 kilometres), and Tijger, south of Celebes (72 × 36 kilometres). Twenty-seven atolls are recorded inthe Caribbean, but perhaps only 10 deserve that name. The problem of the existence of true atolls inthe Red Sea is discussed in section 5 of this chapter (ridge reefs). The most comprehensive study of atolls is by Wiens (1962).

Asymmetry

Atolls are generally more or less asymmetrical, with difference between the windward and leeward sides. Such differences are well=known at Bikini and nearby atolls especially in teh distribution and slope of the spur and groove systems. In the Society and Tuamotu Islands, the low isles on atoll rims are principally or exclusively on

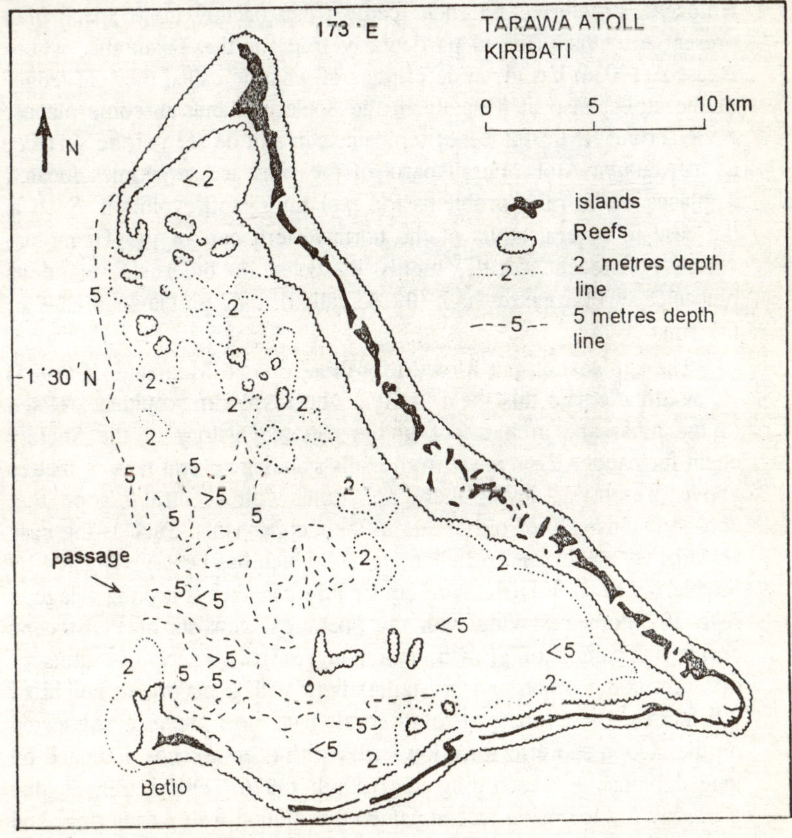

Fig. 14.13. Tarawa Atoll, Kiribati, Micronesia. An asymmetrical atoll, without a rim on the leewar dside, and with a very shallow lagoon.

the windward side, where sediments are thrown up by waves, and such a distribution is considerably exaggerated in Kiribati atolls such as Tarawa Apaiang, Maiana, Abemama, and Butaritari, in Caribbean atolls such as the Turneffe Islands, and even more in the atolls off the shelf of Nicaragua where the leward side is almost abesnt, so that the atoll consists only of an arc facing east. Such atolls are transitional towards the arcuate reefs formignthe Great Barrier Reef. By contrast, low isle are almost continuous around the atoll rim at Diego Garcia in the Indian Ocean.

Aatoll History Kept in Rim Structure

Atoll rims often retain remnants of their Pleistocene or early Holocene structure, and such features can occupy more space than present growths. This is particularly true for the Tuamotus, where Agassiz (1903a) has given descriptios of what he called the 'old ledge' of the atolls; also at Mopelia in the Society Islands in some places, even Tertiary (Mio-Pliocene) remnants can still be seen at the surface, as at Mataiva Atoll. Fossil parts of the rims are sometimes located in places which raise problems for explaining atoll evolution. Such is the case in several atolls of the northwestern part of the Tuamotus, where Pleistocene corals, highly dissected by bioerosio, stand as pinnacles several metres high (the so-called), only on the south side of the rims.

The explanation put forward by Pirazzoli and Montaggioni (1985) is, as already said this local uplift of the fossil rim would be related to the moat and arch effect in the tectonic history of the Society chain formation. Even apart from atolls standing several tens of metres above present sea-level, there are atolls with acentral lagoon that consist exclusively of old corals in their upper rims. Such is the case at Aldabra, southwestern Indian Ocean, which has been invstigated by Stoddart *et al.* (1971). Its rim, cut by thre passages conecting a lagoon 6 to 10 kilometres wide with the open sea, consists of Pleistocene corals in position of growth, belonging to two successive units of different composition and crystallizatiion, with a maximum height of 7.5 metres; these emerged fossil corals havc been intensely dissected on the lagoon shore to form low cliffs with deep notches preceded by jagged pinnacles. According to Trudgill (1976), cliff retreat is due primarily to bioerosion, accompanied by solution, salt weathering, and abrasion. These cliffs arebacked by rasp-like surfaces called 'champignon', while smoother surface with lower relief form the 'platin'. Smal unconsolidated dunes occur, especially on the windward coast.

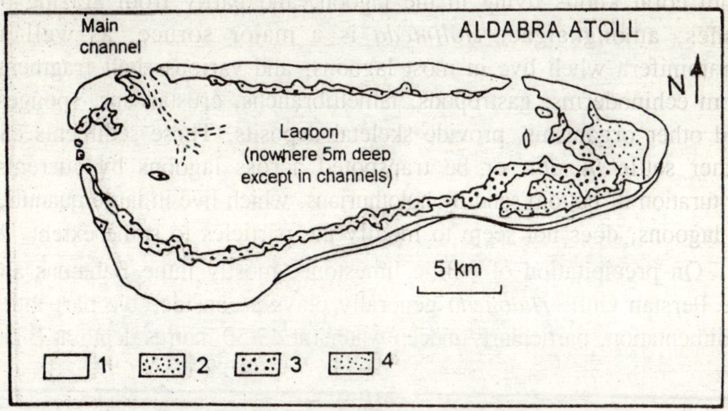

Fig. 14.14. Aldabra Atoll, Western Indian Ocean. A slightly emerged atoll. 1, champignon; 2, platin; 3, mangroves along lagoon shores; 4, beach and dune sand.

The lagoon is less than 5 metres deep, except in channels that open towards the passages, which exceed 20 metres in depth, and are scoured by tidal currents (the tidal range exceeds 3 metres at spring tides). Mangroves grow on the lagoon shores.

Aldabra is thus an excellent example of an atoll combining fossil and modern geomorphological features, the fossil ones playing an important role in a still living structure. Situated in the Mozambique Channel, Europa Atoll, described by Battistini (1966), resembles Aldabra very much, but is smaller (6 × 6 kilometresa). It is also slightly emerged, and probably of Upper Pleistocene age, jaged by karstification, with a lagooon shallower than at Aldabra in which grow extensive mangroves of variouis species.

Lagoon Sediments

Information on atoll lagoon sediments has been given at Bikini, Enewetak, Rongerik, and Rongelap in the Marshall Islands by Emery *et al.* (1954); at Johnston, an isolated atoll-like feature between Hawaii and the Line Islands by Emery (1956b); at Kapingamarangi inthe Carolines by McKee *et al.* (1959); at Mopelia in the Society Islands by Guilcher *et al.* (1969); at Kure and Midway in Hawaii by Gross *et al.* (1969); in the Caribbean by Milliman (973), and by Guilcher (967a) for several areas. In contrast with sediments in lagoons between barrier reefs and high volcanic or sedimentary hinterlands, sediments in atoll lagoons are entirely calcareous, except for a few diatom and spicules of silica embedded in sponges, and sometimes traces of aeolian dust. The sources of supply are coral particles of various sizes, derived from coral knolls living in the lagoon; and partly from grazing by fishes; amongnalgae, *Halimeda* is a major soruce, as well as foraminifera whch live in most lagoons; and various shell fragments from echinoderms, gastropods, lamellibranchs, crustaceans, sponges, and other organisms, provide skeletal deposits. These sediments can either settle *in situ*, or be transpoted across lagoons by currents. Trituration of bottom sand by holothurians, which live in large quantities in lagoons, does not seem to modify the particles to some extent.

On precipitation of oolitic limestone, mostly inthe Bahamas and the Persian Gulf. *Halimeda* generally plays a considerable part inthis sedimentation, particularly indeep water (at 45-50 metres depth at Bikii.

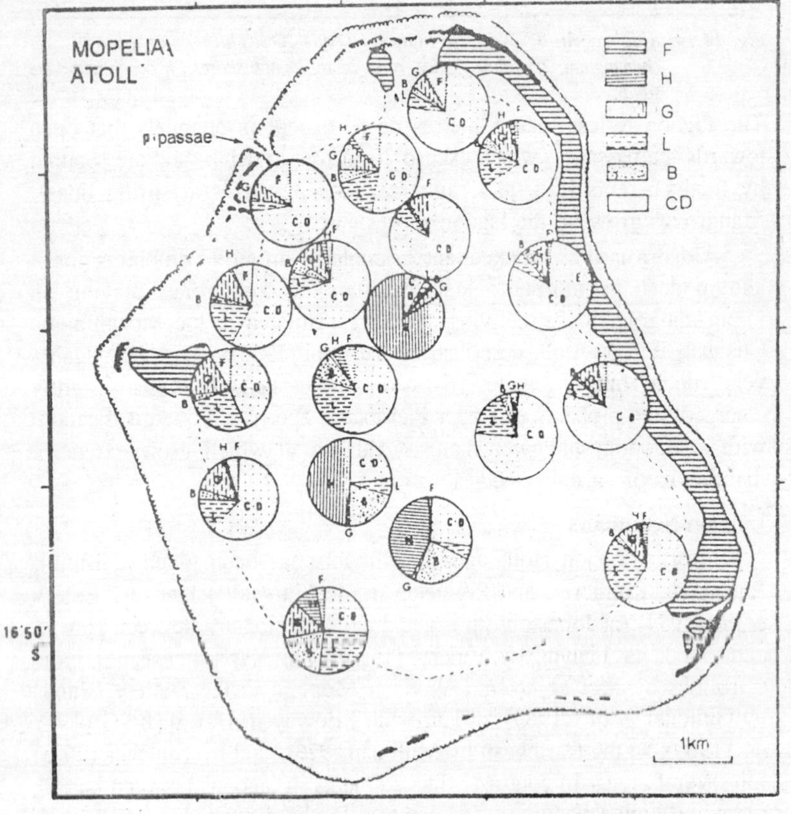

Fig. 14.15. Mopelia Atoll, Society Islands. Sedimentary composition of lagoon deposits. F—Foraminifera; H—Halimeda; G—Gastropoda; L—Lamellibranchiata; B—Bryozoa; CD—Coral debris.

Rongerik, Rongelap, and in the deepest parts at Mopelai where they form 42 to 74 per cent of the sediment), or a ring-like pattern round the central area. By contrast, *Halimeda* debris is rare at any depth at Johnston, where coralline algae play a large part. Foraminifera are mostly rrepresented in beaches and nearby shallow waters (up to 85 per cent of the sand grains in beach sands at Rongerik; more scattered, however, at Mopelia). Foraminifera found in deeper water at Kapingamarangi and Enewetak are of different species to those settling in shallower areas.

Fine coral debris occurs principally near the rim of the four Marshall atolls, decreasing towards the middle, and the distribution of gastropod and lamellibranch shells is the same at Mopelia. Coarse coral debris is found at all depths, even in the central parts where it settles around the coral knolls, at the foot of which talus is visible at Bikini and Mopelia. At Kure and Midway, the world's northernmost atolls, the sediments in lagoons are, in decreasing order of abundance, foraminifera, coralline algae, shell fragments, coral fragments, and *Halimeda*. The scarcity of *Halimeda* may be due to the high latitude (28°25', 28°13'), but as it is also rare at Johnson (17°), it may be a special feature of the Hawaiian region. In Caribbean atolls, *Halimeda* is often abundant in ssediments, as at Serrana, Roncador, and Alacran.

Mangroves

Mangroves are not so frequent in atoll lagooons as on the iner shores of fringing reefs and in 'low wooded islands' in Queensland, Madagascar, the Red Sea, the Caribbean, and Indonesia. They were not reported in the Marshall Islands, and appear to be completely absent from the Society and Tuamotu Islands, in spite to attempts to introduce *Rhizophora* to the Moorea lagoon in 1965. They were not mentioned in the four atolls described by Milliman (1969) off Nicaragua. On the contrary, *Rhizophora* thrives in sheltered parts of the Abemama and Tarawa lagoons in Kiribati, where it grows vigorously on calcareous sands and muds. Mangroves are also present in the Carolines, for example at Losap Atoll, at Aldabra and Europa Atolls in the Indian Ocean, as said previously, and in two of the three atoll lagoons near Belize, Caribbean.

Pinnacles or Knolls in Lagoons

The floors of many lagoons bear numerous pinnacles or knolls, which rise from various depths gives an example from Bellingshausen Atoll in the Society Islands, whcih has no passage and is fed with surface oceanic water only by overwash: in spite of this, coral growth appears to be excellent. Pinnacles rising to the surface are a common occurrence in the Tuamotus, for example at Arutua, Manihi, Ahe, Tikehau, Mururoa. In the Indian Ocean, Diego Garcia is another striking example. The pinnacles generally have living corals on their surfaces.

However, a host of pinnacles in lagoons is not a universal feature. For example, theyappear to be rare at Scilly in the Society Islands, where no knoll is visible at depths shallower than 12 metres. There is apparently no relation between the occurrence or number of pinnacles and the degree of connection between the lagoon and the open sea, and thisis true also in lagoons behind barrier reefs: Bora Bora lagon has goo dconnections with the ocean but no knolls, while Tupai, Tetiaroa, Bellingshausen and Kaukura have knolls and no deep passage; Rangiroa has knolls and two deep passages: and the general situation is the same at Bikini, Rongerik, Enewetak and Rongelap in the Marshall Islands.

Since pinnacles do nto seem especially related to present conditins of vigorous coral growth, another possibility is that they are remnants of karstification during low sea-level phases, a hypothesis favoured by supporters of the karstic saucer theory. But pinnacles can be of composite origin: their core can be Pleistocene, and their surface or upper part Holocene; some can be old and others recent. Only the higher ones, rising to present sea-level, can be taken, if they are Pleistocene, as evidence for an antecedent platform beneath the present atoll lagoon. The significance of the pinnacles will depend on numerous sections in many places, and radiometric dates. A few sections and dates will not be decisive. It adequate investigations are carried out in this way, the clue—or one of the clues—of atoll genesis may be there.

Ava, Koa, Kairua/Passage, Overflow Channel, Temporarily Closed Channel

In the Tuamotus, the communication between atoll lagons and the ocean occurs in two ways, recognized in Paumotu terminology There are a few passages everal metres dee, which aresutiable for navigation, called *ava*. In addition, there are many (several thousand in the Tuamotu archipelago) very shallow passages, scattered around the atolls between the *motu*(s) (islets) but better defined than the low flat areas through which swash penetrates in most cases into the Marshall lagoons. These are known as *hoa*(s), a word which can be translated as overflow or overwash channel; with the intervening elongated motus, they result in teh typical Tuamotu landscape. They have been suposed by Chevalier *et al.* (1968) to result from joints formed by compaction of the reef, and to have been subsequently widened by erosion. They act efficiently as lagoon feeders in water.

Series of hoas with intervening motus have also been observed intheKiribati Group at Tarawa, Abemama and Maiana and even at Majuro in the Marshall Islands, an archipelago where they seem, however, much more exceptional than in the Tuamotus, while the avas are more numerous. But, as the Tuamotu lagoons are often wide, the fetch is large, and large waves are generated when the wind is strong. These lagoon wavers are able to rework the sandy beches on the inner side of the motus, and build long sand spits across the inner ends of the hoas during periods of moderate oceanic swell, hence lack of currents throughthe hoas.

Agassiz (1903a) already observed these spits. Ponds are thus formed in the hoas, which become *tairua*(s) and are more or less persistent.

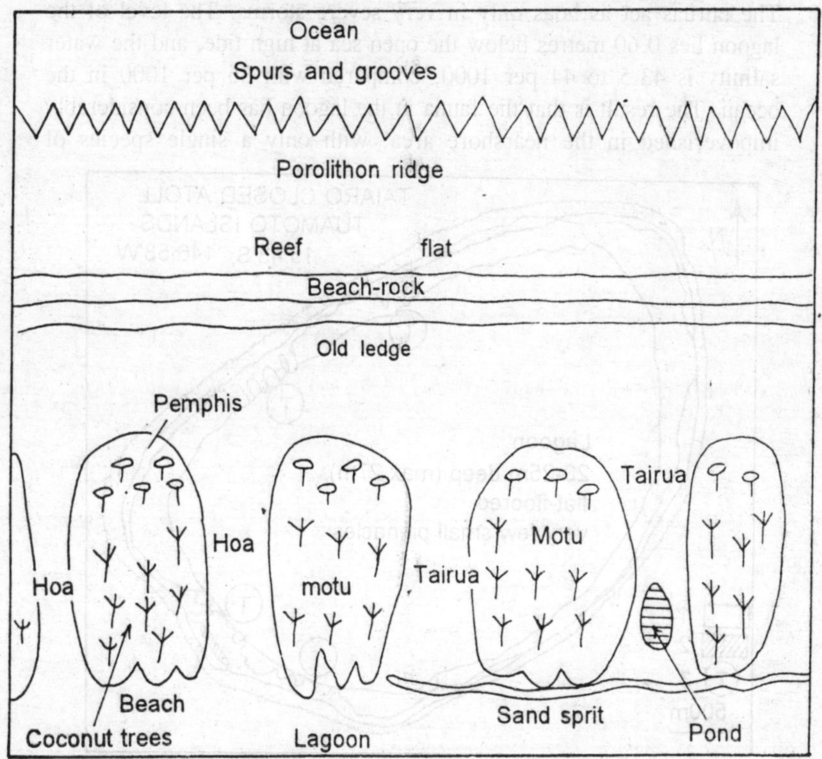

Fig. 14.16. Elements of an atoll rim in the Tuamotus.

Some have been used for aquacultureat Rangiroa Atoll. They also exist in the Bora Bora barrier and at Tarawa, where open hoas are, however, more frequent. Avas, as at Mopelia and Rangiroa, are generally deeper in their central part than at the lagoon entrance, which remains shallow because of the decrease of scour by tidal currents. A number of atolls in the Tuamotu and the Society Islans have no ava, so that the connectiion in surface with the ocean is possible only through hoas, or over the reef flat without hoas, the second case occurring mainly in the Society Islands as at scilly, Tupai, and Bellingshausen.

An extreme situatin has been analysed at Taiaro in the Tuamotus, an atoll 4.5 kilometres longand 3 kilometres wide, which, like some others in the same group, is a closed atoll, the rim consisting of a continuous circular island, built during a Holocene (1000-2000 BP) phase of sea-level 1.5 to 2 metres higher than now. It is breached by 18 tairuas, i.e. hoas which were open when the sea-level was higher. The tairuas act as hoas only in very severe storms. The level of the lagoon lies 0.60 metres below the open sea at high tide, and the water salinity is 43.5 to 44 per 1000, compared with 36 per 1000 in the ocean. The result is that the fauna in the lagoon has been considerably impoverished in the nearshore area, with only a single species of

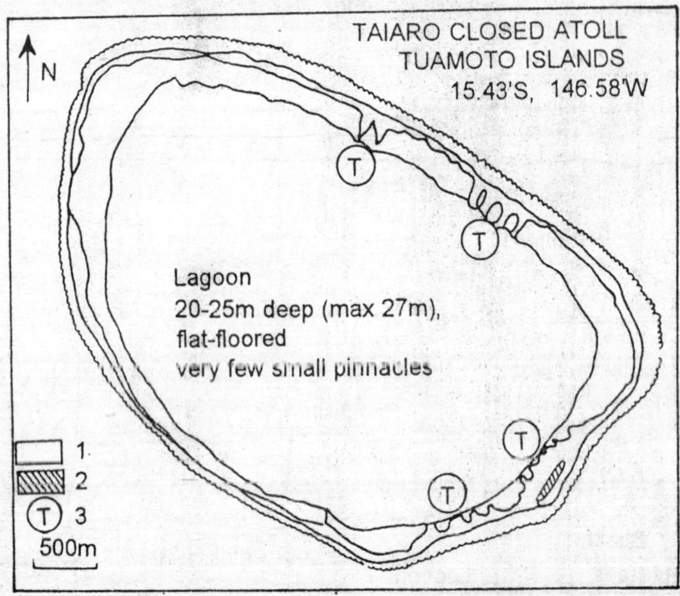

Fig. 14.17. Taiaro close atoll in the Tuamotu Island. 1, terrestrial vegetation; 2, pond 3, groups of tairuas.

hermatypic coral (one *Porites*) and one echinoderm. Various shells, however, live in the lagoon, but it bears no important pinnacle, a faact not favourable tothekarstic hypothesis of the evolution of atolls, since pinnacles should exist as residuals of solution of a former open reef platform. Takapoto, also in the Tuamotus is not so closed as Taiaro: it has a few open hoas, and the lagoon has more than 400 pinnacles of various sizes with living corals on their surfaces.

Water Circulation in Lagoons and Percolation at Depth Through the Atolls

Lagoon circulation has ben studied in severla places, notably by Von Arx (1954) at Bikini and Rongelap in the Marshall Islands. As expected, the main factor is the trade winds, which force swell into the lagon over the windward, eastern side. The water in excess escapes partly over the leeward side, but some comes back to the east along the bottom, so that there is a division into rotating compartments. In addition, tidal currents run to and fro in the deep passages. The circulation is more marked when trade winds are strong, in the winte months of the northern hemisphere; it slackens in summer, when the doldrums settle near the two atolls.

Other observations have been made in the Society and the Tuamotu Islands at the surface and on atoll rims, but not at depth. Except for the tidal currents which alternate through the avas, as inthe Marshalls, the processes are different here. In addition to the trade winds which drive water in lagoons from the east, there is occasional or even frequent heavy swell from the south, generated in the roaring forties. This swell piles water into the lagoons freely, because islands are rare or absent on the southern side, being mostly located in the east, in the Society Islands at least. During such periods in 1963, the sea level in Mopelia lagoon rose 40 to 50 centimetres above normal; the alternating tidal currents in the single deep passage were replaced by a strong and continuous flow out over the leeward side. On Scilly Atoll, such events have been frequent enough to create a morphology of crescentic accumulations of sand on the leeward rim, with convex sides that face the lagoon and not the ocean. The surface circulation is the same at Mataiva Atoll and Tikehau Atoll in the Tuamotus, according to Delesalle (1985, Delesalle *et al.* 1985, and Harmelin-Vivien, 1985): oceanic wate enters through the hoas in the south and east, and flows out through the ava in the northwest or west.

A similar pattern has been observed at Kaukura in the Tuamotus, but as that atoll has no ava, the water goes in and out only through the hoas, thus flowig lagonwards or oceanwards according to their location. In addition to this surface circulation, percolation certainly occurs at depth across the atolls, since they contain may voids, which may be original cavities or holes formed by karstification during low sea-level phases. The existence of a freshwater lens, called the Ghyben-Herzberg lens, at a shallow depth beneath the islets of the rim, has been known for a log time, but it is now known that early descritioins of this lens are oversimplified, although it certainly exists. At a lower level, oceanic water circulates through the karstified limestone. Studies carried out at Rangiroa and Mururoa in the Tuamotus have shown that the content of organic matter in the lagoon water is considerably higher than in the upper oceanic water surrounding the atolls: at Mururoa the biomass is more than ten times larger inside than outside.

Ecologically, these Polynesian atolls thus appear as 'oases' scattered in the desert ocean surface. According to Sargent and Austin (1954), the situation is the same at Rongelap Atoll in the Marshall Islands, in which 'the productivity is considerably hgiher than that of the adjacent waters or any other open marine areas'; these authros speak of 'flourishign coral atolls in an empty ocean'. Rougerie and Wauthy (1986, and in press) suggested that these contrasts result from an 'endo-upwelling' of deep oceanic water, richer innutrients thanthe surface water, flowing across the inner, lower parts of the atolls. The role of this endo-upweling process varies from one atoll to another: it seems rather weak at Taiaro Atoll whil eat Takapoto, a half-closed atoll (same section), it seems stronger. At Enewetak in the Marshall Islands, it has been shown that effective permeability increases with increasing depth, a fact which may help to understand the deep percolation.

Reticulated Atolls

Mataiva Atoll, 10 kilometres long and 5 kilometres wide at the westernmost end of the Tuamotus, has a very fine reticulated or honeycombed lagoon. This lagoon 'is made of about 70 pools of varying sizes (100 m to over 2 km), with an average depth of 8 m (down to 30 m). The shallow "ridges" which separate these pools are under 0.1-0.8 m of water and form a network more than 200 km long'. In a few places, these ridges rise 20-30 cm aove sea-level, and are made of fossil corals of Middle Holocene age in position of growth. This strange pattern, and the rich phosphates which exist inside the cells, are though to derive from a complicated evolution. A reef platform, which outcrops on the north coast of the rim as a strongly dolomitized reef, and dips sothward belo wthe lagoon, developed probably in the Mio-Pilocene. This platform was karstified during period(s) of emergence, possibly in the Late Pliocene, to produce the honeycombed pattern. The origin of the phosphates remains uncertan as in other atolls; they could have formed from bird excrements and bones, or from marine roganic matter, or from drifted volcanic material, or from endo-upwelling according to Rougerie and Wauthy. The last episode was a slight emergence after a Holocene maximum sea level, which appears no ony in some lagood ridges, but also in jagged upper parts of the southwestern outer rim. There are two or three idistinct generations of beach rock in some of the outer beaches, as on the east coast. In the course of this evolution, tropical karstification appears to have occurred. But it is not clear why several uplifts and lowering would have occurred at Mataiva, and apparently none in the other near-by Tuamotus.There are some typical hoas and tairuas.

The lagoon behind the barrier reef between Raiatea and Tahaa in the Society Islands displays a similar honeycombed pattern over a much smaller area. Hao Atoll in the Tuamotus also has a more or less similar structure in its lagoon, and Fakaofo in Tokelau seems to have too. Canton Atoll in the Phoenix Islands, described by Smith and Henderson (1978), has a reticulated framework resembling Mataiva, but not so complete: the ridge network is less continuous and the cells have more interconnections. These ridges are covered near the surface principally be branching corals, but at lower levels the coral growth is poor or nil, because the atoll receives oceanic water only through one passage: there is no real hoa on the windward side, and the southeastern end is a cul-de-sac. No information is available on the internal structure of the ridges, but it is possible that karstifiation has occurred, although that hypothesis has not ben put forward.

At first sight, it would seem that Farquhar Atoll in the western Indian Ocean is of the same type as Mataiva and Canton: its lagoon contains more than twenty ridges, and in the southern and northern corners enclosed subsidiary lagoons occur. But the ridges are largely sandy, and always rectilinear and subparallel, and they do not enclose a set of cells as at Mataiva. The lagoon depth is less than 15 metres. The oldest parts of the atoll are Holocene (3000-3600 BP) and have slightly emerged (2 metres). It may be that Farquhar would not have been a true atoll, but a bank reef until quite recent times, with initially separate parallel sand stripes. Christmas Atoll in the Line Islands is more truly reticulated than Farquhar, but it owes this aspect to some

500 hypersaline shallow lakes and lakelets, so that the lagoon proper is restricted to the northwestern corner.

Christmas is certainly a very old structure, and could be called a dying atoll, like Jarvis Atoll, 420 kilometres to the southwest, which also has a filled lagoon floor, with evaporite deposits. It must be remembered that the Line and Phoenix Islands lie in the dry zone o fthe Equatorial Pacific. In conclusion, reticulated patterns in atoll lagoons seem to be of diverse origin, although the most typical, at Mataiva and Canton, might result from more or less similar evolutions.

Drowned Atolls

Drowned atolls are more numerous tha nreticulated atolls, and their history has been generally simpler. Saya de Malha in the Indian Ocean has been investigated by Russians with a research submarine, and is typical. It forms a ring approximately 40 kilometres in diameter, with a central depression 70 metres deep in the north and 140 metres in the south, a peripheral rim only 8 metres deep at its top with depressions which look to be former passages. This rim stil bears many living colonies of *Acropora*, *Pocillopora*, and *Porites*, as well as *Halimeda* sponges, urchins and other invertebrates at 20-25 metres depth. The asymmetry o the (former) lagon seems due to active sedimentation inthe shallow northern part, which lies on the leeward side, and is thus sheltered from currents geneated by the southerly monsoon. Saya de Malha shows tha a drowned atoll is not necessarily anextinct atoll: although its rim no longer attains the sea surface, active growth and sedimentation continue.

Drowned atolls are feequent in the Carolines, as at McLaughlin Bank and Gray Feather Bank. Velasco Reef. Palau, is another example, with a rim drowned under 10-15 metres of water, enclosing a lagoon 40 to 50 metres deep. Pulap Atoll in the Carolines (*ibid*) is half submerged, the easten side being entirely under water. The Geyser Bank, between the Comores and Madagascar, is a half-drowned atoll, and the nearby Zelee Bank is completely drowned.

On the Mid-Pacific Mountains, north of the Marshall Islands at 22°N, Darwin Guyot, under 1266 metres of water, is very interesting because, in spite of its great depth and age, it has kept an atoll shape with a peripheral rim, and a central depresson approximately 18 metres deeper. Shallow-water rudists of Cretaceous age have been dredged from the rim and the 'lagoon'. As it seems to be the oldest structure of this kind not to have been covered by later deposits, its name, coined by Ladd, Newman and Sohl, is certainly welcome.

Double and Triple-ring Atolls

These frameworks resemble the double and multiple barrier reefs discussed the difference being the absence of central high island. Nukuoro in the East Carolines has an inner ring of lagoon pinnacles, and Greenwich, in the smae group, has at least one and perhaps two inner rings, in addition to the usual atoll rim. Explanations suggested by Tayama resemble two of those proposed here for double barriers: that currents entering the lagoons through passages favour the growth of pinnacles along a course more or less parallel to the rim; or that there have been sudden tectonic movements durig the subsidence. The first explanation if favoured by Kwajalein Atoll, Marshalls and Ant Atoll, Caroline, where swamrs of pinnacles spread laterally from the inner ends of deep passages across the atoll rim.

Giant Atolls

Atolls are large as Kwajalein, Rangiroa and Tijger must have very large basements. It may be that some of them rest upon twin volcanoes instead of ony one. Rangiroa for example has a size adequate for a basement made of volcanoes similar to the pair forming Tahiti Island. This idea receives support from the existence of twin atolls with the shape of an 8 at Woleai in the West Carolines. It is understandable that, in the course of subsidence, the central isthmus has widened gradually. Indeed, at Rangiroa there is a remnant of such an initial shape. On the other hand, Kuenen (1933, 1950) explained Tijger Atoll as 'topping a geanticline like a chimneyot standing on a roof, but it is much broader than the crown of the ridge and has evidenlty spread outward on its own screes.'

Faros (or Faroes)

Although they have more or less similar counterparts in othr eareas,the faros of the west coast of India whence the name comes, are a unique assemblage which must be described first. The Southern Maldives do not show quite typical features, but north of the Veimandu Channel they include:

(a) Two large circular groups of atolls stretching 260-300 kilometres north-south and about 90-100 kilometres east-west.

(b) Each ofthese two groups contains seven or eight large or very large atolls, 30 to 80 kilometres log, the lagoons of which being generally 35-70 metres deep. Between these atolls are channels generally more than 400 metres deep. Comparable depths occur in the centre of the two large circular groups.

(c) Faros occur on the edges of the large atolls. Some are 10 kilometres or more in length, but the majority are smaller. The smallest are almost circular, but others are elongated parallel to the edges of the large atolls. Their lagoons vary in depth between 4-6 metres and almost 40 metres. Three of the large atolls of type b contain many faros in their lagoons in addition to those occurring on their outer edge.

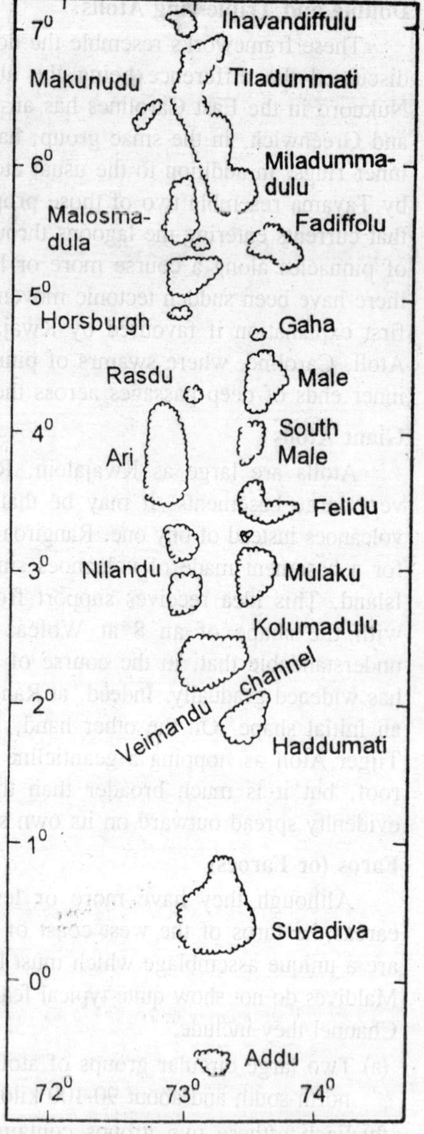

Fig. 14.18. General outline of the Maldives, eastern Indian Ocean.

Agassiz (1903b) and Gardiner (1903) have accurately described the Maldives, but, as Davis (1928) has said, their tentative explanations of the formation of these reefs are inadequate. Gardiner particularly trie to escape the necessity for subsidence, and he supposed an accumulation of deep-sea corals(?) over a pre-coralline basement, cappepd by shallow corals which would have created large atolls of type b. One interesting idea he suggested is, however, that the peripherla growth of these circular groups has prevented the development of reefs in the middle. More recently, Purdy in a brief abstract (1981) deduced, from 'seismic stratigraphy and results of a single well,' that the Maldives developed during a long-continued subsidence of an Eocene volcanic substrate, leading to an accumulation of more than 2100 metres of shallow-water carbonates. He considered that 'the reef development took place in a decidedly non-Darwinian sense,' since it was 'not accompanied bya genetic succession of fringing-barrier-atolls'. Purdy's concept is not explained in sufficient detail to allow a sound opinion. Apart from the prolonged subsidence, which is essential, the problem of the formation of a and b types seems to require further information. The small faros, which remain the most curious features of the Maldives, have horseshoe shapes, with the convex

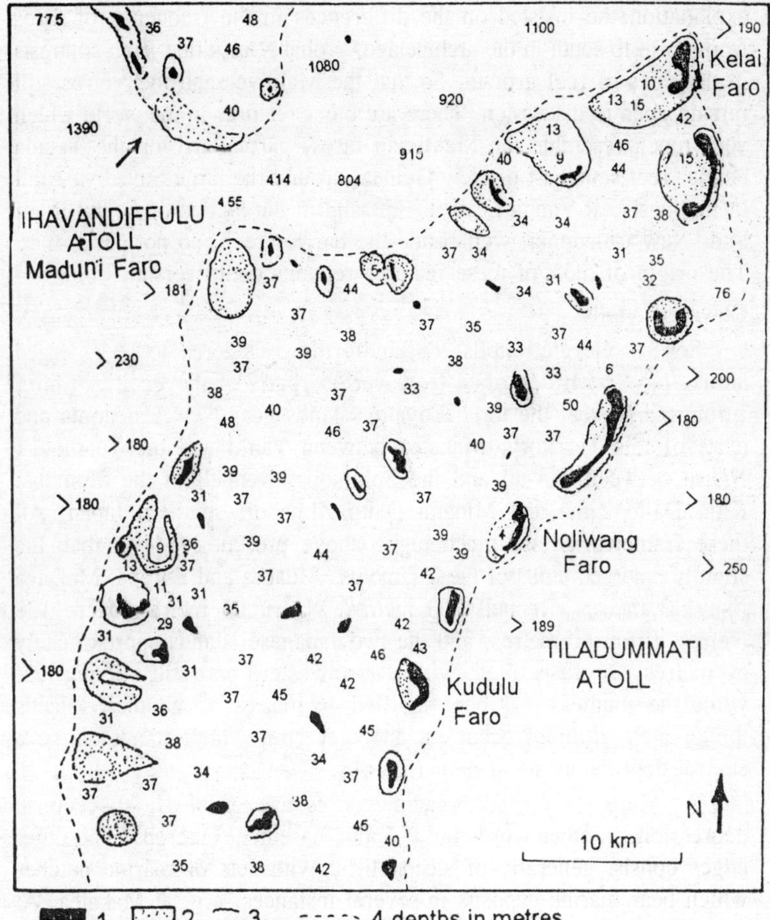

Fig. 14.19. Northern half of Tiladummati Atoll, Maldives, with faros. 1, islands; 2, very shallow; 3, edge of reefs; 4, limit of atolls.

side facing the open sea, and thus seem to be related to the surf generated by the alternating monsoon winds on both sides of the b type atolls, while the faros situated in the middle of these atolls are almost symmetric.

The same author (1963) has also drawn the attention to the similarity of the Maldive faros with the horseshoe-shaped features found on the Mayotte and New Caledonia barrier reefs, and on the outer Great Barrier Reef, which are obviously due to wave refraction on either side of passages. However, Stoddart has not accepted this explanation: he insisted on the differences in the frequency of faros from north to south in the archipelago, probably resulting from contrasts in the rates of reef growth. So that the Maldive sand their faros still remain open to discussion. There are other features in the world which very much resemble the Maldivian faros, particularly on the Tagula barrier reef southeast of New Guinea; around the large Suvadiva Atoll in Indonesia; at Vanua Levu in Fiji; and in the lagoon of Grand Recif Sud, New Caledonia, well inside the barrier reef and not parts of it. The origin of most of these features remains controversial.

Elevated Atolls

Several elevated atolls—or landforms considered to be emerged atolls—have been described from various parts of the Pacific: Mare, Lifou and Ouvea, the three Loyalty Islands near New Caledonia and parts of this Territory; Makatea between Tahiti and the Tuamotus; Nauru between Tuvalu and the Solomons: Rennell in the Slomons: Kita, Daito Zima and Minami Daito Zima off southern Japan. All these frameworks rise much highe above present sea-level than the slightly emerged atolls of the Tuamotus, Aldabra and Europa. Makatea rise 113 metres, Rennell 200 metres, Nauru 65 metres, Mare 138 metres, Lifou 60 metres, and the two Japanese islands approximately 50 metres; Ouvea is tilted, with a northwestern part still a living reef while the southeast has been uplifted 46 metres. In all these islands the greatest altitudes occur on an outer rim, which stands above a central depression, as in modern atolls.

At Mare, two small volcanic cores are visible in the central depression, in other words the volcanic basement outcrops. The outer edges consist generally of steep cliffs, with sets of marine notches which bear marine deposits in several instances, e.g. at Makatea. At Mare the peripheral rim includes smaller, shallower depressions which have been interpreted as fossil faros by Chevalier (1968), following Guilcher (1963). All the raised atolls have been strongly karstified,

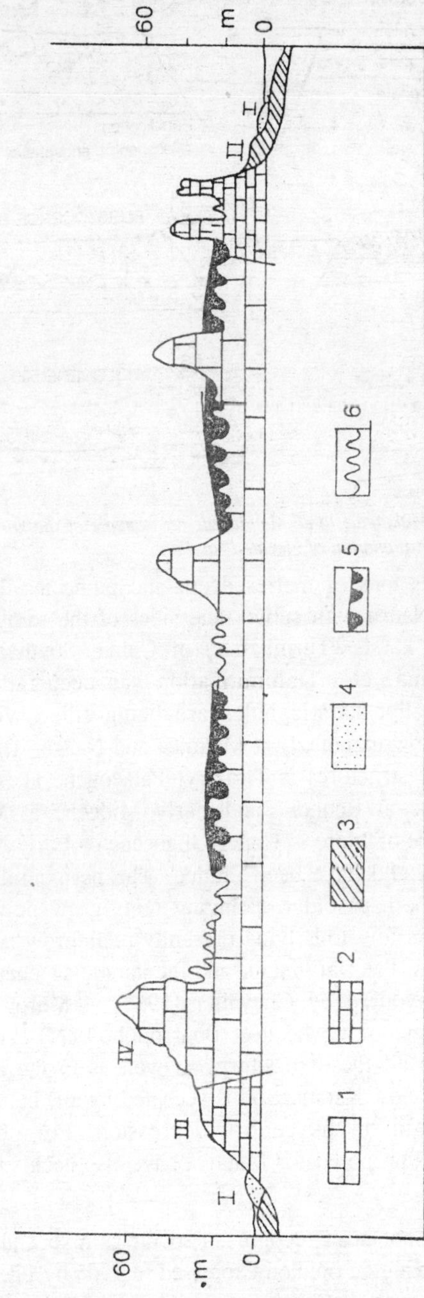

Fig. 14.20. Section across the elevated atoll of Nauru, Central Pacific, according to Leontyev and Medvedev. 1, inner lagoon coral reef; 2, circular fossil rim coral reef; 3, Pleistocene reef; 4, loose Holocene deposits; 5, phosphate-filled karastic holes; 6, karastic holes exposed by phosphate exploitation; I, Holocene terrace; II, III, IV, Pleistocene notches at 10-11, 24-25, and 60 metres above sea-level.

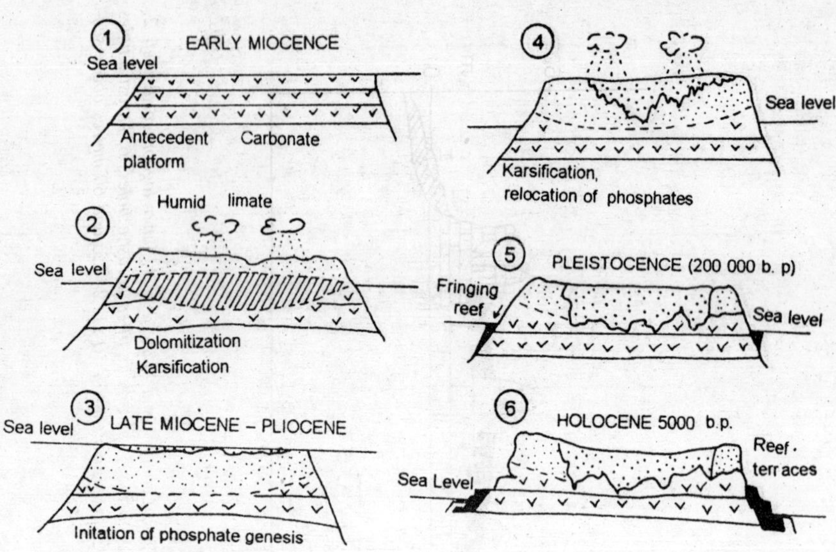

Fig. 14.21. Evolution of Makatea island, simplified. An example of the karstic saucer theory as an explanation of elevated 'atolls'.

with caves sometimes tens of metres deep, and pinnacles that stand particularly high at Nauru with subvertical sides of the samy type as in continental tower karsts (Turmkarsts) of China, Southeast Asia, Cuba, New Caledonia, etc. Dolomitization has occurred. In the depressions, the smaller karstic holes are being filled with thick phosphate deposits (now exploited) at Makatea and Nauru. The age of the older parts of the structures is Tertiary: Palaeogene at Nauru, at least Middle Miocene at Rennell, and Early Miocene at Makatea; while the volcanic core of Mare is Upper Oligocene or Early Miocene. Mare and Makatea would have been faulted. The peripheral notches are Pleistocene and perhaps older. Fringing reefs grow now at sea-level, around them. As for atolls lying presently reefs grow now there are two interpretations: the Darwinian, and the karstic saucuer theory. The first one was favoured by Chevalier (1968) at Mare, and the second one by Montaggioni at Makatea. Bourrouilh (1977) extends the karstic explanation to all these structures as well as to the atoll-like features in the Fiji Islands. Karstication is accepted by all, but opinions differ about the origin of the central depression. For phosphate formation, the theories proposed for Mataiva have also been suggested.

Ridge Reefs

Ridge reefs have not been recognized so far as a special type of coral reefs; but, following an opinion expressed in 1965 by Ph. Kuenen after he read Guilcher's report (1955) on teh Farsan Bank in the Red Sea, it seems that they should be so. It must be remembered first that the Red Sea, in which this type of reefs is found, is an ocean in the making, since it ies between the gradually diverging African and Arabian plates, and, in the north, the Sinai platelet. Recent studies have shown that its evolution consists of progressive oceanization. proceeding from south to north, alogn a central graben originally surveyed by R V *Calypso* through a series of 'cells', located over more or less regularly spaced 'hot points'.

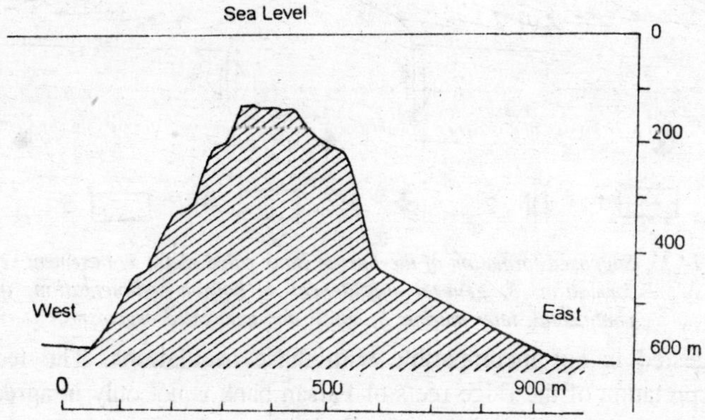

Fig. 14.22. Echo-sounding profile by R V Calypso across Shab Suleim ridge.

The general trend of these cells and of the graben is northwest to southeast, a direction which is also followed by the ridge reefs found on either side of the graben. Farsan Bank, on which the biggest reefs of this type are situated, is a huge area off the Asir coast of western Saudi Arabia, over 580 kilometres long and 100 kilometres wide between the Tihama (coastal plain) and the central graben. It consists of alternations of hollows several hundreds of metres deep and coral reefs rising abruptly to the surface. In the area investigaed off Lith, the reefs consist of long ridges (e.g. Shab suleim, Shab Sahabak) running NW-SE, i.e. in the general direction of the Red Sea, the depth on either side of Shab Suleim being between 600 and 700 metres, beside extremely steep slopes. Similar ridges have been described off Jeddah by Mergner (1967) and called by him outer reefs. These features have been interpreted by Guilcher as coral structures which originated on small horsts, and then grew upward duringthe progressive subsidence of Farsan Bank.

The present state of researchdoes not allow to know what happened during the Quaternary oscillations of sea-level, but the evolution

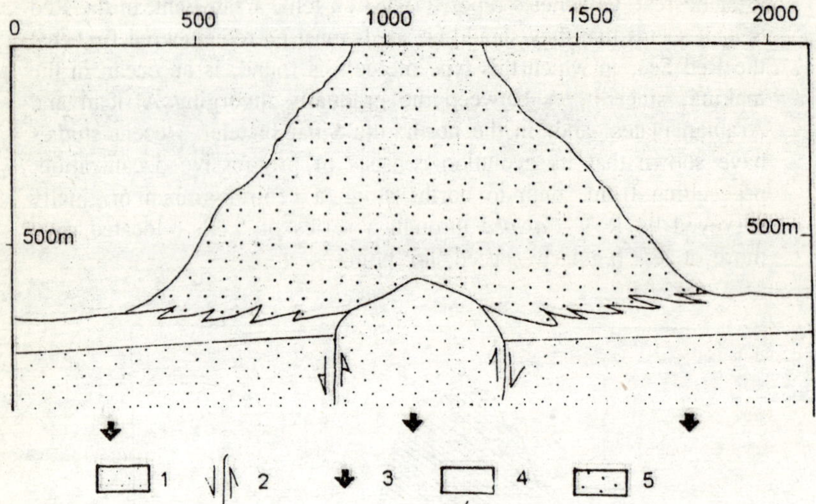

Fig. 14.23. Supposed formation of the Farsan Bank ridge reefs. 1, basement; 2, horst formation; 3, general subsidence; 4, bottom sedimentation, initially continental, later marine; 5, corals and associated organisms.

suggested is not incompatible with these oscillations. The tectonic interpretation of the ridge reefs of Farsan bank is not only in agreement with the general evolution of the Red Sea; it also fits the morphology and structure of Abulat Island, which lies closer to the coast, on a higher step than Farsan Bank, and consists of emerged corals, heavily faulted, again in a NW-SE direction. Abulat is thus a raised, emerged ridge reef. In addition, at the northwestern end of the Red Sea, the southern tip of the Sinai Peninsula, which is situated at the tripele junction between the African and Arabian plates and the Sinai platelet, consists of emerged coral reefs that have been cut into small horsts and grabens, which run in the same direction. These frameworks continue into the shallow area at the entrance of the Gulf of Suez, where some of them have been interpreted as the outcome of tidal currents running to and fro. From a discussion on that point, it results that the ridges are structurally controlled, being thus tectonic ridges, of smaller size than those on the Farsan Bank; the submerged ones may well be influenced in their growth by local tidal currents, which do not exist elsewhee in the Red, Sea, where the tidal range is extremely small, particularly over Farsan Bank.

Ridge reefs also occur in shallow water on the western side of the Red Sea, along the coast of Sudan, where El Rabaa (in Guilcher,

1986) has interpreted them as resulting from slow subsidence of horsts; and this is in agreement with Braithwaite (1982). It is even more interrestin that, like Farsan Bank off Saudia Arabia, the Sudanese ridge reefs extend to the deep sea in front, as shown by Schumacher and Mergner's (1985) excellent study of Sanganeb Atoll. This atoll is elongated, and extends in a north-south direction, parallel to the coastal ridge reefs; it rises from depths of 550 metres in the east and 750-825 metres in the west.

It is interpreted by the German authors as having been built on a horst: the thickness of teh coral reefs is a matter for speculation, but the tectonic interpretation is in total agreement with the stuctural Red Sea context. Sanganeb Atoll, although it resembles other atolls in the world in its shape and depth of lagoon, thus belongs to a Sea. Aanother reef, called Green Reef, southeast of Suakin, Sudan, resembles an atolland was visited by R V *Calypso* in 1952, but this has not yet been studied in detail. Summarizing these resuslts, ridge reefs are certainly widespread in the Red Sea, and include as subtypes the shallow and emerged ridge reefs on the easter, western and northern sides, the huge ridge reefs in deeper water off Arabia and Sudan, and one ridge atoll off Sudan.

As many other features concerning coral reefs, the Red Sea ridge reefs were mentioned by Darwin (1842), on the basis of information from Captain Moresby. He wrote: 'some of them are very long in proportion to their width ... many of them rise to the surface ... they extend parallel to the shore ... the sea is generally profoundly deep close to them as it is near most parts of the coast of the main land .... These data were unfortunately forgotten for a very long time after him. So that Darwin deserves the name of ridge reefs' grandfather

Are ridge reefs represented in other areas? Braithwaite (1982), referring to Heidecker (1973), suggesgted that 'there are ajnalogies between the tectonicsetting of the reefs here (near Port Sudan) and those off the Queensland coast'. But, although Heidecker gave evidence for tectonic and structural control of recent reefs off northeast Queensland, it does not seem that these features really resemble the Red Sea ridge reefs. So far as is known, this type of reefs still lacks counterparts elsewhere.

Special Reefs

There are other reefs, which may be very large, and do not fall into the preceding categories.

Bahaman Reefs

Although the Bahaman Banks lie in an environment where temperature. Salinity and low turbidity are favourable for reef growth, they are not true coral reefs, in spite of the fact that corals grow on their upper edges. They will be included in this general review for comparison, with a selection from the large bibliography on the Bahamas. These banks extend over 10 degrees of longitude off Florida, Cuba and Hispaniola. The size of the Great Bahama Bank (550 × 330 kilometres), and the Little Bahama Bank (250 × 70 kilometres) exceedss the largest atolls, including those in the Maldives and the Great Chagos Bank; they decrease in sizxe in the southeastern banks, where the biggest one, Caicos Bank, is still 100 kilometres long.

Like atolls they bear low islands on their periphery, and their inner part are submerged; but these inner parts are generally much shallower than in atolls, reachaing depths of only a few metres, so that they are submarine platforms rather than lagoons. These inner platforms, particularly on the Great Bahama Bank, bear ooolites of cryptocrystalline aragonite, a sediment which on the western parts of the bank forms, shallow, extremely mobile ridges, in which the asymmerty of the ripples changes with the tide. The other sediments on the same bank are more stable, and include mixed skeletal sands, pellets, 'grapestones' (aggregates), and calcareous ooze or mud in the deeper parts.

True coral reefs are generally restricted to the top of the eastern, windward slope of the bank. The islands, always located on the sides, never in the middle, lie generally on the eastern edge. Although usually low, they sometimes bear Pleistocene aeolianites, e.g. at New Providence Island, where these consolidated dunes reach 25 to 30 metres at Nassau City, standing above and behind Pleistocene beach ridges. The largest island, Andros, displays an extremely complicated pattern of low Pleistocene karstified limestones (including 'blue holes') with multiple long, sinuous consolidated ridges lying just above present sea-level, which are fossil marine ripples. Hurricanes now create lenticular mud bodies called cyclonic trails, which are superimposed over the fossil features. In spite of their differences with 'ordinary' coral reefs, the Bahaman banks have a few counterparts. In the shallow Gulf of Batabano, 7 to 12 metres deep, between mainland Cuba and the Isle of Pines, the distribution of sediment types is, according to Daetwyler and Kidwell (1959), much as on the Great Bahaman Bank: in the north, calcareous muds occur; the central part is occupied by oolites

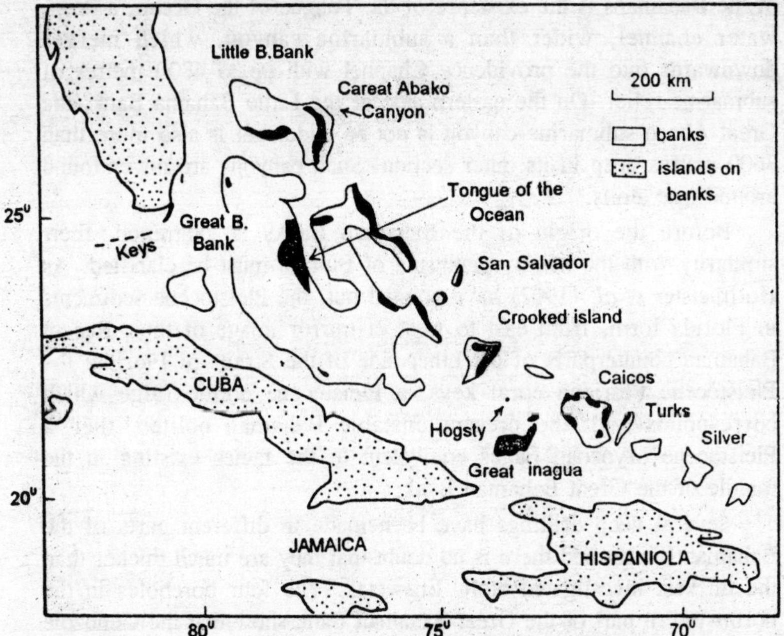

Fig. 14.24. General setting of the Bahaman Banks.

and grapestones; and in the south, skeletal calcareous sediments and a row of coral reefs are found on the edge of the very steep slope descending to the depthss of the Caribbean Sea.

On the other hand, the precipitation of oolites, although infrequent on true coral reefs, is not restricted to the Bahaman Banks and the Gulf of Batabano, since it occurs in the shallow waters of the southern part of the Persian or Arabian Gulf, and also on the western Florida shelf and the northern shelf of Yucatan. A framework resembling an atoll more thean the other Bahaman features is Hogsty Reef, in the Southeastern part of the group, described by Milliman (1967). It has the shape of a small (9 × 5 kilometres) atoll, with a shallow lagoon, open on its leeward side. But its rim appears to be a Pleistocene dune, and hermatypic organisms, which now flourish on it, seem to have been only recently (re?) introduced. Hogsty Reef is thus an atoll in terms of configuration, but quite different in structure from the usual type.

The submarine outer slopes of the Bahaman Banks are extremely steep, and drop precipitously (commonly 28 to 40°) to great depths (more than 3500 metres), as around atolls. But the most curious aspect of the Bahamans is the existence of the Tongue of the Ocean, a deep-water channel, wider than a submarine canyon, which merges downwards into the providence Channel with up to 4800 metres of submarine relief. On the eastern side of the Little Bahama Bank, the Great Abaco submarine canyon is not so wide, but is also more than 3600 metres deep in its outer section. Such canyons are never found around true atolls.

Before the origin of the Bahaman Banks is examined, their similarity with the nearby peninsula of Florida must be clarified. As Hoffmeister *et al.* (1967) have pointed out, the Pleistocene sediments in Florida form, from east to west, a mirror image of their Recent Bahaman counterparts of the other side of the Straits of Florida: the Pleistocene Floridan coral keys, a Pleistocene oolitic ridge which corresponds with the present unstable Bahaman oolites, then a Pleistocene bryozoan facies equivalent to the facies existing in the middle of the Great Bahaman Bank.

Several deep drillings have been made in different parts of the Bahaman Banks, and there is no doubt that they are much thicker than the thickest investigated atoll, Enewetak. The four boreholes in the northwestern part of the Great Bahaman Bank show that the Cenozoic and Cretaceous sediments are at least 4800 metres thick; the pre-Jurassic basement being at -6000 or -7000 metres. These figures require a long continued subsidence, which was initiated at a time when the Atlantic Ocean was just opening between America and Africa. Deep channels such as the Tongue of the Ocean are certainly very old, but whether they are as old as the opening of the Atlantic remains a matter of discussion. Several models have been proposed, including one by Mullins and Lynts (1977), according to which the depressions were initiated as grabens when the platform was still part of Africa; the platform would have later drifted and rotated during the Cretaceous and Early Tertiary, with subsidence testified by the thick sedimentation on the banks.

In the Great Abaco canyon, the steep walls consist of Cretaceous massive limestones, and the orientatioin of the canyon is structurally controlled. The absence of filling by sedimentatioin in that canyon may be due to episodic erosion along the axis during the pliocene and the Pleistocene, but erosional activity is presently small. In the Florida Straits, on the western edge of which coral growth now plays a larger part than in the bank environment, the absence of filling is more understandable than in the Tongue of the Ocean and other similar troughs, because of scour by the Gulf Stream. Leg 101 of the *Joides Resolution* Ocean Drilling Program (ODP) in 1985 jsought to collect firm data on the Bahaman structure and its evolution, but techincal problems fhampered drilling in the Florida Straits and oter troughs, so that the interpretations are still hypothetical.

Houtman's Abrolhos

The southernmost ones in the Indian Ocean, are not only unique in their location off a western continental coast (apart from some poor reefs off the west coast of Africa); they are also unusual in their shape, as they are neigher atolls nor banks nor fringing reefs. Darwin (1842) left their classification in doubt, and Davis (1928) hesitated between 'bank reefs' and 'bank atolls'. More recently they have been described accurately by Teichert (1947) and Fairbridge (1946-1947), and their ecology defined by Hatcher (1985). The reason for their good development at 28°30'S-29°S is the warm Leeuwin current. Houtman's Abrolhos, situated some 70 kilometres offsore, rise near the outer edge of the Australian shelf from depths of 50-55 metres. They comprise four groups of islets: the two northernmost are somewhhat shapeless, while Easter Group and Pelsart Group, in the south, are V shaped, with the apex in the south, and a wide opening with many coral patches or knolls in the north. The structure is complex. The foundation in each case is a Pleistocene coral reef limestone, rising up to 1 or 2 metres above high tide level in its highest parts. It is covered in many places by a shelly limestone, above which a shingle limestone is found, and finally a dune limestone reaching 10 metres in thickness at West Wallaby Iland. Subaqueous shingle depositss, beach ridges of coral shingle, small dunes in the Wallaby Islands, and guano, represent the modern sediments.

The growth of corals is luxuriant in many places, and the algal flora is extremely diversified, with both tropical and temperate components. From a geomorphological point of view, the most interesting group is certainly Pelsart, in which the eastern branch of the V forms an extremely narrow (30 to 600 metres wide) emerged island 11 kilometres long, in which various sedimentary component outcrop, wereas the western branch is an outer reef at low water level with a few remnants of a higher, former reef. It is surprising to find uch a complicated reef on the outskirts of the coral world. Generally the reefs in such a situation are simply fringing reefs, as at Lord Howe in the Tasman Sea; but Bermuda reefs, also at the limits of coral reef growth in the North Atlantic, are equally somewat

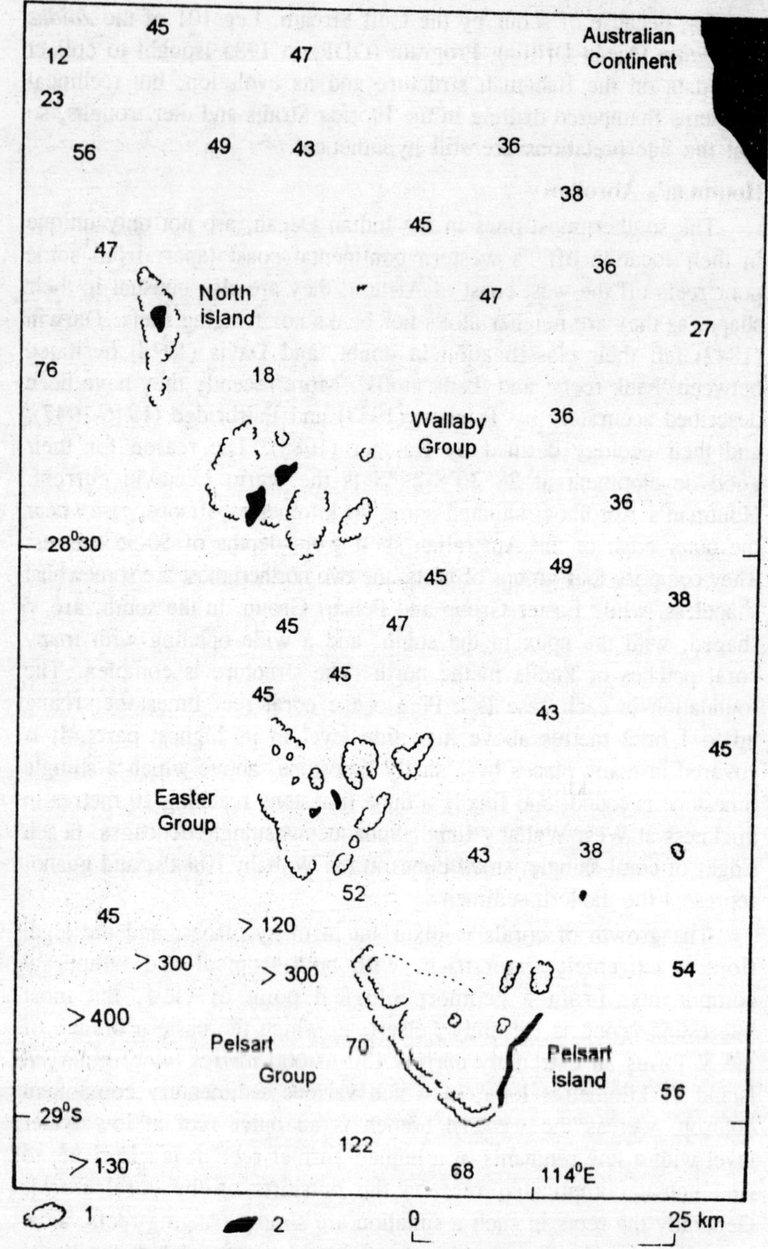

Fig. 14.25. Houtman's Abrolhos, Western Australia. Depths in metres. 1, reefs; 2, islands and continent.

complex, being clasified as a bank-atoll by Davis (1928), althoug this is questionable. The clue to this problem may lie in the excellent hydrological envvironment in both cases, with warm currents favouring coral growth in spite of the latitutde.

Comparison between the Indo-pacific and Atlantic Coral Worlds, and problems of the Brazilian Reefs

The main contrast between the Indo-Pacific and the Atlantic coral worlds lies in the number of hermatypic coral genera in the two areas. The heart of the coral kingdom is situated in Melanesia, Indonesia and the Eastern Indian Ocean. From that area the number of genera decreases outwards, especially in Polynesia, and, as a whole, the Western Atlantic is much poorer. However, the Caribbean reefs are by no means pale images of their Indo-Pacific brothers. On their outer, windward sides, they often display vigorous growth ot *Acropora palmata*, with *A. cervicornis* a little lower, which can match similar partss of reefs in the Indo-Pacific. But, as Adey and Burke (1976) and Laborel and Bouchon (1986) have pointed out, such patterns are replaced in other sites by fleshy algal pavements, also associated with strong wave energy, as in Martinique where they seem to ave recently replaced *A. palmata*.

Invasion of soft algae, which change the nature of the reefs, may be related to an increase in turbidity, resulting, in several cases at least, from modern erosion on the nearby subaerial slopes. Yet in the heart of the coral kingdom, the reefs of souteast Asia have to withstand heavy sedimentation from large rivers, so that the contrast in this respect is far from being complete. On the other hand, many Caribbean reefs are subject to mor severe hurricanes than a large majority of Indo-Pacific reefs. Although present in the Caribbean, there are very few atolls in comparison with the Indo-Pacific (17, or even 10, instead of more than 300), and often not so typical as they are open on their leeward sides; this shape is also represented in the Pacific, e.g. at Tarawa, but is rather rare there. A reason for the scarcity of atolls in the Caribbean may be that migration of a plate over hot spots, whic appears to be the origin of atoll chains in the Pacific and possibly also in the Indian Ocean, has not been effective in the Caribbean.

The movement of the Caribbean plate between the Nazca, Cocos and American plates has not resulted in atoll chains, apparently because of a lack of melting anomalis in the asthenosphere. Milliman (1977) remarked that, instead of subsidence, much of the Caribbean area had uplift and active volcanism during the Tertiary; but the same author and Supko (1968) expressed the counter opinion that atolls such as Courtown and Albuquerque, off the Nicaraguan shelf, were probably built on a subsiding basement as in the Pacific examples. Information on the deep structure of that area is not conclusive in this respect. The Venezuela basin rests upon an oceanic-type crust, at least as old as Late Cretaceous, with a thick pile of Tertiary sediments above it—a structure apparently favourable to attoll genesis; but no typical atoll has been reported from that area.

Conversely, the Columbia basin contains atolls off Nicaragua, but, in the Deep Sea Drilling Project, Leg 15 failed to obtain useful structural data in that basin: there was only one drilling, n°154, which penetrated into thick volcanic sands, silts and clays, and stopped at 270 metres in them. It may be that the karstic sau cer theory finds more support in the Caribbean, particularly in Belize barrier reef

Fig. 14.26. Coral reefs and their environment in Brazil. 1, warm current; 2, cold current; 3, heavy sedimentation from rivers; 4, main reefs.

than generally in the Pacific; although the evidence from Fiji appeared convincing. Outside the Caribbean, coral reefs are represented in the Atlantic along northeastern Brazil, where they have been described in detail by Laborel. The development of the Brazilian reefs has been hindered by two obstacles in the north the Amazon River, at the mouth of which the coral growth is prevented by the freshwater and the suspended sediments supplied by the largest river in the world, so that reefs are absent north from Sao Luis de Marahao; in the south, a double thermal obstacle: an upwelling in Cabo Frio area, and cold water deriving from the Falkland current.

The southernmost boundary of the reefs tus lies between Rio de Janeiro and Santos. In the intermediate area, there is an impoverishment of coral in both directions: towards the nort between Cabo Sao Roque and Maranhao, and towards the south from Cabo Sao Thome near Rio. In addition, coral growt is still restricted by the sediment and freshwater supplied by Rio Sao Francisco and Rio Doce: so that, finally, the favourable areas are limited to the Natal-Recife coast on one hand, and the Sao Salvador—Abrolhos region on the other. In the Abrolhos region (not to be confused with Houtman's Abrolhos in Western Australia), which includes the Abrolhos archipelago proper and the nearby Paredes reefs, coral growth is rich. The reefs are everywhere at least 20 metres thick and include a large number of flourishing pinnacles (the so-called chapeiroes) which may coalesce laterally. Their surface is covered by thriving colonies of coral, with lateral overhangs, as in the best reefs of the Indo-Pacific region.

In other Brazilian areas, a common peculiarity of the reefs is their association with *arrecifes*, which are subparallel to the coastline and parly submerged lines of Holocene beach rock, dated at between 7000 and 2000 BP. This type of reef occurs mostly in the Natal-recife area, whre corals form fringing reefs attached on the outer side of the seaward arrecife. Particularly near Joao Pessoa in Paraiba, the hermatypic corals are only scattered over the arrecife, whereas algae (e.g. *Caulerpa*, *Halimeda*) are predominant, so that the name of algal reef would be appropriate. The calcareous (95-98 per cent of carbonates) sediment consists essentially of *Halimeda* particles in many places. The turbidity of the water is high during the rainy season, which lasts for seven or eight months; mud thus carried to te reef gives a grey colour to dead *Halimeda* fragments, which strongly differ in colour from similar particles encountered on Pacific and Indian Ocean atolls. This turbidity is probably one of the causes of the poor development of hermatypic reefs. The coral reefs exising along the Brazilian coast never go beyond the stage of fringing reefs, although they had ancestors during Pleistocene low ea levels. Laborel also notes that, on islets lying far off the coast such as Fernando do Noronha and Rocas, there are not true coral reefs, but only scattered coral colonies which have not been able to build real reef frameworks. In spite of the name 'Atol das Rocas', and the circular shape of this structure with a very shallow central depression, less than one metre deep, it is not a corallian atoll since it is almost exclusively made of calcareous algae. In these sites, water turbidity cannot be cause of limited coral growth, so that the poor development of reefs in Brazil remains some what unsolved.

Reef Degradation

In recent years, coastal geomorphologists have been increasingly interested in problems of coastal management, because coasts have been more and more subject to damage by man's activities, and these changes have been investigated by geomorphologists. In a similar way, the vigour and even the existence of coral reefs have been threatened by man, so that in the International Congresses at Manila (1981) and Tahiti (1985) such topics were widely discussed. A book dealing with coral reef geomorphology must therefore include a review of these various modifications, and of the efforts made to reduce or halt them, since without such management there will be no more reef geomorphology to study. In a few areas, scientists considered that the situation is not dangerous.

Such is the case at Madagascar, where Rabesandratana (1985) expressed the view that reefs are useful because they attract a large number of tourists, provide lime from broken shells, are a source of food with their fishes, gastropods, crustaceans and holothurians, and thus play a role in the socioeconomic activities of the country. No mention is made excess in exploitation nor danger of disappearance. Similarly, Mallik (1985) considered that Kavaratti and Kalpeni Atolls, Lakshadweep (formerly Laquedives, Indian Ocean) constitute 'an enormous potentiality ... for industrial use (in) the manufacture of cement, calcium carbide, etc. ... It is expected that exploitation of 1 or 2 metres of sand from the lagoon floor for industrial purpose will not affect the equilibrium, and attempts should be made to achieve an optimum depth of dredging'.

In the South Pacific, the coral reefs of the Solomon Islands 'are still close to their natural state except perhaps in localized areas', and in Vanuatu 'there has probably been little human impact'. On a world scale, the general opinion is, however, that reefs and their environment have deteriorated. A first cause of dstruction has been extensive illegal dynamiting as a fishing technique since the Second World War. This has affected many South Pacific Islands that were occupied by the US Army and navy during the war, since, when the departed in 1945, American troops left behind them considerable amounts of explosives. They were particularly used by the Wallisians, who have, in the following years, completely devastated the lagoon between their island and its barrier reef. Some parts of the reefs of new Caledonia, Fiji, and Samoa have also been damaged in the same way.

Even outside the South Pacific, dynamiting has been reported from a wide range of reefs, such as those in Mauritius, Tanzania, Indonesia, and Sri Lanka. In several instances, over exploitation of living resources, without much dynamiting, is sufficient to create a hazard. Such is a the case in the Maldives where traditional fishing, which exploited moderately reef fishes, turtles, molluscs, crustaceans, and octopi, has recently improved its catches considerably by the use of mechanized vessels and expansion through the archipelago to face the demand of local tourism and world market, taking a much wider range of marine products than the former Maldivian output. On the Chinese island of Hainan, where extensive coral reefs were reported in the 1950s, they have now largely disappeared or been replaced by algal covers. The damage is due mainly to the use of fine meshed nets which have 'removed all the fish', thus destroying the coral community.

Yet the impact on Pacific reefs and shores of the war itself has not proved, as a whole, very durable. The Japanese fleet sunk in the depths of Truk lagoon in the Carolines has been buried under a growth of marine organisms; some remnants of American lorries persist on the shores of Tarawa and other military landing sites; silhouettes of wrecked Japanese ships variegate the marine skylines of Guadalcanal and new Britain; and the American quays built on fringing reefs on the inner reaches of the Bora Bora lagoon have been quietly invaded by the near shore vegetation.

Another potential threat is the possibility of establishing military bases in atolls. Such has been the case on Aldabra, Indian Ocean, a slightly emerged atoll of Sangamon age known to be the only undisturbed habitat of the Giant Land Tortoise in this ocean. Its bird population and flora are also most interesting. It was considered, in the middle 1960s, as a possible site for the development of a British military airfield, which would certainly have completely modified the environment.

A cause of coral reef destruction is the mining of coral and dredging of sand and gravel for construction and industrial use. In French Polynesia, the main source of construction material, mainly for road and urban development, has always been the coral reefs, which have been exploited by dredging. The extraction of these materials has led to a multiplication of exploited sites, mainly in the vicinity of Papeete, Tahiti, but also around Moorea and the Leeward Islands such as Raiatea. In the Tahiti alone there were 36 sites of reef mining in 1985. Extraction was generally by hydraulic dredges working along dredging paths. The results are a series of excavations with a chaotic aspect and an irregular geometry, completely disturbing the structure of the reefs. In the lagoon of Grand Cul de Sac Marin on Guadeloupe Island in the French Caribbean, dredging of coral sand has led to a considerable proliferation of halophilous bacteria in waters over the dredged areas, related to the suspension of fine particles: an outcome

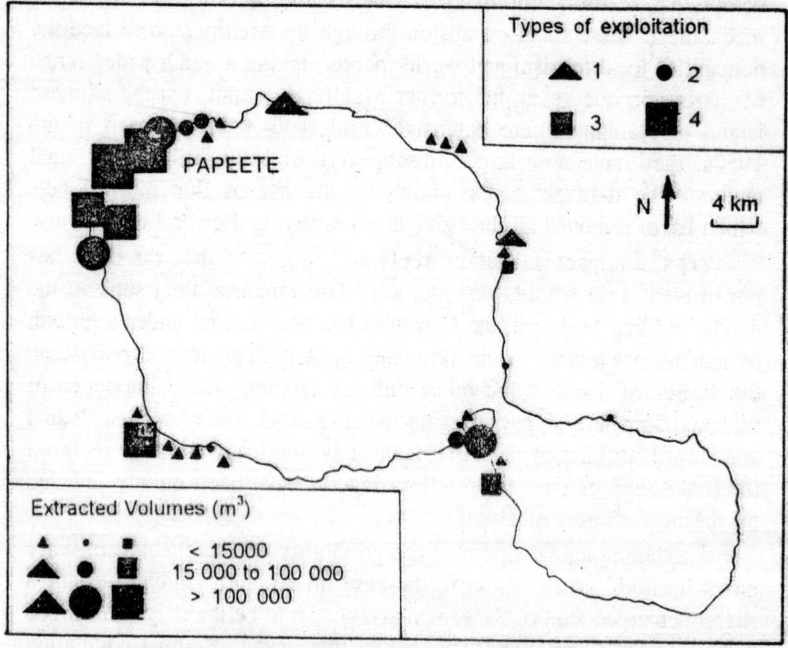

Fig. 14.27. Reef dredging sites at Tahiti. 1, channels and small harbours; 2, coral mining; 3, coastal development; 4, extensive coastal development grouping several sites.

is the disappearance of 20 of the 29 fish species living around the reefs. At Phuket Island, Thailand, the coastal fringe and near shore area have been mined for tin for 70 years by dredging, where it has resulted in heavy damage, the lower parts of neighbouring slopes being has resulted in heavy damage, the lower parts of neighbouring slopes being covered with fine sediment 2.5 centimetres thick. Nickel laterites have been exploited in many coral reef areas in Central American and the Caribbean, Papua New Guinea, Indonesia, Queensland, Philippines. The effects on reefs as generally little known; but investigations in New Caledonia where opencast mining for nickel is considerable, have shown the resulting deposition of red muds near the mouths of several rivers and the modification of coastal sedimentation affecting some fringing reefs; other New Caledonian coral reefs lie generally farther from the mainland, and have so far escaped the effect of clays washed down from hilltop mines, but parts of the lagoon are not so secure.

When reefs exist in front of expanding cities, urbanization may encroach upon them. This has recently happened at Jeddah, Saudi Arabia, where the author has observed in 1981 the deterioration of the fringing reef that in the nineteen fifties was living all along that part of the Red Sea coast. From 1950 to 1980, Jeddah grew from 50,000 to more a million inhabitants. A part of its expansion occurs on the coastal side, to the detriment of the former boat channel, upon which a maritime 'corniche' (sic) of wealthy residences and palaces with swimming pools has been built. The outer edge of the reef has become adjacent to the built-up area, in front of a wall made of huge blocks extracted from the crystalline basement and from the fossil reef existing farther inland: in such a position corals are unlikely to survive. In front of the downtown skyscrapers, the reclamations are partly used for car parking: and there too, the reefs has been destroyed.

Another danger derives from soil erosion and use of pesticides of agriculture on lands behind coral reefs. Such effects are common as results of tremendous growth and modernisation of agriculture in developing countries. A typical example has been reported from Mayotte Island, Western Indian Ocean, a volcanic island surrounded by a large barrier reef. Investigations made in the lagoon in 1959 had shown that the calcium carbonate content of sediments was higher than 95 per cent in the outer and middle parts; but more recent sampling in 1983 by Thomassin's team pointed, however, to a decrease in calcium carbonate in the eastern area from 67-99 per cent to 50-71 per cent.

At the same time, a whitening and a massive death of corals, and a burying of corals under mud, were observed in the same area. This increase in terrigenous supply to the lagoon is undoubtedly related to the rapid increase of Mayotte population, which grew from 23,364 inhabitants in 1958 to 52,035 in 1980 (62 to 139 per square kilometre), following a moderate increase in previous years, and led to an extension of the track network, cultivation on steep slopes, and consequent very heavy soil erosion.

In the Ryu Kyus, Japan (Muzik, 1985), intensive development for industry and tourism has had damaging effects on the marine environment. 'Natural vegetation has been intensively removed for dams and pineapple fields. River banks and shorelines have been paved with cement, leading to extreme loss of topsoil with every rainfall. Terrigenous sediments are measurably accumulating on the beaches, in lagoons, and even on the reef slope. Living coral cover, measured in % in 1972 and in 1984 in 67 stations around Okinawa Island, has passed in places from 50-90 to 2, from 20-70 to 1-5, from 50-100 to 2, from 50-80 to 2, etc. Not only the amount of living coral cover has decreased, but also the number of species (for example, from 78 to 18 in one station)'. Although original devastation of the coral reefs can be attributed to extensive predation by *Acanthaster planci*, there can be no doubt that the problems of coral reef survival have been compounded by man, as a result of agriculture, tourism and industry.

Discharge of sewage is also responsible for large losses on coral reefs. Along the northwestern coast of Tahiti, where 74 per cent of the island population lives, observations have been made by the Laboratory of Environmental Study and Survey in Tahiti, which reported that 'various forms of chemical aggression have been identified and quantified from analysis of waters and/or sediments: detergents, hydrocarbons, pesticides, metals'. Three rivers flowing to the lagoon have particularly contributed metals, and faecal germs from urbanization, Papeete harbour, and local industries are responsible for further damage.

Oahu in Hawaii is another island with coral reefs where the impact of tourism by sewage discharge may be very large. It has been studied in Kaneohe Bay for several years by biologists, geographers and engineers. This bay, which is rather confined, with a poor circulatory pattern in its southeastern part, had flourishing coral reefs before 1939. Since the end of the war, the watershed about the bay has been subject to urbanization, the population increasing almost ten-fold. The result was that the waters of the bay were marked by eutrophication by massive sewage discharge, increased run-off from the watershed in storms, and increased siltation from raw lands left in the curse of development of new subdivisions: so that, by 1974, no coral way left growing in one third of the bay, in another one third almost all coral had been overgrown by green algae, and only in the last third (northwestern corner), where the circulation was better, the coral reefs were reasonably intact.

Collection of specimen by and for visitors is another cause of degradation. Corals and marine shells have attracted men for a very long time. African populations have used cowries as money since remote periods, oyster shells from lagoons were and are still used as collars in Papua New Guinea, and paintings of the sixteenth century show the appreciation of marine shells by Italian ladies at that epoch. But the damage to reefs remained minor until the recent development of tourism allowed crowds of American, European and Japanese people to visit the reefs, walk on them, dive and collect find specimens. The habit of selling shell collars for farewell at airports, in Tahiti for example (several dozens of collar for those who have many friends in the island), is now a profitable industry for local people, who collect the material from the reefs of the surrounding islands. And, for amateurs not able to collect their own specimens from reefs, shops around the world are largely supplied by contractors.

The international trade in ornamental corals and shells has thus known a considerable increase in the 1970s. Taiwan and Japan have been the main centres of the precious coral industry; Philippines were the main exporter of stony corals; Philippines, Mexico, Haitian have been among the major suppliers of ornamental shells; the main consumers being the US, Japan and several European countries. From the early 1970, Japanese and Taiwan coral fishing fleets have been organized, operating in waters off Haiti, the Philippines and Australia as their own waters were rapidly depleted. Exports from the Philippines grew from 631 tons in 1976 to 1962 tons in 1978; for Taiwan, from 26 to 206 in the same period. The main shells taken are *Strombus gigas*, *Cyprea tigris*, *Nautilius pompilius*, and species of Volutidae, Conidae. It has been mentioned already that large collections of *Charonia tritonis* may be responsible for the outbreak of *Acanthaster planci* on the Great Barrier Reef and at Mauritius, and hence the destruction of coral colonies. In the Maldives, small scale collection of precious corals and shells had not damaged the reefs over a long period, but with recent big increases depletion has become disastrous there too.

In many areas, there is combination of the above mentioned causes of destruction, which makes the situation quite dangerous. Martinique Island in the Caribbean, for example was an area where 45 coral species had been identified. There has been damage by heavy siltation, especially in Fort de France Bay and on the southwestern coast, where it derives from sewage tied to urbanization, industrial development, deforestation of hillsides, and road works in the vicinity of the coast. Deposition of mud has spread from the centre of the colonies, particularly the small foliated ones, while the meandroid forms are not so easily invaded. Several areas, that were still very rich a few decades ago, are now buried under a general muddy mantle, where only the shape of the former colonies still appear as ghosts. Similarly in Thailand mining activities, dynamite fishing, bottom trawling, coral collecting, industrial and domestic pollution occur at the same time. Pollution and destructive fishing techniques have also combined in Tonga where corals are deliberately broken by women and children during fishing activities and for tourists.

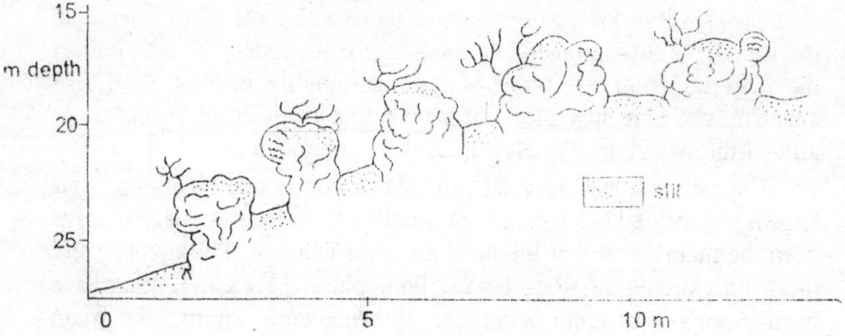

Fig. 14.28. Siltation in progress on reefs in Fort de France Bay, Martinique, Caribbean.

In Fiji, there has been over fishing, owing to population pressure, and impacts from tourism development and pollution in urban areas. In American Samoa, the best documented country in the South Pacific, a considerable decline in reef quality has been observed, and very serious degradations have been reported around Pago Pago harbour. In Paonape in the Carolines destruction of habitat from reef flat dredging, terrestrial sedimentation from construction activities, degradation from sand extraction, increased fishing pressure, decreased water quality have all occurred. As a whole, 'most of the accessible reefs in Polynesia are in various stages of decline'. Also Caribbean reefs have significantly deteriorated since 1975, damaged by a combination of large hurricanes

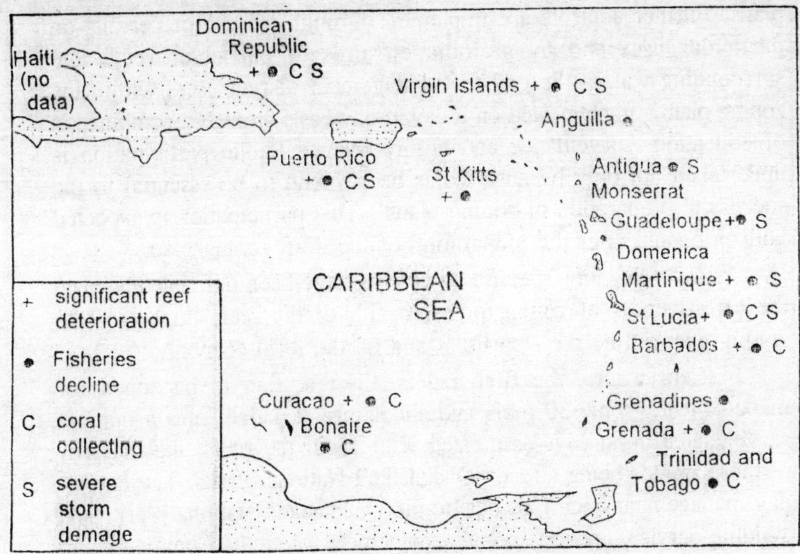

Fig. 14.29. Reef degradation in Eastern Caribbean.

and human activities. Runoff following deforestation of hillsides is a major problem in Guadeloupe, Costa Rica and the Virgin Islands. Careless anchoring and boat grounding are other causes of destruction, as are dredging, use of explosive in fishing, coral collecting, agricultural chemicals.

Efforts Against Degradation

The main efforts in this way have been on the Great Barrier Reef of Queensland, which include the widest area of coral reefs in the world. They led to the creation of the Great Barrier Reef Marine Park (GBRMP), which is expected to be completed and under management in 1988 over a total area of 343,800 km^2.

The creation of this marine park was decided by the passing of the BGRMP Act by the Australian Government in 1975. It was preceded by many small national parks of Queensland, such as those listed for the northern sector by Saunders (in GBRMP Authority, 1978,). This huge area has been divided into five sectors.

'Firstly, a framework for management is constructed through a zoning process which partitions the uses of the Marine Park. This process is analogous to methods developed for town and country planning and similar in some respects to techniques used in the British National Park system. Zoning is complemented by site-specific management plans. Further controls are implement through the use of permits on particular uses and by periodic closures of particular reefs and surrounding waters. Second the management of reef users within the zoning plan is implemented on a day-to-day basis through a combination of cducation, surveillance and enforcement... Public participation is undertaken not only because it has been found to be essential to the successful preparation of zoning plans'. This participation is expected both prior and after the preparation of the draft zoning plan.

As a whole, four specific objectives have been defined related to the effectiveness of zoning plans, health of the reef, impact of the zoning plan on the reef, on users and on the local economy.

In Capricornia, the first and smallest section to be zoned for management, 'some 30 reefs and shoals are included, plus a number of vegetated coral cays associated with 13 of the reefs, the majority of these islands being already Queensland National Parks. The day-to-day management is carried out by the Queensland National Parks and Wildlife Service, under a special arrangement where the Commonwealth Government provides initial capital funding, and 50 per cent of operating costs'. The management has been in operation since 1982. Education/information activities are widely included to increase public awareness of the marine park concept. Surveillance uses aerial and surface patrols. Management includes deterrence and detection of infringements, monitoring of natural resources and control of allowed activities.

In the Cairns and Cormorant Pass sections, the second ones to be organized, the allowed uses are different according to areas; in some places, most activities are accepted including fishing, shipping, defence operation, collecting and research (both requiring permits), while in other is only scientific research and traditional hunting and fishing are admitted with permits, recovery of minerals and pear fishing commercially or with SCUBA are banned in all zones.

As to the commercial coral collecting on the Great Barrier Reef (Oliver and McGinnity, 1985), approximately 45 tonnes of scleractinian coral are harvested annually, and almost exclusively sold domestically as curious or specimens for marine aquaria. Collecting is restricted to licensed areas of reef fronts, each approximately 400 metres long. In 1983 there were only 12 active collectors, two of them accounting for over 60 per cent of the annual harvest. *Pocillopora damicornis* makes up over 70 per cent of the total amount collected. So far, exports amount to less than 150 kilograms per year. It might be, however, that the ban of coral exports decided by the Philippines would result in enquiries in Australia by large scale American and European importers: in which case the Marine Park Authorities will have to monitor even more closely the coral harvest. In any case, the absence of a strong social or economic pressure in Australia for an expansion of the coral trade is certainly a very favourable factor for the wealth of the Great Barrier Reef.

Beside these Australian achievements, Salvat (1981) listed 16 other countries who had at that time 55 coral parks and reserves: Israel (1), Djibouti (2), Kenya (3), Malaysia (3), Indonesia (1), Philippines (14), Japan (11), Palau (1), American Samoa (1), Tonga (3), French Polynesia (2), USA (8), Mexico (2), Colombia (1), Costa Rica (1), Brazil (1).

These parks were generally small, but a special mention must be made of those created in the Philippines, the host country of the 4th International Reef Symposium, which has paid a special attention to his reefs, subject to various stresses, perturbation and overexploitation. A protection of sites has been made by creation of parks. But management on conventional modes appeared difficult if not impossible: so that a formula of municipal coral reef park was conceptualized. This system consists of the creation of small protected areas by local governments, which are include in the national scheme for marine parks envisaged for the whole country. The brief description of two such parks by Castneda is interesting. On the other hand, the protective management of coral reef fishes is also under study in the Philippines at a station on steep outer slope of Sulimon Island, in connection with the very high fishing pressure on most reefs of that country.

Another point concerning the Philippines is that, after the very large exports of corals in the nineteen seventies (see above), a total ban on coral collection was considered necessary until more investigations of population biology. But, since the collecting industry employed over 1000 individuals, and brought in millions of dollars in export earnings, the ban has been difficult to enforce.

Among other countries, the efforts made in the United States are noteworthy, the comparatively small area covered by coral reefs being very carefully managed. In Florida, where total prohibition of commercial coral colleting has been in force in state waters since 1974, two marine sanctuaries exist on reefs at Key Largo and Looe Key. Since the anchors of boats greatly damage fragile corals, a system of mooring buoys has been introduced at Key Largo, where the number of visitors is large, and has proved effective, other Caribbean and Pacific countries have sought advice on setting up such buoys to conserve their reefs. Also at Key Largo, the grounding of a small freighter in 1984 caused substantial damage: coral colonies abraded or overturned, reef crushed and covered with coralline debris. A multifaced program for reef recovery was developed and the results will be synthesized into a management document that discusses the extent of damages associated with ship grounding and identifies strategies for reef recovery following such a disturbance. The wreck may thus be used as an experiment useful for other areas.

In Kaneohe Bay on Oahu, Hawaii, where damage through sewage discharges has been previously described here, pressure from the pubic and scientific community compelled the local government and military to terminate large sewage discharged by 1978 in the southeast lagoon, which was most threatened. New surveys in 1983 revealed a remarkable recovery of corals in the southern and middle lagoon, and continued coral abundance in the northern lagoon where no decline had been previously observed. Green algae, on the contrary, which had proliferated through eutrophication, declined greatly. Here too, the results of studies may be applicable elsewhere in management.

In Tahiti the abovementioned observations on lagoon pollution have led to recommendations for improvement. Eighteen investgated stations have been classified according to the quality of the environment, in order to be used as guidelines for further work. The process is, however, ten years behind relative to Oahu. Concerning the dredging of coral material around Tahiti and Moorea, it is planned to limit the number of coral reef areas to be dredged, to restore damaged sites (restoration already being undertaken at Moorea), to use improved techniques in places where dredging continues (techniques already tested, that limit the flow of fine sediment), and develop the use of substitute material, such as basaltic rocks and alluvial deposits exploited in closed excavations on Tahiti Island.

Achievements for reef protection have been recorded in the Caribbean, especially at Barbados, where works were undertaken in 1979, and a reserve 2.5 x 1 kilometres opened in 1980 for environmental monitoring, local people and tourist education concerning conservation of coral reef ecosystems. In the Virgin Islands, the United States manage a national park where the role of biosphere reserves in conserving the marine resources is explored. There too an emphasis is made on local people and their involvement in conservation and management.

In New Caedonia in spite of the fact that the reefs themselves have not yet been much affected by open cast hilltop mining of nickel orc, the damage too vegetation, slope and valley landscape had led the Association Néo-Calédonienne pour la Sauvegarde de la Nature to promote an interest in nature conservation and landscape protection, especially to preserve the remaining *Araucaria* forests. It is not too late to start restoration, since in areas where mining ceased three or four decades ago, vegetation has revived on the river banks and the red clay has been washed away downstream; it continues, however, to flow towards the river mouths and out into the neighbouring sea, where it increasingly threatens the coral reefs. But it may be hoped that efforts in managing the inland landscapes will help to prevent its deleterious impacts.

At aldabra, plans for a military base were finally abandoned during the economic crisis of 1967, for financial rather than scientific reasons; however, a large campaign in England, supported by the Royal Society, added its weight for the preservation of the area. A big naval American base has since been created in the same ocean at Diego Garcia Atoll. It is probably impossible to reconcile all imperatives in coral areas; it was probably more important to preserve nature at Aldabra than at Diego Garcia since its fauna showed unique features.

Perspectives, Hindrances and Philosophy

A crucial point in the improvement of ref management ins the awareness of officials and the general population of the importance of coral reefs and their conservation. Such an awareness is comparatively easy to obtain in developed countries, but it is necessary that it grows in developing countries, where a very large number of coral reefs are found. It has already been observed in states such as the Sudan, which has a very fine collection of reefs on its coast. The Sudan Marine Conservation Committee, comprising all departments which deal with the sea, has since 1980 been the driving force in marine conservational activities in that country. It has proposed projects such as beaches for recreation, protected diving sites, and the Port Sudan Marine National Park. A series of talks, accompanied by slides and movies about the conservation of marine environment, were given to a wide spectrum of Sudanese communities in Arabic and English languages. The reaction of the audience showed that the message was well received. Public education was also developed by the direct observation of suitable sites, with the help of invited foreign marine scientists. Support from ICUN and the World Wildlife Fund was obtained. A visit of the Duke of Edinburgh to Sanganeb Atoll and reefs contributed to the spread of the idea of establishing a marine national park in that offshore area.

Foreign financial support was obtained to strengthen the national marine conservational institutions, and to repair the Sudanese survey vessel for oceanographic investigations. So that within a few years the basic ideas were promoted, and public understanding has been growing.

Various perspectives have appeared in other developing countries, such as Indonesia, where more than 40 potential sites have been under study for the establishment of a marine park system, with assistance by the World Wildlife Fund, FAO, UNESCO, and UNDP. Such assistance is quite necessary in the Third World, since the lack of money in these countries is obviously a major problem.

On the other hand, the specificity of developing nations in this respect has been well stressed upon for Ponape Island, in the Carolines, where Holthus (1985) noted that 'the luxury of delaying or removing reef resources from use while studies are conducted or preserves are set aside is unrealistic. Increasing number of indigenous people depend upon coral reefs for their livelihood, and resources for short-term economic gains. Reef management must reflect these economic, social and political realities in order to remain viable and achieve its goals'. He thus pointe out the main problem in coral reef management in the Third World, and his remarks help us to understand what has happened in Sri Lanka, for example, where legislation is adequate to prevent the degradation of coral reefs, under several acts and ordinances, but 'lack of enforcement and political considerations and priorities of a developing nation have rendered this legislation ineffective. A few reefs which have been declared reserves or sanctuaries have not functioned as such due to lack of follow-up regulations and a clear policy on an enforcement authority' (in so far as such a policy would be feasible in this social environment). The case of Sudan has demonstrated that education is an essential tool, but education cannot solve everything. Local scientists and administrators who have such an education in charge must learn the constraints of the society in which they work, and modify accordingly their goals.

Even in developed countries, rational management of coral reefs probably cannot obey exactly the same rules everywhere. Bradbury *et al.* (1985) maintain that good policy in this field 'should be qualitative, local, adaptive, and concerned with the whole ecosystem'. These principles derive from the fact that coral reef ecosystems are characterized by their spatial and temporal heterogeneity. 'Beware the reef manager who claims to the answer'. It is perhaps not surprising that such a statement has been expressed by Australian scientists, who have worked in the huge area of the Great Barrier Reef, where the environments are necessarily heterogeneous over 14 degrees of latitude from Torres Strait to Gladstone. Several years of precise management and planning in such diversified reefs may be considered as an insurance of the value of this opinion.

A seminar held on coral reef fisheries during the Tahiti Congress has similarly concluded that 'a given management plan will very from country to country and between areas within a country'; but in the various solutions adopted 'it cannot be overemphasized that conservation of the reef environment is an essential prerequisite to managing the fisheries'. The main administrative management potions should be to restrict harvests with size limits, catch quotas and seasonal closures, to create permanent closed reserves, and selective fishing.

Coordination between countries which have reefs is also necessary. Attempts on the international scale have been made, particularly in the South Pacific Regional Environment Programme (SPREP), implemented by the South Pacific Commission in Nouméa, New Caledonia, under the guidance of a coordinating group comprising representatives of several international organizations. Meetings have been held to finalize a convention on environmental protection of coral and lagoon ecosystems.

These considerations will perhaps seem to have extended outside the scope of coral reef geomrphology. But the shape and structure of the coral reefs are in the dependence of all their living environment. Molluscs and fishes need suitable water and sedimentary conditions to thrive, and these are closely associated with the reefs. Coral gardens would be dull without schools of fish swimming to and from and grazing the colonies. Is there a fatal trend toward the impoverishment of the reefs through an ineluctable increase of human pressure? Or may we still hope, as Dahl and Pichon did when opening the symposium on protection and conservation of the reef environment at Tahiti in 1985, that 'a concerted effort by everyone interested in coral reefs can help to stem the tide of reef destruction'? The future will give the answer.

INDEX

D